BASEMENT TECTONICS 9

Proceedings of the International Conferences on Basement Tectonics

VOLUME 3

The titles published in this series are listed at the end of this volume.

BASEMENT TECTONICS 9

Australia and Other Regions

Proceedings of the Ninth International Conference on Basement Tectonics, held in Canberra, Australia, July 1990

Edited by

M. J. RICKARD

Department of Geology, The Australian National University, Canberra, Australia

H. J. HARRINGTON

Tasmanian Hardrock Pty. Ltd., Canberra, Australia

and

P. R. WILLIAMS

Bureau of Mineral Resources, Geology & Geophysics, Canberra, Australia

SPRINGER SCIENCE+BUSINESS MEDIA, B.V.

Library of Congress Cataloging-in-Publication Data

International Conference on Basement Tectonics (9th : 1990 : Canberra, Australia)
Basement tectonics 9 : Australia and other regions : proceedings of the Ninth International Conference on Basement Tectonics, held in Canberra, Australia, July 1990 / edited by M.J. Rickard, H.J. Harrington amd P.R. Williams.
p. cm.
Includes bibliographical references and index.
ISBN 978-94-010-5173-6 ISBN 978-94-011-2654-0 (eBook)
DOI 10.1007/978-94-011-2654-0
1. Geology, Structural--Australia--Congresses. 2. Geology, Structural--Congresses. 3. Cratons--Europe, Eastern--Congresses. I. Rickard, M. J. (Michael John), 1932- . II. Harrington, H. J. (Hilary James), 1924- . III. Williams, P. R. (Peter Roddick), 1950- . . IV. Title. V. Title: Basement tectonics nine.
QE636.I58 1990
551.8--dc20 91-42480

ISBN 978-94-010-5173-6

CONFERENCE COMMITTEE

ORGANIZING COMMITTEE

M.J. Rickard, Department of Geology, The Australian National University, Canberra ACT 2601, Australia.

H.J. Harrington, 16 Hobbs Street, O'Connor ACT 2601, Australia.

P.R. Williams, Bureau of Mineral Resources, GPO Box 378, Canberra ACT 2601, Australia.

P. Wellman, Bureau of Mineral Resources, GPO Box 378, Canberra ACT 2601, Australia.

SYMPOSIUM CONVENERS

J. Bain, Bureau of Mineral Resources, GPO Box 378, Canberra ACT 2601, Australia.

M. Duggan, Bureau of Mineral Resources, GPO Box 378, Canberra ACT 2601, Australia.

M. Muir, CRA Exploration Pty Limited, 139 Canberra Avenue, Fyshwick ACT 2609, Australia.

M.J. Rickard, Geology Department, The Australian National University, GPO Box 4, Canberra ACT 2601, Australia.

C. Simpson, Bureau of Mineral Resources, GPO Box 378, Canberra ACT 2601, Australia.

SPONSORS

International Basement Tectonics Incorporated.

Geological Society of Australia Incorporated and its Specialist Group in Tectonics and Structural Geology.

The Department of Geology, The Australian National University.

The Bureau of Mineral Resources Geology & Geophysics.

ARCO Oil and Gas Co, USA.

BHP-UTAH Minerals International.

Carpentaria Exploration Co Pty Ltd.

CRA Exploration Pty Ltd.

Western Mining Corporation Ltd.

ANSETT Airlines.

Qantas.

Commonwealth Bank.

FOREWORD

The Ninth International Conference on Basement Tectonics was held at the Australian National University in Canberra 2-6 July 1990. The opening keynote address was given by Prof. R.W.R. Rutland, Director of the Bureau of Mineral Resources. Other keynote speakers were E.S.T. O'Driscoll, an Australian consultant, and Prof P. Bankwitz, Central Institute for Physics of the Earth, Potsdam, GDR.

Technical sessions were arranged by session conveners on the following five topics:-

i) The structure of the Australian craton and cover basins;
ii) Basement structure of continental regions;
iii) Structural patterns and mineral deposits;
iv) Techniques for analysing basement structures;
v) Structural patterns in oceanic crust.

The arrangement of papers for this Proceedings Volume has been simplified. Part 1 deals with Australia, Part 2 with other areas and Part 3 lists the titles of all the papers read at the conference. Abstracts of these papers are available in Geological Society of Australia Abstracts No 26 and may be purchased for $A10 from the Geological Society of Australia Office, ANA House, 301 George Street, Sydney NSW 2000.

Field trips to view aspects of the Lachlan Fold Belt and the Sydney Basin were assisted by H.J. Harrington, D. Branagan, D. Wyborn, B. Drummond and M.J. Rickard. A longer field trip, aborted through low enrolments, was organized by H.J. Harrington with assistance from W. Preiss, N. Cook, R. Glenn, A. Grady, and P. James; this assistance is gratefully acknowledged.

The Organizing Committee wishes to thank ACTS for their helpful work as Convention Secretariat, L. Moore and W. Crowe who projected the slides and the following who chaired sessions at the meeting: M. Duggan, H.J. Harrington, R. Korsch, M. Muir, C. Simpson, W. Mayer and P. Wellman. We also thank J. Braun and H. McQueen of the Research School of Earth Sciences at ANU for displaying their computer models of basin development. Assistance with typing and editing was provided by M. Coldrick. Drs E. Heidecker and J. Leven assisted the Editors with refereeing manuscripts. We thank all the delegates, especially those from overseas who made the long journey 'down under' and who, by attending, helped make the conference a success. Finally we thank the Trustees of International Basement Tectonics Incorporated for entrusting the Conference to Canberra.

M.J. Rickard
Convener

FOREWORD

The Fourth International Conference on Basement Tectonics was held at the Australian National University in Canberra 2-6 July 1979. The opening keynote address was given by Prof. R.W.R. Rutland, Director of the Bureau of Mineral Resources. Other keynote speakers were L.S.T. O'Driscoll, an Australian consultant, and Prof. E. Bankwitz, Central Institute for Physics of the Earth, Potsdam, GDR.

Technical sessions were arranged by session convenors on the following five topics:-

i) The structure of the Australian cratons and cover basins
ii) Basement structure of continental regions
iii) Structural patterns and mineral deposits
iv) Techniques for analysing basement structures
v) Structural patterns in cratonic cases

The arrangement of papers for this Proceedings Volume has been simplified. Part 1 deals with Australia, Part 2 with other areas and Part 3 lists the titles of all the papers read at the conference. Abstracts of those papers are available as Geological Society of Australia Abstracts No. 1 and may be purchased for $A10 from the Geological Society of Australia Office, AMA House, 301 George Street, Sydney, NSW 2000.

Field trips to view aspects of the Lachlan Fold Belt and the Sydney Basin were assisted by H.J. Harrington, G. Birmingham, D. Wyborn, B. Drummond and M.J. Rickard. A longer field trip, aborted through low enrolments, was organized by H.J. Harrington with assistance from W. Preiss, N. Cook, R. Oliver, A. Grady, and P. Jones; this assistance is gratefully acknowledged.

The Organizing Committee wishes to thank ACTS for their helpful work as Convention Secretariat, L. Moore and W. Crowe who projected the slides, and the following who chaired sessions at the meetings: M. Duggan, H.J. Harrington, R. Korsch, M. Mayo, C. Simpson, W. Mayer and P. Wellman. We also thank J. Braun and H. McQueen of the Research School of Earth Sciences at ANU for displaying their computer models of basin development. Assistance with typing and editing was provided by M. Chadwick. Drs B. [illegible]

M.J. Rickard

CONTENTS

PART I

AUSTRALIAN CONTINENT

BASEMENT TECTONICS IN AUSTRALIA: AN INTRODUCTORY PERSPECTIVE

R.W.R. RUTLAND
Bureau of Mineral Resources
Geology and Geophysics
GPO Box 378
Canberra ACT 2600
Australia

ABSTRACT. As defined for this Conference, Basement Tectonics appears to be nearly synonymous with Intra-plate Tectonics i.e. with the tectonics of regions after a basement has been developed. This overview attempts to provide a general framework of the main themes of the conference with respect to Australia. Two major phases of reactivation of the craton, in the Cenozoic and the mid-Proterozoic are considered. The elucidation of these two main phases of reactivation is important for the deciphering of the primary geometric relationships between the Archaean and early Proterozoic basement provinces in Australia.

The present topography of Australia is largely the consequence of Cenozoic tectonics. There is a broad relation to the Cenozoic plate tectonic system, rather than to the orogenic structure of the basement rocks. This is illustrated by the passive margin development of the Great Dividing Range, and by the broad flexure of the Australian Shield on an E-W axis. However, the Cenozoic basement tectonics does show a dramatic change across the lines of the Torrens Hinge Zone and the Diamantina River, thus reflecting the boundary between the Precambrian craton and the Tasmanides. The Proterozoic reactivation structures were also significant influences on the expression of Cenozoic reactivation.

On the western edge of the Tasmanides, the Flinders Ranges are a remarkable example of reactivation of older structural trends within a broad NNW trending region of Tertiary subsidence. Within the Eromanga Basin the Tertiary reactivation follows northeasterly Permo-Triassic and older trends.

The neotectonics, through its control on erosional and depositional history has exercised important controls on the distribution of mineral deposits, as well as on the quality of soils.

In the Precambrian Shield of Australia it is clear that the main features of the basement blocks and of their cover basins are the consequence of rather well defined episodes of basement reactivation which are reflected in the pattern of dipole gravity anomalies and in aeromagnetic data. Thus the Archaean Pilbara and Yilgarn provinces have had essentially cratonic character since 2400 Ma but much of the present boundaries are the results of reactivation during the period 1400-1000 Ma. This same period of reactivation has also largely obscured the primary relations between Proterozoic orogenic provinces in the Amadeus Transverse zone, where there has also been younger reactivation.

The Kimberly region provides an example of repeated basement reactivation in nearly orthogonal fault zones which have controlled the emplacement of diamond pipes. Kimberlite intrusions also occur in the region of the Delamerian fold belt where the Cambro-Ordovician fold trends are controlled by reactivation of Proterozoic basement trends.

Major faults are important in controlling mineralisation in the Archaean Yilgarn province and in the Proterozoic provinces developed between 1800 and 1500 Ma. In both cases the faults affect sequences which have been subjected to extensional and compressional episodes and which apparently overlie on older basement. It is not clear that the faults reflect the structure of the older basement but they are important elements in discussions of lineament tectonics.

M. J. Rickard et al. (eds.), Basement Tectonics 9, 3–21.

Introduction

Basement tectonics means very different things to different people as the varying content of the previous eight international conferences demonstrates.

As defined in the circular for this Conference, Basement Tectonics appears to be nearly synonymous with Intra-plate Tectonics, ie. with the tectonics of regions after a basement has been developed. That is, Basement Tectonics is not concerned with the processes which produced the initial architecture of the basement (as might be inferred from the term eg. Dennis 1967) but with the tectonic processes which subsequently modified the basement and its cover. This overview attempts to provide a general framework of the main themes of the conference with respect to Australia.

Perhaps the core of basement tectonics has been in the study of reactivation of the basement on the scales of field geology - and more particularly in the study of lineaments.

Most often we are concerned with the reactivation of old structures within the basement, and with the control that these exert on structures in overlying cover sequences. Of course the reactivation may be due to a new geodynamic system, the structures being reactivated because they represent weaknesses in an inhomogeneous basement. Their orientations may not then be an accurate guide to the controlling geodynamic system.

The basement or craton as a whole can also be considered to be reactivated even if a completely new set of structures due to a new geodynamic system is superimposed since the new structures will displace old structures and surfaces, and modify the enveloping surface of the basement. The basin and swell structure of the African crust provides a well known example where the largest structures reflect a new geodynamic system but where in detail, there is reactivation of old structures. The historical development of Basement Tectonics reflects not only the tectonic styles involved (as summarised in the Conference brochure) but also philosophical approaches to the subject. Thus in a broad way it has tended to reflect fixist rather than mobilist views, and vertical tectonic movements rather than horizontal. Similarly it has tended to reflect the pragmatic and empirical approach of the field geologist, concerned with geological patterns at various scales, rather than the modelling approach of the geophysicist.

In Australia the pioneer studies of Sherbon Hills and others have their roots in notions of fundamental rhegmatic patterns in the crust. More recently, as a fuller geophysical understanding of crust and mantle rheology and dynamics have begun to emerge, and as the role of extensional tectonics has been more fully appreciated, there has been increasing integration of basement-tectonic phenomena into the overall plate-tectonic scheme. In Australia, for example, this has been exemplified by the development of ideas concerning the uplift of the Eastern Highlands (see below).

However, in general the nature of intraplate (or intracratonic) tectonic movements on scales of hundreds or thousands of kilometres appears not to be well enough established geologically for satisfactory geophysical modelling to be achieved—in order to relate the mechanical processes in the lithosphere to the thermal and mechanical processes below (eg. Lambeck et al. 1986). Consequently many features of the basement tectonics of the older areas of the earth's crust are not well integrated into global geodynamic schemes.

This paper will therefore briefly review general topographic, geological and geophysical data for Australia in order to examine the relation of intra-plate phenomena to the Cenozoic geodynamic scheme and to the basement structure of the continent.

Secondly it will examine the role of reactivation by intraplate tectonics in establishing that earlier basement structure.

Cenozoic Basement Tectonics

In a broad way the whole of Australia has acted as a craton since the early Mesozoic and some topographic features may be inherited from Cambrian or even Precambrian times (eg. Stewart et al. 1986; Ollier et al. 1988a). But the present relief of Australia is largely the consequence of Cenozoic tectonics, developed since the widespread planation and marine invasion of the continent during the Cretaceous.

A key question is how far the Cenozoic intra-plate phenomena are directly the effect of the Cenozoic plate tectonic pattern, and how far they reflect the earlier basement structure within the craton. Important data sets therefore are topography, geological structure, and gravity (see eg. BMR Earth Science Atlas of Australia).

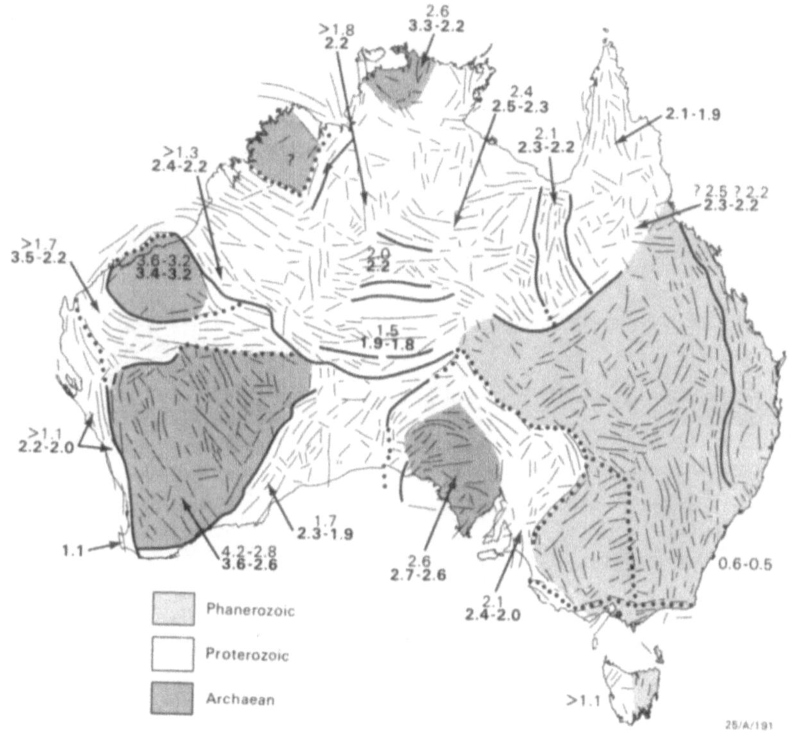

Figure 1. Relationship of pattern of gravity trends and 'dipole' gravity anomalies to ages of primary crustal formation. Trends (thin lines) from gradients on the Bouguer gravity-anomaly map (Bureau of Mineral Resources, 1976). Boundaries of crustal blocks shown as thick lines where marked by major gravity gradients ('dipole' gravity anomalies), and by a line of dots where inferred from trends, from gravity-anomaly level and from geology. Ages (Ca) derived from Pb and Sr isotopes (upper values) (Page et al. 1984), and from Sm-Nd model ages (lower values, bold) (McCulloch 1987). From Wellman (1988).

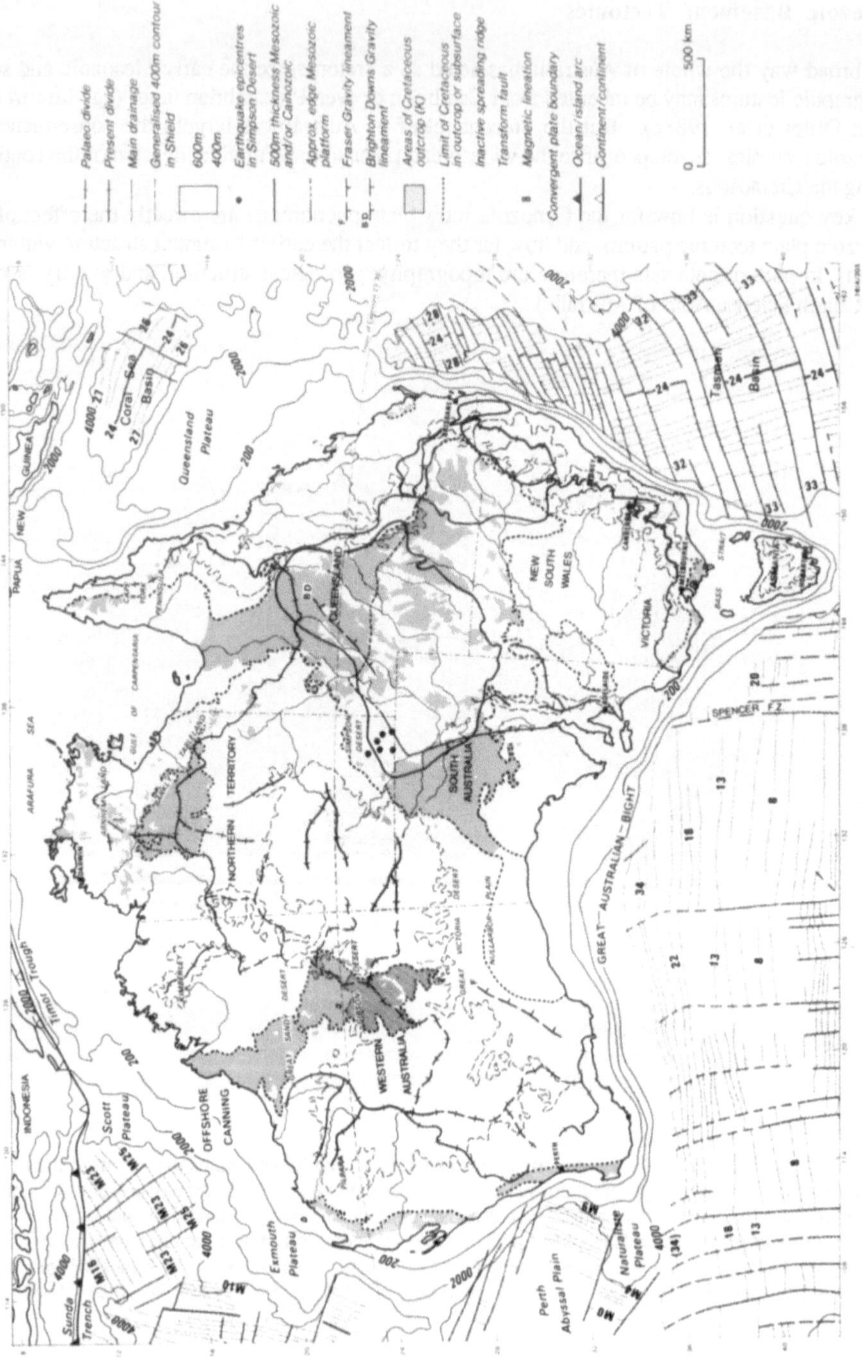
Palaeo divide
Present divide
Main drainage
Generalised 400m contour in Shield
600m
400m
Earthquake epicentres in Simpson Desert
500m thickness Mesozoic and/or Cainozoic
Approx. edge Mesozoic platform
Fraser Gravity lineament
Brighton Downs Gravity lineament
Areas of Cretaceous outcrop (K)
Limit of Cretaceous in outcrop or subsurface
Inactive spreading ridge
Transform fault
Magnetic lineation
Convergent plate boundary
Ocean/island arc
Continent/continent
0
500 km
INDONESIA
PAPUA NEW GUINEA
ARAFURA SEA
GULF OF CARPENTARIA
Coral Sea Basin
Queensland Plateau
Timor Trough
Sunda Trench
Scott Plateau
Exmouth Plateau
OFFSHORE CANNING
NORTHERN TERRITORY
WESTERN AUSTRALIA
SOUTH AUSTRALIA
QUEENSLAND
NEW SOUTH WALES
VICTORIA
SIMPSON DESERT
GREAT SANDY DESERT
GREAT VICTORIA DESERT
GREAT AUSTRALIAN BIGHT
SPENCER F.Z.
Tasman Basin
Perth Abyssal Plain
Naturaliste Plateau

Gravity and magnetic data are usually used at the province scale where the anomalies closely reflect major variations in rock composition and structure. At this scale gravity is useful in defining trends within provinces and also defining province boundaries where these are marked by paired ('dipole') anomalies (eg. Wellman 1976, 1978; see Fig. 1). At the continental scale the areas of higher topographic elevation and of higher negative Bouguer anomalies over the Precambrian Shield area do not show a close relation to the basement structure. There is a broad E-W topographic swell between 21° and 28° South which reflects uplift during the Cenozoic (Fig. 2). Cretaceous sediments are preserved in both north and south Australia where they dip beneath sea level (eg. Hays 1967) but the Cretaceous outcrops increase in elevation inland so that at about 26°S latitude in the shield area they are at about 500 metres above sea level. This broad E-W crustal swell may be complementary to the crustal sag behind the northern (collisional) margin of the Australian Plate but its geometry and history of development need to be much better understood in order to constrain possible models for its development. Thus on this continental scale topography and gravity appear to reflect the Cenozoic geodynamic scheme rather than the structure of the old basement rocks.

It is notable however that the main dipole free-air gravity anomalies which reflect major middle Proterozoic crustal structures are generally parallel to the regional topographic contours. These structures are, locally, also closely related to the boundaries of the main topographic swell. The areas of highest elevation in Central Australia are also closely related to major crustal structures marked by dipole anomalies. These structures were strongly reactivated in the Alice Springs Orogeny (Devonian- Carboniferous) and there was also minor reactivation in the Cenozoic. However, the ENE trend of the MacDonnell Ranges strikes obliquely across the E-W trend of the geological structure and of the gravity anomalies and is clearly not directly related to them.

The broad E-W topographic high in the shield in Central Australia between 23° and 27° South, is abruptly terminated eastwards across a NE trending line beyond which Cretaceous sediments dip beneath the Tertiary sediments in the Simpson Desert. This topographic lineament crosses the Amadeus Transverse zone in an area close to the boundary between the main Tasmanide orogenic belt and its paratectonic extension into the Amadeus Basin (eg. Rutland 1976). This NE-SW topographic lineament also corresponds broadly with an abrupt change in the Magsat anomaly pattern (Johnson et al. 1986). From this line a broad tectonic and topographic depression extends SSE from the Simpson Desert to the Murray-Darling Basin across the Phanerozoic tectonic elements of the Tasmanides. Within this broad NNW-SSE trending region of Cenozoic subsidence, the Flinders Ranges are a remarkable example of positive reactivation of older structural trends.

Thus there is a broad contrast in continental topography between the Precambrian Shield and the Tasmanides. However, the main NE-SW topographic lineament referred to above extends to the southwest where it is clearly *not* the boundary between the Shield and the Tasmanides but appears to correspond most closely to the edge of the platform at the end of the Palaeozoic (indicated on Fig. 2 by the limit of the main Cretaceous outcrop), east of which a number of Late Palaeozoic basins were developed. (See eg Totterdell 1990). It may also be related in part to the older basement structure indicated by the dipole gravity anomaly northwest of the Gawler Province. Thus part of the Shield — the Archaean to Proterozoic Gawler Province, with its Cretaceous cover — lies to the southeast of this line. This may be a consequence of modification of this segment of

Figure 2. Generalised topography and drainage of Australia related to Cretaceous outcrops and main areas of Mesozoic and Cenozoic deposition onshore; and to the record of ocean-floor spreading offshore.

the lithosphere during late Paleozoic extension.

The transition from the Gawler Province, with its local flat-lying Lower Palaeozoic cover to the paratectonic folded Late Precambrian and Cambrian rocks of the Flinders Ranges, takes place across the Torrens - Lake Eyre lineament which was also a very important feature in Cenozoic basement tectonics (see below) and which is aligned with the Spencer Fracture zone offshore (Fig. 2).

Northwards from about 23° North the NE-SW topographic lineament does not persist and the limit of the main Cretaceous outcrop on Figure 2 turns eastward so that it closely follows the boundary of the Tasmanides. However there is a suggestion in the gravity anomalies and the gross topography of a significant crustal linear feature extending northwards beneath the Georgina Basin (Fig. 2). This line may reflect Proterozoic tectonics. Thick fault-bounded Adelaidean sedimentation appears to be restricted to the east of the line (eg. Tucker et al. 1979) and the older crust east of this line including the Mt Isa Inlier with its strong positive Bouguer as well as Free Air gravity anomalies appears to have behaved differently from the rest of the Shield. The line is reflected in a local present-day drainage divide. However the Cenozoic uplift history appears to have been broadly similar east and west of this line (Hays 1967) and Cretaceous outcrops occur at about 300m elevation at the northern end of the Mt Isa Inlier. Indeed Cretaceous rocks are preserved at over 200m elevation along the main divide linking the Mt Isa region to the Eastern Highlands. Southeast of the Mt Isa Inlier and southwest from the Eastern Highlands in Queensland the Cretaceous sediments dip beneath Tertiary sediments into the Cenozoic tectonic depression (Fig. 2). It is evident from the preservation of Cretaceous sediments along the divide in northern Australia that uplift and erosion are largely post-Cretaceous so that the topography reflects the Cenozoic tectonics. However it is also clear in eastern Australia that the pattern of Cenozoic tectonics also reflects the underlying structure. The areas of tectonic subsidence show correlations with the areas of late Palaeozoic basins and also with the margins of the Tasman orogenic province. Moreover, the areas of thicker Mesozoic deposition (Fig. 2) also reflect the basement structure (Fig. 1).

East of the broad NNW-SSE region of Cenozoic subsidence overlying the western Tasmanides, the elevation rises to segments of the Great Dividing Range broadly parallel to the Queensland and New South Wales coasts.

The gross relief of the Great Dividing Range has been explained in a number of ways (eg. Smith 1982; Lambeck & Stephenson 1986). The maximum uplift as indicated by the trend of the divide (eg. Young 1989) is oblique to the trends of the main Palaeozoic orogenic zones and it is clear therefore that the relief is not inherited from the Palaeozoic structure. There is abundant evidence that the uplift is largely Cenozoic and the parallelism of the divide with the extensional rift structures offshore provides a compelling argument for relating the uplift to the passive margin development (eg. Falvey 1974; Ollier 1978, 1982; Moore et al. 1986; Lister & Etheridge 1989; Weissel 1990).

The earthquake pattern (see eg. BMR Earth Science Atlas; Gaull et al. 1990) is also of considerable potential in showing whether current tectonic activity reflects the Neogene geodynamic pattern or whether it is controlled by older basement structures (Fig. 3).

In eastern Australia the earthquake activity broadly parallels the passive margins and presumably reflects the subsequent adjustments to that passive-margin development. The earthquake activity appears to be rather stronger in the NNE trending belt parallel to the New South Wales margin, where it crosses the older orogenic trends obliquely, than it does in the NNW trending belt parallel to the Queensland margin, where it is more nearly parallel to the orogenic trends.

Nevertheless when the drainage regimes west of the passive margin uplifts are examined (eg. Fig. 2; and BMR Earth Science Atlas) it appears that the New South Wales uplift, although of greater magnitude, has a more localised regional tectonic effect than the Queensland uplift. In the broad topographic depression previously referred to, the predominant drainage direction is from NNE to SSW ie. from the Queensland passive-margin uplift. WNW drainage from the New South Wales passive-margin uplift is relatively more localised. This may indicate greater tectonic control by the northern-plate boundary rather than by the Tasman opening or the N-S hot-spot traces (see eg. Johnson 1989). Jones & Veevers (1983) suggested that in Queensland the area of the Great Dividing Range was an active margin during the Mesozoic, and this may also be a factor in controlling the SSW tectonic slope.

It should also be noted that the predominant SSW drainage is also following Tertiary, Permo-Triassic and older structural trends through much of the topographic depression. This SSW drainage extends with remarkable constancy throughout the broad topographic depression but there are abrupt changes of drainage direction across both its northeastern and western boundaries. Its northeastern boundary is marked by abrupt topographic, gravity, and underlying basement changes in a zone parallel to the Brighton Downs lineament (Fig. 2). An abrupt change of drainage

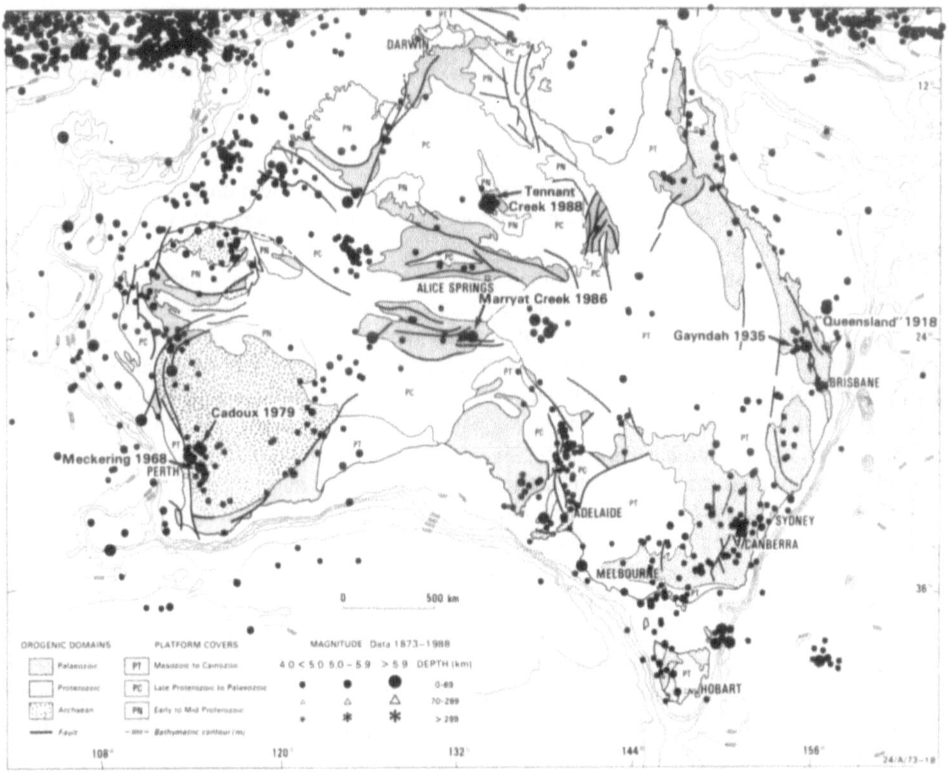

Figure 3. Earthquake epicentres of Australia with Richter magnitudes equal to or greater than 4 for the period 1959-1988, superimposed on the general geology. From Gaull et al. (1990).

direction from SE to SSW takes place in this zone which is clearly one of the most important Cenozoic reactivation features in the Australian crust.

The SSW drainage extends all the way to the deepest part of the topographic depression on its western margin which is marked by the NNW trending Lake Torrens-Lake Eyre lineament (see eg. Wopfner & Twidale 1967; Wopfner 1985). Tertiary fault structures, mainly down throwing to the east, have been mapped northwards from Lake Torrens to the western side of Lake Eyre. At about 28° South, close to Oodnadatta they give way to Tertiary N-S folds in the area of the Emery Range, where a N-S negative gravity anomaly also appears to truncate the ENE gravity trends of the Proterozoic Musgrave Ranges to the west. The main earthquake activity in South Australia (Fig. 3) marks this trend (and the anomalous positive reactivation of the Flinders Ranges within the broad area of subsidence), while the concentration of earthquake activity in the southeast corner of the Northern Territory (Fig. 3) appears to mark the intersection of the Torrens-Eyre and Brighton Downs margins of the subsiding area ie. it marks the corner of the subsiding slab against the more positive areas to the west and northwest — notably the eastern termination of the Musgrave complex.

Thus the Torrens-Eyre and Brighton Downs lineament zones mark the borders of a subsiding slab which was tilted to the SSW in Cenozoic time, but both lineament zones also represent reactivation of older zones on the boundary of the Tasmanides with the Precambrian Shield.

This broad pattern of a broad NNW trending region of Tertiary subsidence emphasises the anomalous character of the Flinders Ranges as an uplifted zone in the Tertiary. It poses the question as to why this segment of the margin of the Tasmanides should have behaved differently from those to the north and south. The ranges are essentially coincident with the paratectonic Delamerian fold belt, whereas the areas to north and southeast were areas of basinal deposition during the Adelaidean and early Palaeozoic and presumably had different crustal characteristics. The difference in tectonic behaviour during the Cenozoic is presumably related to the difference in basement characteristics.

As already noted, the Precambrian Shield, with its broad E-W uplift has behaved quite differently from the Tasmanides. The main E-W trend is in fact segmented into Archaean and Proterozoic elements and there is evidence of an old drainage divide trending NE between these elements (Fig. 2; since modified by a shift of the divide to the east and by local reversal of drainage). This old divide is distant from the northwest coast of Australia but is broadly parallel to it and to the offshore extensional basins developed in the Mesozoic (Fig. 2). In the Northern Territory the old divide parallels a broad positive gravity anomaly and may reflect a Proterozoic hinge zone between platform to the southeast and deep basins to the northwest.

Earthquake activity is broadly associated with the passive margins of western and northwestern Australia (Fig. 3). The pattern of earthquake activity in the shield area generally reflects older structural zones especially in zones of Proterozoic reactivation around the Yilgarn craton. There is also a belt of activity parallel to the Fitzroy Trough on the south margin of the Kimberley plateau and a strong concentration west of the western termination of the Arunta complex which appears analogous to that east of the eastern termination of the Musgrave complex referred to above. However, although there are younger N-S fault trends in the Phanerozoic cover which laps onto the Yilgarn craton these N-S trends of younger tectonic activity are *not* clearly reflected in the earthquake pattern (Fig. 3). Current analyses of earthquake risk make some assumptions about the relationship of seismicity to crustal structure (Gaull et al. 1990) but no analysis with specific reference to the history of Cenozoic tectonic activity has been attempted. The general relation of the earthquake activity to the block structure determined in the Proterozoic suggests that Cenozoic reactivation has been largely focussed on these older inhomogeneities. The earthquake activity on the Meckering line (Fig. 3) shows a good correlation with the topographic divide. On the other hand, there is little seismic activity along the topographic divide in Northern Australia although there is some in both the Tennant Creek and Mt Isa inliers, presumably reflecting Cenozoic reactivation.

If the broad features of the topography of both eastern Australia and the Shield area can be related to the Cenozoic geodynamic scheme, albeit modified in detail by the basement structure, the absence of passive margin uplifts along the southern margin of the Shield clearly requires explanation. The main phase of ocean floor spreading south of the Shield overlaps with that off the Victorian and New South Wales margins (Fig. 2; Veevers 1990) but the topographic expression on land is quite different.

In the case of the New South Wales and Victorian (and Queensland) segments of the passive margin in eastern Australia, there is a strong marginal uplift so that the main drainage is inland away from the margin. In the case of the south (and west and northwest) Australian margins there

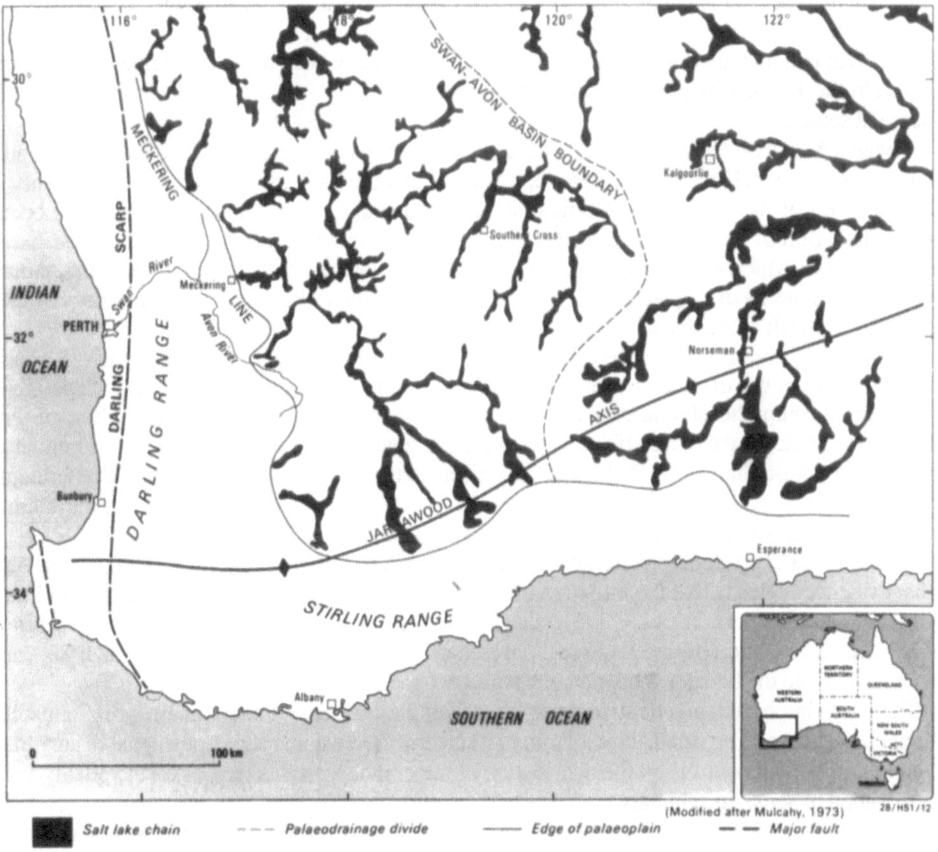

Figure 4. Major physiographic and tectonic features of southwestern Western Australia. From Ollier et al. (1988b), modified after Mulcahy (1973).

is a much broader continental shelf, and there is generally drainage towards the margin from much of the inland. Broadly the change occurs across the boundary between the Shield and the Tasmanides which appears to have controlled the development of a major zone of transform faulting on the ocean floor to the south (the Spencer Fracture Zone, Fig. 2). There appears to be an a priori argument that the contrast in margin development reflects the different basements involved but there is also a ready explanation in terms of the detachment model of continental extension which predicts different geometries for upper-plate (east Australian) and lower-plate (south Australian) margins (Etheridge et al. 1989). The contrast in basement geology may therefore merely have controlled the switch, across the major transform zone, from upper-plate to lower-plate geometry.

Since their formation however these margins have also been subject to Tertiary compressional stresses which have reactivated the earlier extensional structures, and which have also had effects inland.

In Western Australia and South Australia there was significant Eocene marine transgression which extended several hundred kilometres inland from the Great Australian Bight. Since that time there has been post-Eocene differential uplift of about 300 metres across the Jarrahwood Axis (Cope 1975; Ollier et al. 1988b; Fig. 4). Reference has already been made to changes of drainage pattern in the Tertiary in central Australia; and minor Tertiary faulting and basin formation are known in some areas.

Models which treat the continent as a stable craton, subject only to epeirogenic movements, are evidently not applicable. As noted above, it is clear that the E-W topographic high represents a gross arching of the Precambrian Shield since Cretaceous times. Given that there has been differential uplift of more than 500 metres, it is to be expected that there are many more localised tectonic effects which have affected the palaeo-drainage pattern. It therefore appears that much more attention needs to be given to differential Tertiary tectonics in elucidating the history of the regolith over much of Australia.

In particular careful attention needs to be given to dating the differential movements especially through the wider use of fission-track dating techniques. In this way it may eventually be possible to develop a comprehensive history of Cenozoic basement tectonics and to relate it both to the phases of extensional tectonics offshore and to the subsequent collisional tectonics on the northern margin of the Australian plate. Collisional tectonism along the New Guinea portion of the northern margin was initiated in the Oligocene at about 30 Ma (Pigram et al. 1989, 1990). On the Timor margin it is usually supposed that collision occurred late during the Neogene (eg. Daly et al. 1991) but there is evidence of compressional tectonic activity in the Oligocene (Hillis 1990). This with other evidence suggests that the initiation of collision on the Timor margin could have been Late Eocene-Oligocene (pers. comm. C.J. Pigram, 1991). Much remains to be done before the features of Neogene reactivation of the craton (eg. Glaessner 1953; Hays 1967; Ollier & Taylor 1988) can be integrated into the broader geodynamic scheme.

There are important practical reasons for developing a better understanding of Cenozoic basement tectonics. The neotectonics, through its control on erosional and depositional history has exercised important controls on the distribution of mineral deposits, as well as on the quality and preservation of soils. In summary:

> Tertiary tectonics controls the generation and preservation of mineral deposits within the regolith — from bauxite to uranium stream channel deposits and from opals to mineral sands. Evidently Tertiary uplift also controls the areas of well exposed basement rocks which host most metalliferous mineral deposits. For example, Broken Hill and Mt Isa lie close to drainage divides. Key questions therefore are whether old basement structures are important in

controlling the Tertiary uplift and whether the covered areas have similar character to the exposed areas (as discussed above for the Flinders Ranges). Studies of the Cenozoic cover may help to answer these questions and can provide clues to the underlying basement. Tertiary reactivation is also an important control on petroleum accumulations both onshore and offshore and this needs to be analysed on a continent-wide basis.

It is also necessary to examine all this evidence very closely in order to see through the younger tectonic history and to fully understand the relations of the Archaean and Proterozoic orogenic provinces and their middle Proterozoic reactivation. Clearly a multidisciplinary approach is required and there is a particular need for 1:1m neotectonic maps.

Proterozoic Basement Tectonics

Proterozoic and Archaean provinces of Australia have acted as a craton at least since ca 1000 Ma. At that time the Adelaidean cover sequences were deposited and have been relatively little disturbed since, except in the Amadeus Transverse Zone which bears an aulacogen-like relationship to the Tasmanides.

But large parts of these provinces have also had cratonic character since 1800 Ma. That time marked the termination of the continent-wide Barramundi Orogeny and deposition of Carpentarian cover sequences. However there was strong local reactivation of Carpentarian cover sequences between ca 1800 and 1500 Ma and a new pattern of reactivation between 1400 and 1000 Ma. These reactivation events are essentially responsible for the present block and basin structure of Australia (Rutland 1981). One consequence of this intracratonic tectonic history is that the relations between the earlier (Archaean and early Proterozoic) orogenic provinces are obscured and it is very difficult to establish their original relationships. Major strike-slip movements may have occurred for example, but conclusive evidence of this has not yet been produced.

As a consequence of these reactivation events and the deposition of younger cover, Archaean-early Proterozoic relations are only well exposed in Western Australia between the Pilbara and Yilgarn Archaean provinces in the Capricorn orogenic zone. Elsewhere the Archaean provinces are limited by younger reactivation boundaries.

Three examples of Proterozoic reactivation are briefly described here as a background to the more detailed discussion elsewhere in this volume.

(1) *The Albany-Fraser Zone* (see eg. Gee 1979; Wellman 1988, 1990). The gravity and total magnetic intensity maps of this area show dramatically how the Archaean trends have been truncated by the younger orogenic belt (Fig. 5). The Archaean rocks have clearly been reworked in the margins of the younger belt which also displays significantly deeper crustal levels as a result of upthrusting (Fig. 6). This zone is seen in some interpretations as the margin of the Archaean crust against newly formed Proterozoic crust (eg. Wellman 1988). This implies earlier rifting and drifting of the previous extensions of the Archaean crust. While rifting probably did take place it is also possible that the Proterozoic crust incorporates earlier, thinned, Archaean crust (Etheridge et al. 1989). Reworked Archaean basement underlying Proterozoic rocks is in fact widely distributed in the Gawler Craton to the east. In any case, the geochronology of the zone indicates various phases of reactivation and as already noted the zone also appears to control some present-day earthquake activity.

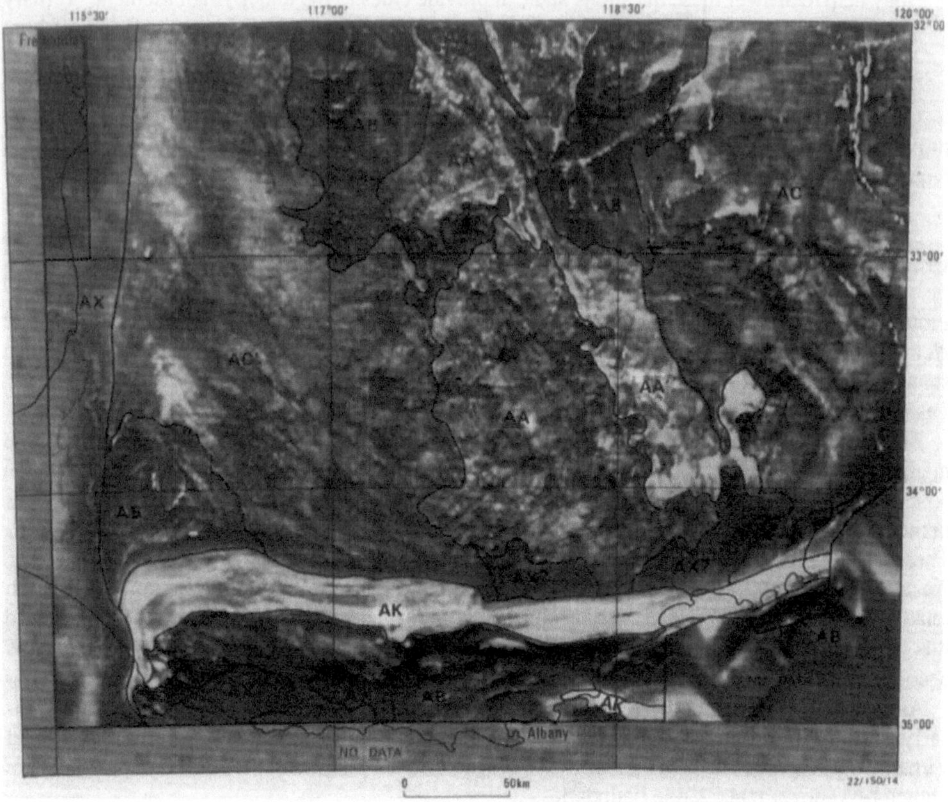

Figure 5. Total magnetic-intensity pixel map of southwestern Australia. Symbols refer to main magnetic domains distinguished by textural characteristics and anomaly widths. AK is the main deformed zone of felsic and mafic granulites in the Proterozoic Albany orogenic belt, with the Yilgarn Archaean orogenic province to the north. From Tucker & D'Addario (1986). Note the bend in the Albany belt at the intersection of E-W and N-S reactivation trends, indicating sinistral movement in the N-S zone.

(2) *The Kimberley Region.* This region provides evidence of repeated basement reactivation in nearly orthogonal zones (eg. Rutland 1981; Wellman 1988).It has been argued by Hancock & Rutland (1984) that the initial orogenic belt is a consequence of A-subduction and did not involve a Wilson cycle. Irrespective of this conclusion it is clear that all later phases of tectonic activity in the belt represent intra-cratonic reactivation. The zone also appears to have localised the intrusion of kimberlitic rocks during both the Proterozoic and the Neogene (Jaques et al. 1986, 1990; Fig. 7). The Proterozoic lamproites may be related to rifting which preceded strong NNE trending, folding and faulting dated at about 1050 Ma (Plumb & Gemuts 1976). The Miocene lamproites may be related to a N-S trend crossing the Fitzroy trough and the adjacent King Leopold zone of Proterozoic reactivation (Fig. 7).

Other reactivation events at approximately 1500 Ma and 600 Ma have been tentatively identified in the King Leopold domain (Plumb & Gemuts, 1976).

(3) *Arunta Block.* The large Bouguer gravity anomaly gives a clear indication of the major N-dipping crustal structure at the southern margin of the Arunta block which has now been interpreted in detail on the basis of seismic reflection studies (Goleby et al. 1989) Shaw et al. describe the structure in more detail elsewhere in this volume.

The folds of the Devonian-Carboniferous Alice Springs Orogeny reflect a late phase of reactivation in the zone. The zone is mainly due to earlier reactivation before 1100 Ma, and this

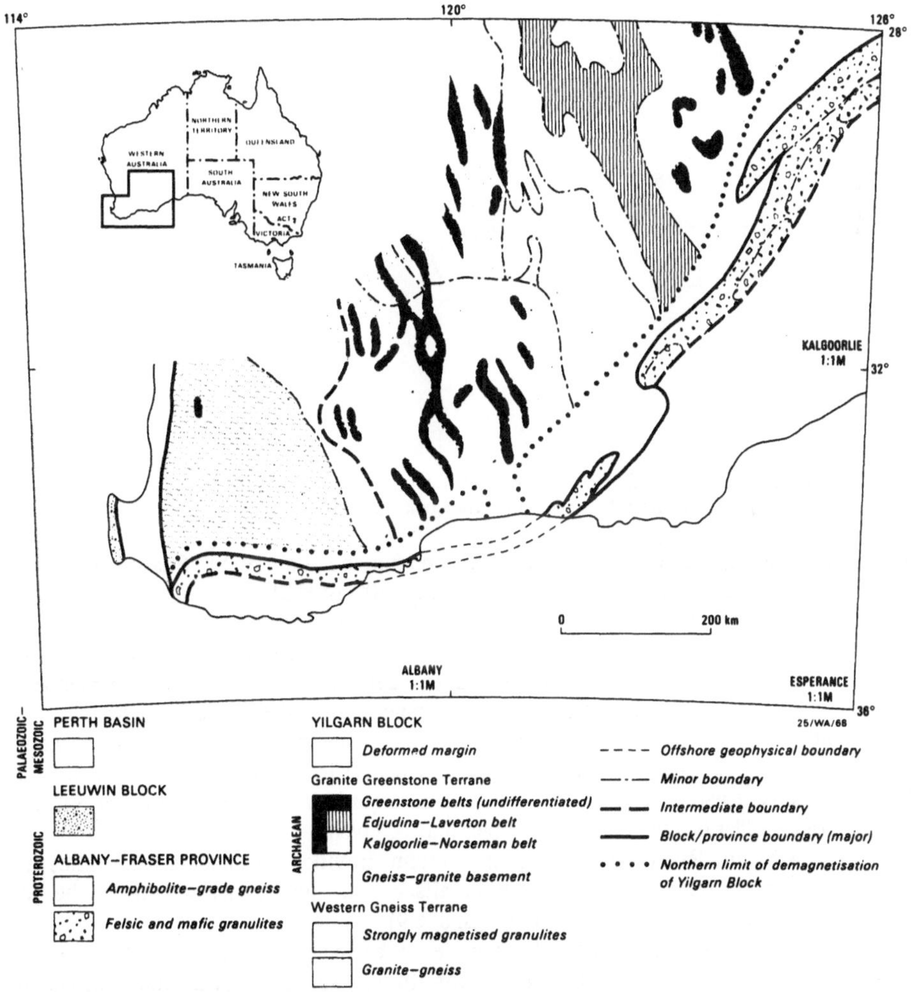

Figure 6. Basement geology of the Albany, Esperance and Kalgoorlie 1:1,000,000 sheets, as inferred from interpretation of regional aeromagnetic data. From Whitaker A.(1990).

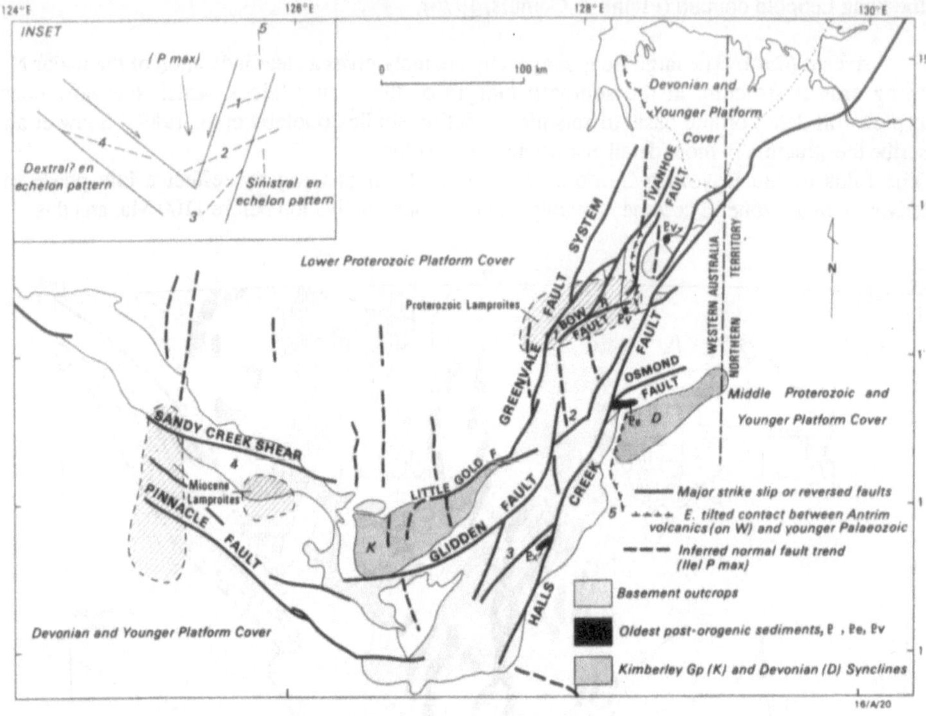

Figure 7. Areas of Proterozoic and Miocene lamproites in relation to fault systems in the reactivation zones flanking the Kimberley Basin. Modified after Hancock and Rutland (1984).

was possibly superimposed on Barramundi collision (Rutland et al. 1988). Shaw & Black (this volume) date the main period of Proterozoic shearing at 1500-1400 Ma.

Amongst other examples of Proterozoic reactivation the Proterozoic Olary province displays shear zones first developed at around 1600 Ma which were reactivated so as to exercise control on Delamerian (ca 500 Ma) folding of the Delamerian fold belt (Glen et al. 1977). These phenomena can now be observed because of the Cenozoic uplift and erosion of the Flinders Ranges.

Major faults are also important in controlling mineralisation in the Proterozoic sequences developed between 1800 Ma and 1500 Ma. These sequences have been subjected to extensional and compressional episodes and overlie an older basement. The mineralisation is related to reactivation of the older basement. Two examples may be given:

(1) In the Alligator Rivers Uranium/gold/platinum province recent detailed studies have established the relationship of this mineralisation to shear zones which cut the post-Barramundi Orogeny unconformities (Figs. 8 & 9).

(2) In the Mt Isa orogenic domain the complexity of the fault structure and evidence of extensional tectonics elsewhere in the region strongly suggests that the main Mt Isa fault structure is due to compressional reactivation of earlier extensional structures.

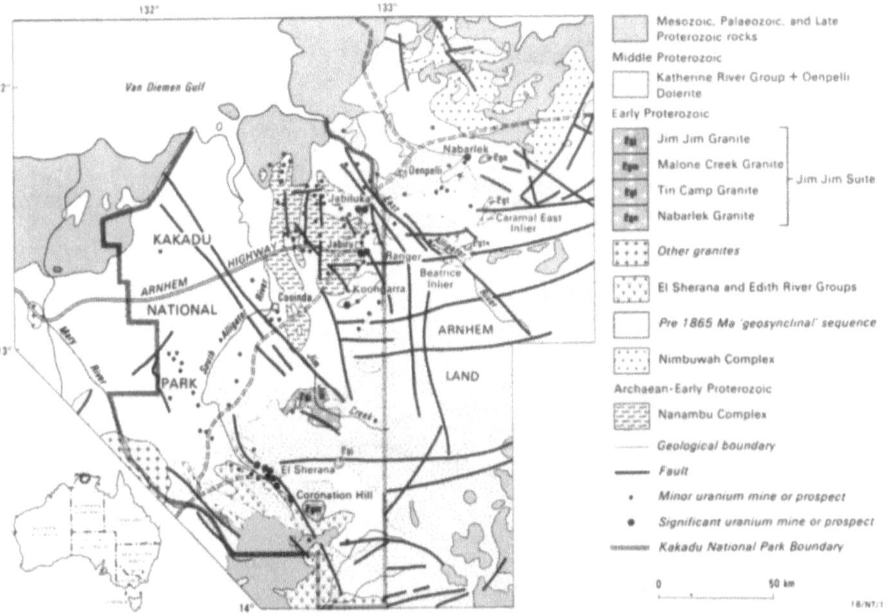

Figure 8. Distribution of major granite suites and uranium deposits in the Kakadu and southwestern Arnham Land region. Note the fault zone along the South Alligator River in the region of the unconformity below the Middle Proterozoic sequence. From BMR 90.

Similarly major faults are important in controlling mineralisation in the Archaean Yilgarn province. It is not clear that the faults are inherited from structures in the older basement to the greenstone sequences, but they are important elements in discussions of lineament tectonics.

Conclusion

Within the limits of the definition adopted, two main phases and types of basement reactivation in Australia have been briefly reviewed. The Cenozoic reactivation exercises control on present day topography and exposure and was itself controlled by the Cenozoic geodynamic scheme and by basement structure. The broad Flexure of the Precambrian Shield on an E-W axis, and the passive-margin development at the Great Dividing Range are most clearly controlled by the Cenozoic plate-tectonic framework. In relation to the broad contrast between the Precambrian Shield and the Tasmanides it seems likely that the crustal responses to Cenozoic reactivation have been influenced by various earlier episodes of lithospheric extension, notably those at the

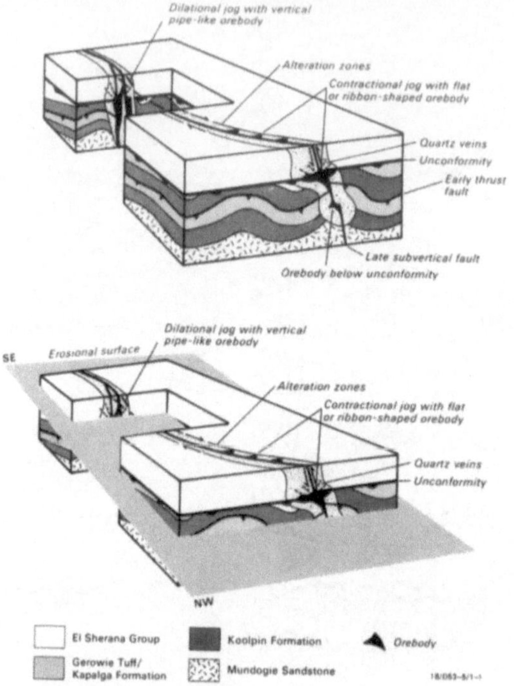

Figure 9. Schematic simplified representation of geological features controlling mineralisation in the South Alligator Mineral Field. The unconformity is between the little deformed felsic volcanics and sandstones of the El Sherana Group and the underlying highly folded and deformed South Alligator Group (both Early Proterozoic). From BMR 90.

beginning and end of the Paleozoic. The areas modified by these episodes, in the Tasmanides and in the Gawler craton have responded to Cenozoic reactivation differently from the Precambrian areas to the west.

The study of the Cenozoic reactivation has been neglected but modern data sets and technologies offer the prospect of rapid progress. The earlier Proterozoic reactivation was responsible for the main block structure of Australia and was also presumably due to a changed geodynamic scheme. The Proterozoic reactivation structures were significant influences on the Cenozoic reactivation. The understanding of primary relationships between basement provinces in Australia depends to a significant extent on the elucidation of these two major phases of reactivation of the craton.

References

BMR Earth Science Atlas of Australia. Various sheets. Bureau of Mineral Resources, Geology and Geophysics, Canberra, Australia.

BMR 90, 1990. Yearbook of the Bureau of Mineral Resources, Geology and Geophysics. 1990. p. 9-15.

Daly, M.C., Cooper, M.A., Wilson, I., Smith, D.G., & Hooper, B.G.D. 1991. Cenozoic plate tectonics and basin evolution in Indonesia. Marine and Petroleum Geology **8**, 2-21.

Etheridge, M.A. 1986. On the reactivation of extensional fault systems. Philosophical Transactions Royal Society of London **A317**, 179-194.

Etheridge, M.A., Symonds, P.A., & Lister, G.S. 1989. Application of the detachment model to reconstruction of conjugate passive margins. In, Tankard, A.J. & Balkwill, H.R. (eds) Extensional Tectonics and Stratigraphy of the North Atlantic Margins. American Association Petroleum Geology Memoir **46**, 23-40.

Falvey, D.A. 1974. The development of continental margins in plate tectonic theory. Australian Petroleum Exploration Journal **14**, 95-106.

Gaull, B.A., Michael-Leibe, M.D., & Rynn, J.Y.W. 1990. Probabilistic earthquake risk maps of Australia. Australian Journal of Earth Sciences **37**, 169-187.

Glaessner, M.F. 1953. Some problems of Tertiary geology in southern Australia. Journal & Proceedings Royal Society New South Wales **87**, 31-45.

Glen, R.A., Laing, W.P., Parker, A.J., & Rutland, R.W.R. 1977. Tectonic relationships between the Proterozoic Gawler and Willyama orogenic domains. Australia. Journal Geological Society of Australia **24**, 125-150.

Goleby, B.R., Shaw, R.D., Wright, C., Kennett, B.L.N., & Lambeck, K. 1989. Geophysical evidence for 'thick-skinned' crustal deformation in central Australia. Nature **337**, 325-330

Hancock, S.L., & Rutland, R.W.R. 1984. Tectonics of an early Proterozoic geosuture: the Halls Creek orogenic sub-province, northern Australia. Journal of Geodynamics **1**, 387-432.

Harrington, H.J., Simpson, C.J., & Moore, R.F. 1982. Analysis of continental structures using a digital terrain model (DTM) of Australia. Bureau Mineral Resources Journal of Australian Geology and Geophysics **7**, 68-72.

Hays, J. 1967. Landsurfaces and laterites in the north of the Northern Territory. In, Jennings, J.N., & Mabbutt, J.A. (eds) Landform Studies from Australia and New Guinea, Cambridge University Press, 182-210.

Hillis, R.R. 1990. Post-Permian subsidence and tectonics, Vulcan sub-basin, NW Shelf, Australia. In, Australian Institute Mining and Metallurgy Pacific Rim Congress **2**, 203-211.

Jaques, A.L., Lewis, J.D., & Smith, C.B. 1986. Kimberlites and lamproites of Western Australia. Geological Survey Western Australia Bulletin **132**.

Jaques, A.L., O'Neill, H.St.C., Smith, C.B., Moon, J., & Chappell, B.W. 1990. Diamondiferous peridotite xenoliths from the Argyle (AKI) lamproite pipe, Western Australia. Contributions Mineralogy and Petrology **104**, 255-276

Johnson, B.D., Mayhew, M.A., O'Reilley, S.Y., Griffin, W.L., Arnott, F., & Wasilewski, P.J. 1986. Magsat anomalies, crustal magnetization and heat flow in Australia. Summary of poster given at 4th Kimberlite Conference, Perth.

Johnson, R.W. (ed.) 1989. Intraplate volcanism in Australia and New Zealand. Cambridge University Press, 408p.

Jones J.G., & Veevers J.J. 1983. Mesozoic origins and antecedents of Australia's Eastern Highlands. Journal Geological Society of Australia **30**, 305-322.

Lambeck, K., McQueen, H.W.S., Stephenson, R.A., & Denham D. 1984. The state of stress within the Australian continent. Annales Geophysicae **2**, 723-742

Lambeck, K., & Stephenson R. 1986. The Post-Palaeozoic uplift history of southeastern Australia. Australian Journal of Earth Sciences **33**, 253-270.

Lister, G.S., & Etheridge, M.A. 1989. Detachment models for uplift and volcanism in the Eastern Highlands, and their application to the origin of passive margin mountains. In, Johnson, R.W. (ed.) Intraplate Volcanism in Eastern Australia and New Zealand, Cambridge University Press p.297-313.

McCulloch, M.T. 1987. Sm-Nd isotopic constraints on the evolution of the Precambrian crust in the Australian continent. American Geophysical Union, Geodynamics Series **17**, 115-130.

McGowran, B., Rutland, R.W.R., & Twidale, C.R. 1977. Discussion:.Age and origin of laterite and silcrete and their relation to episodic tectonism in the mid-north of South Australia. Journal Geological Society of Australia **24**, 421-425.

Moore, M.E., Gleadow, A.J.W., & Lovering J.F. 1986. Thermal evolution of rifted continental margins: new evidence from fission tracks in basement apatites from southeastern Australia. Earth and Planetary Science Letters **78**, 255-270.

Mulcahy, M.J. 1973. Landforms and soils of southwestern Australia. Journal Royal Society of Western Australia **56**, 16-22.

Ollier, C.D. 1978. Tectonics and Geomorphology of the eastern highlands. In, Davies, J.L., & Williams, M.A.J. (eds) Landform Evolution in Australasia, ANU Press, Canberra. p 5-47.

Ollier, C.D. 1982. The Great Escarpment of eastern Australia: tectonic and geomorphic significance. Journal Geological Society of Australia **29**, 13-23.

Ollier, C.D., Gaunt, G.F.M., & Jurkowski, I. 1988a. The Kimberley Plateau, Western Australia. A Precambrian Erosion Surface. Zeitschrift für Geomorphologie **32**, 239-246.

Ollier, C.D., Chan, R.A., Craig, M.A., & Gibson, D.L. 1988b. Aspects of landscape history and regolith in the Kalgoorlie region, Western Australia. Bureau Mineral Resources, Journal of Australian Geology and Geophysics **10**, 309-313.

Ollier, C.D., & Taylor, D. 1988. Major geomorphic features of the Kosciusko-Bega region. Bureau Mineral Resources Journal of Australian Geology and Geophysics **10**, 357-362.

Packham, G.H., & Falvey, D.A. 1971. An hypothesis for the formation of marginal seas in the Western Pacific. Tectonophysics **11**, 79-109.

Page, R.W., McCulloch, M.T., & Black, L.P. 1984. Isotopic record of major Precambrian events in Australia. Proceedings 27th International Geological Congress **5**, 25-72.

Pigram, C.J., Davies, P.J., Feary, D.A., & Symonds, P.A. 1989. Tectonic controls on Carbonate Platform evolution in southern Papua New Guinea; passive margin to foreland basin. Geology **17**, 199-202.

Pigram, C.J., Davies, P.J., Feary, D.A., Symonds, P.A., & Chaproniere, G.C.H. 1990. Controls in Tertiary carbonate platform evolution in the Papuan Basin: New play concepts. In, Carmon, G.J., & Carman, Z. (eds) Proceedings 1st Papua New Guinea Petroleum Convention, p.185-196.

Plumb, K.A., & Gemuts, I. 1976. Precambrian geology of the Kimberley region, Western Australia. 25th International Geological Congress Excursione Guide No.44C, Sydney. 69p.

Rutland, R.W.R. 1981. Structural framework of the Australian Precambrian. In, Hunter, D.R. (ed.) Precambrian of the Southern Hemisphere, Elsevier. p.1-32.

Rutland, R.W.R., Goleby, B.R., Shaw, R.D., & Wright, C. 1988. Tectonic significance of Proterozoic thrusting in Australia: evidence from deep seismic reflection profiling. Geologiska Föreningens i Stockholm Förhandlingar **110**, 410-416.

Shaw, R.D., & Black, C.P. (this volume). The history and tectonic implications of the Redbank Thrust Zone, central Australia, based on structural, metamorphic and Rb-Sr isotopic evidence.

Shaw, R.D., Korsch, R.J., Wright, C., Goleby, B.R., & Collins, C.D.N. (this volume). Thrust tectonics in Central Australia based on the Arunta-Amadeus seismic-reflection profile.

Smith A.G. 1982. Late Cenozoic uplift of stable continents in a reference frame fixed to South America. Nature **296**, 400-404.

Stewart, A.J., Blake, D.H., & Ollier, C.D. 1986. Cambrian River Terraces and Ridge tops in Central Australia : Oldest Existing landforms? Science **233**, 758-761.

Totterdell, J.M. 1990. Notes to accompany a 1:5,000,000 scale Permian structure map of Australia. Bureau of Mineral Resources Australia Record 1990/40.

Tucker, D.H., & D'Addario, G.W. 1986. Albany 1:1,000,000 map sheet. Magnetic Domains. Bureau Mineral Resources Australia.

Tucker, D.H., Wyatt, B.W., Druce, E.C., Mathur, S.P., & Harrison, P.L. 1979. The upper crustal geology of the Georgina Basin region. Bureau Mineral Resources, Journal of Geology and Geophysics **4**, 209-226.

Veevers, J.J. 1990. Antarctic-Australia fit resolved by satellite mapping of oceanic fracture zones. Australian Journal of Earth Sciences **37**, 123-126.

Weissel, J.K., & Karner, J.D. 1989. Flexural uplift of rift flanks due to mechanical unloading of the lithosphere during extension. Journal Geophysical Research **94(BIO)**, 13919-13950.

Weissel, J.K. 1990. Long term erosional development of rifted continental margins: towards a quantitative understanding. Australian Institute Mining and Metallurgy, Pacific Rim Congress **3**, 63-70.

Wellman, P. 1987. Eastern Highlands of Australia, their uplift and erosion. Bureau Mineral Resources, Journal of Australian Geology and Geophysics **10**, 277-286.

Wellman, P. 1988. Development of the Australian Proterozoic Crust as inferred from gravity and magnetic anomalies. Precambrian Research **40/41**, 89-100

Whitaker, A. 1990. In, BMR 90, Yearbook of the Bureau of Mineral Resources, Geology and Geophysics, 1990.

Wopfner, H., & Twidale, C.R. 1976. Geomorphological history of the Lake Eyre Basin. In, Jennings, J.N., & Mabbutt, J.A. (eds) Landform Structures from Australia and New Guinea, Cambridge University Press, p.118-143.

Wopfner, H. 1985. Some thoughts on the post-orgenic development of northeastern South Australia and adjoining regions. Special Publication South Australian Department of Mines and Energy **5**, 265-372.

Young, R.W. 1989. Crustal constraints on the evolution of the continental divide of eastern Australia. Geology **17**, 528-530.

FAULT PATTERNS DURING NORMAL AND OBLIQUE RIFTING AND THE INFLUENCE OF BASEMENT DISCONTINUITIES: APPLICATION TO MODELS FOR THE TECTONIC EVOLUTION OF THE PERTH BASIN, WESTERN AUSTRALIA

D.R. BYRNE AND L.B HARRIS.
Department of Geology
University of Western Australia
Nedlands 6009
Australia.

ABSTRACT. Rift fault patterns are reviewed, and comparisons are made between the expected fault patterns associated with pure extension and transtension, and those found in the Perth Basin, Western Australia. The orientation of the faults in the Perth Basin can be attributed to reactivation of basement structures during sinistral transtension associated with NE-SE extension. A change to dextral transtension is suggested immediately prior to break-up in the Lower Cretaceous. Other features may be explained by permutations of maximum and intermediate stresses, leading to strike-slip movements on NNW striking faults, and basin inversion.

Introduction

In the formulation of tectonic models for the evolution of sedimentary basins, there often appears to be a lack of understanding of the geometry of faults expected in different deformation regimes. Whilst a great deal of emphasis is put on the interpretation of geological cross-sections (eg. in the understanding of detachment faults and the geometry of flower structures in transtensional or transpressional regimes), the pattern of faults in plan view is often rated to be of lesser importance, leading to the formulation of tectonic models which, when viewed in 3D, are incompatible with known deformation responses.

One basin where several models have been invoked to explain the structural evolution, but where discrepancies and contradictions with theoretical data remain, and where the rôle of basement structures in controlling faulting in the cover sequence has not been adequately investigated, is the Perth Basin of Western Australia. The N-S trending Perth Basin extends along the west coast of Western Australia adjacent to the Archaean Yilgarn Craton (Fig. 1). Previous workers have interpreted the Perth Basin as either a typical half-graben along a passive continental margin (Playford et al. 1976) or the product of dextral transtension (Marshall et al. 1989; Stein 1989; Middleton 1990).

Basin formation began in the Silurian and culminated in the Early Cretaceous with the separation of "Greater India" from Australia (Veevers et al. 1975; Powell et al. 1988). A detailed description of the broad geometry and depositional history of the Perth Basin is presented in Playford et al. (1976), and a summary is also given in Marshall et al. (1989). The eastern margin of the Perth Basin is marked by a narrow, 1000km long, generally N-S trending fault zone referred to as the "Darling Fault" along which up to 15 km of normal movement has been reported (Glickson & Lambert 1973; Playford et al. 1976).

High-grade Proterozoic gneisses of the N-S trending Darling Mobile Belt (Trendall & Peer

M. J. Rickard et al. (eds.), Basement Tectonics 9, 23–42.

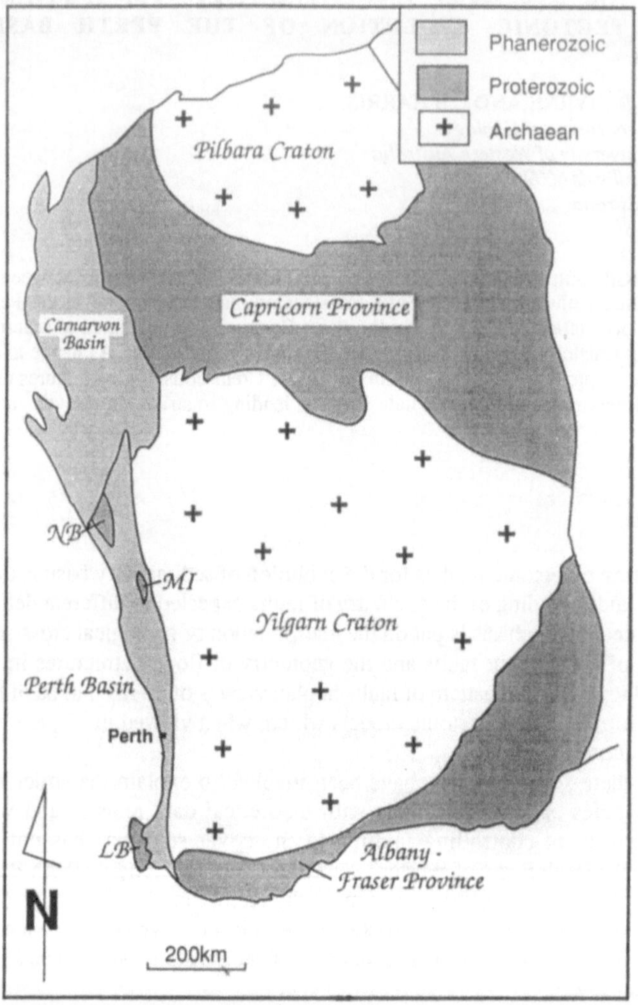

Figure 1. Location map,Western Australia. Features shown by abbreviations are the Northampton Block (NB), Leeuwin Block (LB), and Mullingarra Inlier (MI). (Adapted from Harris 1987).

1975; Fletcher et al. 1985; Harris 1987) underlie the Perth Basin. Basement outcrops in three fault-bounded blocks; the Northampton Block, Leeuwin Block and Mullingarra Inlier (Fig.1). Shear zones (in Archaean rocks) controlling the location of the Darling Fault crop out near the western margin of the Yilgarn Craton.

This paper reviews the two-dimensional geometry of faults in plan view expected during different styles of rifting, and considers also the features resulting from basement reactivation. The compatibility of fault patterns within the Perth Basin with these models, and the influences of basement structure on the localisation and orientation of faults, are then examined, leading to a re-evaluation of the broad tectonic evolution of the Perth Basin.

Structures developed during rifting

Rifts are down-faulted blocks or grabens which involve crustal and/or lithospheric thinning due to horizontal extension (Park 1988). Whilst intra-continental rifting may mark the first stage in the formation and separation of two lithospheric plates, and the formation of new oceanic crust, some rifts may be arrested early and not develop beyond an "immature" stage without oceanic crust involvement. These 'immature' rifts, such as the Baikal, East African and Oslo Rifts (Park 1988), provide information on geometries of faults under known boundary conditions. Rifts vary in style from simple fault-bounded troughs to multi-segmented grabens with numerous junctions and dog-legs, such as the East African Rift (Lowell 1985). The structures associated with rifts and grabens have been described in detail by Lowell (1985), Park (1988) and Etheridge et al. (1988), thus only a brief summary will be presented here. As we are only concerned here with the two-dimensional (plan view) pattern of faults, lithospheric models for extension are not considered.

The orientation and sense of movement on faults during rifting is a function of the following factors:

(i) the orientation of the general plate movement vector with respect to the rift zone, i.e. whether plate separation is orthogonal or oblique to the rift axis;

(ii) changes in the direction of extension;

(iii) whether there has been a permutation between principal-stress components s_1 and s_2 (where $s_1 > s_2 > s_3$) with continued, constant extension direction leading to changes in the deformation response;

(iv) whether the fault sets have monoclinic symmetry resulting from a stress-field where $s_1 > s_2 > s_3$ (eg Anderson 1951), or orthorhombic symmetry where either $s_1 \approx s_2 > s_3$ (extension) or $s_1 > s_2 \approx s_3$ (compression) (Reches & Dieterich 1983; Reches 1983; Krantz 1988).

(v) reactivation of basement structures and local perturbations from the regional stress field due to pre-existing mechanical anisotropies (i.e. previous faults or ductile shear zones or distinct differences in basement lithology).

GEOMETRY OF FAULTS RESULTING FROM EXTENSION ORTHOGONAL TO THE RIFT AXIS

The simplest case of graben formation is where the general trend of a graben is perpendicular to the direction of plate separation, ie. perpendicular to the regional extension direction. Figure 2a summarises the geometrical relationships between the various types of faults expected in this situation. An example of a rift system involving orthogonal extension is the Gulf of Suez, where the fault geometry involves relay faults trending NNW, parallel to the gulf, and dog-leg faults trending NW and NNE (Angelier & Bergerat 1983; Lowell 1985; Chénet et al. 1987). Relay (i.e. en relais, stepping), or longitudinal faults, are the main faults developed during orthogonal rifting. These show normal movement and strike perpendicular to the regional extension direction, and therefore parallel to the axis of the graben (Lowell 1985).

Transverse, or transfer, faults form as accommodation structures which link, and generally trend perpendicular, to the relay faults (Lowell 1985). The transverse faults are analogous to transform faults in oceanic rifts, as they transfer movement between normal relay faults via displacements involving a dominantly transcurrent component (Lowell 1985; Etheridge et al. 1988). Transverse faults are particularly common where listric normal faults have developed (where rotational or scissor movements may occur), and may separate domains where normal faults have opposing dips.

Dog-leg, or oblique, faults are also accommodation structures, but usually strike about 30° to relay faults. Although they form obliquely to the extension direction they may show normal dip-slip movement (Lowell 1985), or oblique-slip movement. Chénet et al. (1987) found that the pitch of slickenside striae on dog-leg faults in the Suez Gulf varied greatly, even where those on the relay faults were essentially 90°. Their formation is not fully understood, with a number of models being presented in Lowell (1985), but they are probably formed during axially symmetric extension as orthorhombic fault sets where s_1 and s_2 switch axes (Reches & Dieteric 1983; Reches 1983; Krantz 1988).

Extension, without any transcurrent movement, usually produces a series of relay faults parallel to the overall trend of the graben. It was found from experimental modelling by Lowell (1985) that the extension mechanism determined whether or not dog-leg faults would form. Arching of a clay model, with the axis of the arch parallel to the graben, produced relay faults without any dog-leg faults, whereas pull-apart with subsidence usually resulted in the formation of dog-leg faults. Dog-leg faults link segments of normal faults to form faults of greater length, in the equivalent manner to the linking of extensional fractures by shear fractures, as commonly observed during rock failure experiments.

Gentle to open folds may form during extension usually as rollover anticlines associated with (and parallel to) listric normal faults (Gibbs 1984; Lowell 1985; Cooke & Harris 1987). Broad synclines may develop to accommodate movement on ramps where faults describe a ramp-flat style.

FAULT GEOMETRIES DEVELOPED DURING TRANSTENSION

When the overall trend of the rift zone is oblique to the direction of relative plate motion (ie. oblique to the regional extension direction), transcurrent movement along the rift zone accompanies extension. Possible orientations and movement senses of faults developed during transtension are shown in Figures 2b and 2c.

Transtension may be accommodated by either of two methods. Firstly, composite structures, generally parallel to the overall graben trend, may be formed by the linking of short segments of orthogonal relay (T shear in Figs 2b & 2c) and transfer faults (eg. Rosedale Fault in Etheridge et al. 1985). The density of transfer faults increases with decreasing angle between the extension direction and the main graben trend. Transfer faults form due to difficulty in developing continuous normal faults along the whole rift length during crustal transtension (Etheridge et al. 1985). This may also result in a series of smaller grabens, each with their axes perpendicular to the extension direction, arranged in an en échelon manner along the overall trend of the main rift.

The more common geometry comprises two oblique-slip fault sets, which are not perpendicular, and are geometrically asymmetric about the graben axis (Etheridge et al. 1988). This situation is best described by wrench geometry (Tchalenko 1968), where the fault set parallel to the major graben axis is parallel to the principal displacement (D) orientation, and the obliquely trending faults correspond to the Riedel (R) orientation (Figs 2b & 2c). Faults with only normal movement can form in the T orientation.

Experimental modelling has provided a better understanding of factors controlling the development of faults in transtension. Experiments by Richard (1990) show that in the case of dextral transtension where there were equal components of normal and dextral transcurrent reactivation of a 45° dipping normal fault, normal- and strike-slip movement took place both along structures striking parallel to the controlling basement fault and also along structures akin to Riedel shears (R), clockwise to the main graben trend. En échelon structures clockwise to the main trend

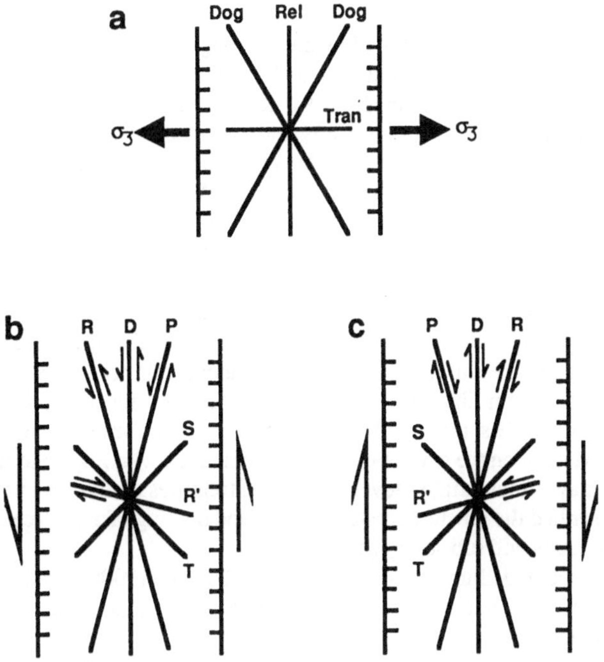

Figure 2. (a) Fault geometry within rifts associated with pure extension: Rel=relay fault, Dog=dog-leg fault, Tran=transverse fault, s_3=extension direction (Adapted from Lowry, 1985); and with (b) sinsitral and (c) dextral transtension: D=principal displacement shears parallel to boundaries of rift zone, R & R'=Reidel shears, P='pressure' shear, T=normal faults, S=fold axes and reverse faults (Nomenclature from Tchalenko 1968).

are best developed where a lower ductile layer in the cover sequence is thin or absent.

Transtensional rifts are commonly formed over divergent basement wrench faults. In cross-section, negative flower structures may develop in the overlying basin sediments (Harding 1985). Development of helicoidal R faults in the basin sediments can result in sub-grabens forming obliquely to the main graben (Naylor et al. 1986). Pull-apart basins in dilational jogs may also form sub-grabens, such as the Dead Sea, which formed due to stepping of a sinistral wrench fault (Chénet et al. 1987). The width of the zone affected in a cover sequence, due to movement on a divergent wrench fault in the basement, depends on the amount of decoupling along sedimentary layers, such as evaporites; the width increases as decoupling increases (Richard 1990).

Some instances of transtensional rifting have involved reactivation of faults in the basin, which formed during orthogonal extension in an earlier stage of graben development. The Upper Rhine Graben, for example, originated during orthogonal extension, and was later subjected to a sinistral shear couple as indicated by the en échelon stepping of grabens, and overprinting of horizontal on dip-slip slickensides on some of the faults (Park 1988). The result is a predominance of faults parallel and asymmetrically oblique, by about 25° anticlockwise, to the rift (Park 1988).

As with pure extension, folds can develop due to roll-over on normal faults, and a central synform may characterise negative flower structures above a divergent wrench fault.

REACTIVATION

Many rifts are at least partially controlled by pre-existing structural weaknesses in the crust, such as those in Africa where both normal and strike-slip reactivation of Precambrian basement structures have been observed (Daly et al. 1989). Rifts in east Africa commonly parallel the trends of Precambrian mobile belts and follow the boundaries of the older cratons (Park 1988). In the Suez Rift, Chénet et al. (1987) concluded that the orientation of longitudinal and dog-leg faults was controlled by older fractures in the Precambrian shield. The sinistral fault associated with the Dead Sea Basin is also a reactivated structure (Chénet et al. 1987).

Reactivation of pre-existing faults in a cover sequence and reactivation of basement faults or shear zones both induce fault patterns which are atypical of either transtensional or extensional rift systems. Angelier and Bergerat (1983) have shown that many "classical" rifts have undergone an early stage of strike-slip faulting, and that the "zig-zag" pattern of structures is due to normal reactivation of former strike-slip faults. As described above, this occurs with a constant orientation of the maximum extension direction. This idea was further developed by Masse (1983) who concluded that whilst rifts may represent reactivation of former transcurrent faults (with reactivation dependent upon the stress field and the orientation with respect to heterogeneities in the basement), rifts may preferentially develop at the extremities of former transcurrent structures. Masse (1983) described the greatest crustal thinning and doming taking place at the intersection of divergent wrench faults and subsidiary grabens.

From experimental modelling on fault reactivation, Richard (1990) concludes that:

(i) reactivation of faults in the cover occurs more easily when the reactivated basement fault has a wrench component;
(ii) given a basal ductile layer, the reactivation of faults in the cover can occur far from the generating basement fault;
(iii) faults produced during initial dip-slip deformation strongly influence patterns of superimposed strike-slip deformation at depth;
(iv) dip-slip faults are commonly reactivated as strike-slip faults, however, at shallower depth, reactivation is less common due to lower cohesion.

CHANGES IN THE MODE OF FAULTING DUE TO STRESS PERMUTATIONS DURING RIFTING

In the discussions above, it is apparent that normal faults are usually thought to be the dominant structures developed in rifting due to regional crustal extension, especially in orthogonal rifting. Normal faults develop in a stress field where s_2 and s_3 are horizontal (with faults striking perpendicular to s_3) and s_1 vertical. However, a common (though often overlooked) feature of many rifts is the permutation of s_1 and s_2, during which the orientation of maximum extension, s_3, remains constant (Angelier & Bergerat 1983). This stress permutation leads to a change from normal faulting to conjugate strike-slip faulting mode. Pre-existing faults oblique to the extension direction may therefore be re-activated with oblique-slip displacements depending on their angle to s_1. Permutation of s_1 and s_2 may also occur during the development of orthorhombic fault sets where $s_1 \approx s_2$ (Reches & Dieterich 1983; Reches 1983; Krantz 1988).

Permutation between s_1 and s_3 may also take place during rifting even though the plate movements are still indicative of regional extension. This is seen only at a late stage in rifting where oceanic crust is being formed at a new spreading centre: the thermally derived topographic elevation of a young ridge induces compression perpendicular to the spreading direction (termed the "ridge-push" force). This has been observed by Bergerat et al. (1990) in Iceland where both normal and strike-slip faulting developed under a maximum horizontal compressive stress perpendicular to the axis of spreading.

Faulting within the Perth Basin

It is beyond the scope of this article to provide a detailed review of the evolution of the Perth Basin; only salient points are summarised, with reference to previously published reviews. Additional information is provided in the environs of the Northampton Block as this area provides indications as to the relative timing of fault and joint sets; the importance of reactivation of basement features in this terrain on the orientation of faults in the basin is discussed below.

Sedimentation in the northern Perth Basin began in the Silurian with the deposition of the Tumblagooda Sandstone (Playford et al. 1976; Hall 1989; Marshall et al. 1989). This unit is usually in faulted contact with the basement of the Northampton Block, but is also seen to overlie the block with an angular unconformity, as at Malarang Park (Fig. 5). At this stage, sedimentation in the Perth Basin was coeval with that in the Carnarvon Basin to the north. The main phase of development of the Perth Basin occurred from the Permian to the Cretaceous, with the deposition of non-marine and marine sediments (Playford et al. 1976; Marshall et al. 1989). NW separation of the Indian sub-continent from Australia is believed to have taken place during the Early Cretaceous at about 120Ma-135Ma (Powell et al. 1988; Marshall et al. 1989), thus marking the onset of oceanic crust formation.

Due to poor exposure and weathered outcrop, there is a lack of detailed structural information regarding the faults in the basin, and no conclusive evidence for components of strike-slip movement. Playford et al. (1976) assumed that movement on the faults was normal. The continuity and trends of the offshore faults are uncertain because most of the information regarding tthem has been obtained from seismic records with widely spaced traverse lines (eg. Marshall et al. 1989).

FAULT PATTERN IN THE PERTH BASIN

Figure 3a shows the fault pattern in the Permian to Early Cretaceous sediments of the Perth Basin to be asymmetric about the basin axis, being dominated by N-S striking faults, parallel to the trend of the basin, and NNW trending faults. Exceptions to this pattern include two E-W trending faults south-east of the Northampton Block, and the NE striking Hardabut Fault, which forms the northern boundary of the Perth Basin. Figure 3b also shows this asymmetry in terms of the number of straight-line fault segments, although it does show symmetry about the N-S basin axis in relation to the variation in trend of the faults. The Darling Fault, along the eastern boundary of the basin, generally trends N-S with flexures varying from 340°-005°. Transform faults in the oceanic crust off the west coast of Western Australia, trend NW (Fig. 3b) rather than NNW (as shown by Veevers & Cotterill 1978; Powell et al. 1988), which suggests that they are probably not genetically or geometrically related to the NNW trending faults in the basin sediments.

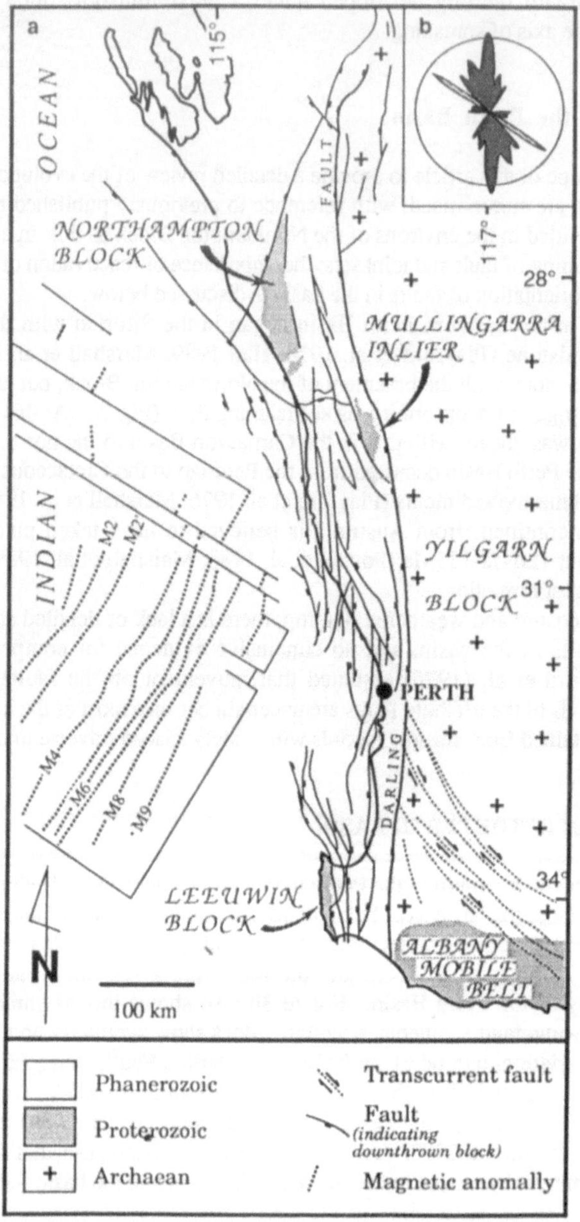

Figure 3. (a) Map of the Perth Basin showing faults and basement blocks. (Compiled from Jones 1976; Warris 1988; Powell et al. 1988; Hall 1989; and Marshall et al. 1989); (b) Rose diagram showing variations in the trends of faults in the Perth Basin, measured from (a); dark grey: basin faults, n=116, interval=5°, radius=15%; light grey: transform faults, n=9, interval=5°, radius=95%.

PREVIOUS MODELS FOR RIFTING IN THE PERTH BASIN

The style of rifting involved in the Perth Basin is uncertain; while Playford et al. (1976) assumed orthogonal extension, Warris (1988), Marshall et al. (1989), Stein (1989) and Middleton (1990) have presented models involving components of dextral movement on N-S faults. Warris (1988) believed that when Greater India separated from Australia, a minor dextral wrench component reactivated faults, especially those trending NNW, producing roll-over anticlines in the hangingwalls. (It should however be noted that it is not necessary to invoke dextral transtension to form the roll-over anticline, since normal movement is sufficient).

The underlying phenomenon suggestive of dextral transtension is the NW separation of India and Australia as shown by the pattern of magnetic anomalies and positions of transform faults (Fig. 3) following break-up at 120Ma. There are still problems in establishing when such a tectonic regime may have commenced since the transform faults and magnetic anomalies give no indication of the stress regime prior to the Early Cretaceous (120 Ma). Marshall et al. (1989) present seismic data from three sub-basins in the offshore northern part of the Perth Basin, which they believe have been in dextral transtension throughout most of their history since the Permian. The sub-basins, and faults within them, trend NNW, and are separated by NW trending wrench faults, presumed to be dextral. They have interpreted negative and positive flower structures in some of the seismic sections, which they attribute to wrench movements on some of the original basin faults previously formed by ENE extension in the Permian. They also recognise numerous narrowly spaced transform faults south of the Wallaby-Perth Scarp (Zenith-Wallaby Fracture Zone), which they consider to be the result of oblique extension of the crust prior to sea-floor spreading.

The general pattern of faults in the Perth Basin, shown in Figure 3, is not consistent with purely orthogonal rifting, and is opposite to that predicted from models of faulting during dextral transtension based on experimental modelling and comparison with other basins as described above.

In attempting to apply a model for classical orthogonal rifting to the Perth Basin, Figure 3b suggests that the symmetrical nature of the trends about N-S in the fault pattern (ignoring the transform faults) may have resulted from E-W extension producing N-S relay, NNW and NNE dog-leg, and E-W transverse faults. However, the E-W faults in the basin do not show signs of transcurrent movement and are depicted by Playford et al. (1976) as southward dipping normal faults in which case they may not be true transverse faults.

N-S dextral transtension, according to Figure 2c, should result in NNE trending faults rather than the NNW trending faults shown in Figure 3a. Both of these models fail to explain the dominance of NNW trending faults and the occurrence of the E-W faults that do not appear to be transform faults. N-S sinistral transtension can explain the dominance of NNW over NNE trending faults, but not the presence of the EW trending faults.

FAULTING EVENTS IN THE NORTHERN PERTH BASIN ADJACENT TO THE NORTHAMPTON BLOCK

Two main faulting events, followed by another extensional event, are evident from examination of the sediments of the Perth Basin around the Northampton Block. The first event involves tilting and down-faulting of the Silurian Tumblagooda Sandstone into the basement. Normal faults associated with this event mostly strike NE (Fig. 4a), such as those at Malarang Park and the Hardabut Fault, whereas the Yandi Fault trends N-S (Fig. 5). The faults at Malarang Park appear to be planar with steep dips, whereas the Hardabut Fault (as described by Prider 1958), is a listric normal fault, and not a thrust as proposed by Playford et al. (1976). The gentle anticline in the Tumblagooda Sandstone by the Hardabut Fault represents a roll-over structure developed in extension and does not indicate a compressive phase. The angular unconformity at Malarang Park between the Tumblagooda Sandstone and Lower Jurassic Greenough Sandstone suggests that the faulting predates the Early Jurassic, and nearby flat-lying Permian sediments reported by Playford et al. (1976) indicate a pre-Permian faulting event.

A second faulting event is inferred from a set of normal faults occurring in the railway and road cuttings at Bringo (Fig. 5). These faults displace basement gneisses, and terminate upward in the Middle Jurassic Bringo Shale, indicating syn-depositional movement during the Middle Jurassic. They strike NW to NNW and E-W (Fig. 4b), and have normal displacements of about 1m. Predominantly normal movement is indicated by striations on one of the faults (Fig. 4b).

The strikes of joints in the Tumblagooda Sandstone and Lower Jurassic Greenough Sandstone have been plotted in the rose diagrams shown in Figs 4c and 4d. There is a wide range in the trends of joints in the Tumblagooda Sandstone, but poorly defined E-W and NNW trends are

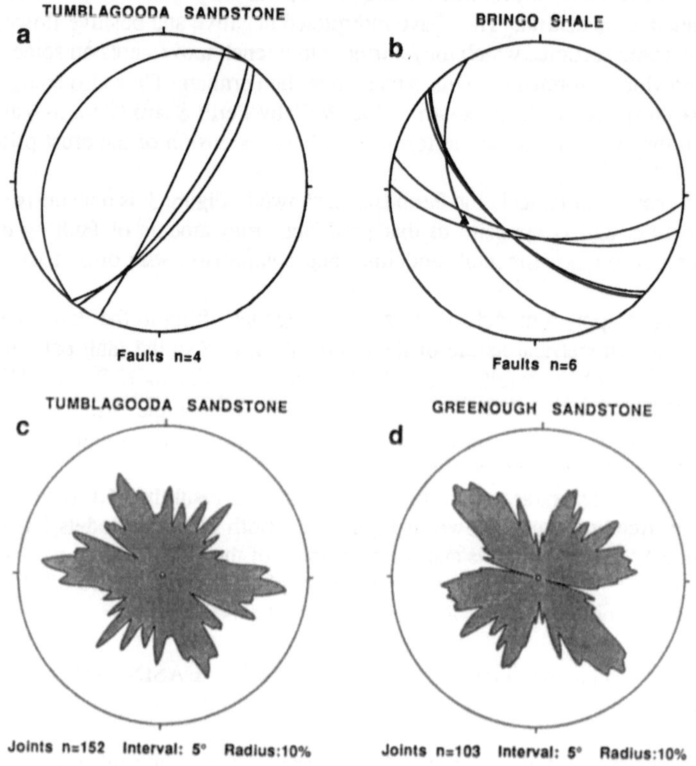

Figure 4. Equal area stereographic projections of faults and striations, from around the Northampton Block, in (a) Silurian Tumblagooda Sandstone and (b) Middle Jurassic Bringo Shale. Moving-average rose diagrams of joints in (c) Tumblagooda Sandstone and (d) Lower Jurassic Greenough Sandstone (dips are subvertical).

modally significant. These joint sets are parallel to the E-W and NNW trending faults in the Bringo Shale (Fig. 4b), which suggests that the joints formed during the Middle Jurassic.

The Greenough Sandstone is dominated by a NW trending set of joints (Fig. 4d), which may be related to the NW trending transform faults formed during rifting in the Early Cretaceous (cf. Fig. 3b). The NNE and ENE trending joint sets in Figure 4d, which are obtusely bisected by the NW joint set,and acutely bisected by a NE joint set, may be a conjugate set also related to this third event.

These three events indicate that the development of the Perth Basin is more complex than previously thought. In summary, there appear to have been (i) NW extension after, and probably during, the Silurian, (ii) NE extension during the Middle Jurassic, and (iii) NW extension during the Early Cretaceous. However, while the E-W faults appear to have been active during the Middle Jurassic, NE extension would not normally produce faults in such an orientation. Therefore, NE extension alone cannot explain their formation.

Influence of basement structures on faulting within the Perth Basin

The inconsistencies between observed and expected fault geometries for the various styles and models of rifting for the Perth Basin, indicate that another factor may be involked to adequately explain the fault pattern. That factor is is the effect of basement structures on the overlying basin.

REACTIVATION OF THE PROTO-DARLING FAULT

The Darling Fault, a fault zone 1000km long and dipping 50°-80°W, is a primordial feature of Perth Basin architecture, and is generally considered to have dominated the development of the Perth Basin, forming its eastern margin with the Archaean Yilgarn Block. The location of the Darling Fault, however, is controlled by pre-existing structural features, referred to by Blight et al. (1981) as the Proto-Darling Fault. The current western margin to the Yilgarn Block has been active since the Archaean with movements documented throughout the Late Proterozoic, Palaeozoic and Mesozoic (Veevers & Johnstone 1974; Bretan 1985; Harris 1987). The long, linear nature of the Darling Fault reflects pre-existing mylonite zones developed during transcurrent ductile shearing: early dextral movements in the Archaean are preserved in the south, with subsequent reactivation and overprinting by Archaean sinistral movements. The importance of this zone as a major crustal feature and an idea of the magnitude of transcurrent displacements along it are indicated by Late Proterozoic deformation. Sinistral transcurrent movements along the Proto-Darling Fault have bent E-W trends of the Albany Mobile Belt (Fig. 3a) into a N-S orientation indicating a minimum of 30km of transcurrent displacement (Bretan 1985). High-grade Late Proterozoic to possibly early Cambrian gneisses in the Leeuwin Block also developed within the N-S sinistral transcurrent regime (Harris 1987). The southern sector of the Darling Mobile Belt therefore appears to represent a sinistral transcurrent shear zone indicating a southwards movement of Greater India relative to the Yilgarn craton at this time. Reverse faulting along this zone has also been recorded by Bretan (1985) and is thought to have taken place during an ESE compression event at about 500Ma (Harris et al. 1989). Basement to the southern part (at least) of the Perth Basin therefore contained a strong N-S structural anisotropy and it can be seen that the western margin of the Yilgarn Block represented a major crustal weakness prior to rifting. Veevers & Johnstone (1974) also suggest that the NW trending transform faults, such as the Wallaby-Perth Scarp, reflect original lines of weakness in the Precambrian crust.

BRITTLE -DUCTILE STRUCTURES IN THE NORTHAMPTON BLOCK

The basement in the northernmost sector of the Perth Basin is exposed in the Northampton Block (Figs 1 & 5). An analysis of structures has been carried out in this block using low level, digitally processed aeromagnetic data (1:25000 scale, recording height 70m, with 150m line spacing) and aerial photography (1:50,000 and 1:40,000 scales) combined with field mapping and structural analyses.

The Northampton Block consists mainly of granulite facies paragneisses, which were folded and metamorphosed at about 1000Ma, and a syntectonic porphyritic granite which has been dated by Rb-Sr methods at about 900Ma to 1150Ma (Prider 1958; Compston & Arriens 1968; Playford et al. 1970; Warren 1973; Hocking et al. 1982; Richards et al. 1985). These are cross-cut by a 550Ma to 750Ma (K-Ar dating by Embleton & Schmidt 1985) dolerite-dyke swarm described by Prider (1958), Warren (1973), Gibson (1974) and Mauger (1978). The dykes are displaced by a set of NNW and E-W trending shears. Hydrothermal Cu-Pb-Zn mineralisation overprints the dolerite dykes (Blockley 1971,1975; Marston 1979) and predates deposition of the Silurian Tumblagooda Sandstone (Fletcher et al. 1985). Alteration of the dolerite dykes by the mineralising fluids has been dated at 434±16Ma (Silurian) by Richards et al. (1985).

The Dolerite-Dyke Swarm : NE striking dolerite dykes, usually about 15m wide and up to at least 28km long, constitute a swarm across the Northampton Block (Fig. 5). Spacing between the dykes is irregular varying from 60m to 2000m with a modal average of 375m. On close examination of aerial photographs and aeromagnetic data, the dykes can be seen to change strike and bifurcate, with segments and splays varying in trend from NNE to ENE (Fig. 6a). The splays usually contain smaller dykes. The larger width of the NE striking dykes, compared to the NNE and ENE trending splays, indicates that they probably correspond to tension fractures resulting from NW-SE extension, whereas the smaller dykes appear to occupy a conjugate set of shears acutely bisected by the larger NE trending dykes.

NNW and E-W trending Shear Zones: NNW striking lineaments (first recognised by Prider 1958) occur at sinistral, horizontal offsets of the dolerite dykes, and are accompanied by a less well-developed E-W trending set showing dextral horizontal offsets (Figs. 5 & 6b). The NNW trending lineaments, visible on aeromagnetic images and aerial photographs, are up to 27km long and are spaced 2km-5km apart. They characteristically bifurcate, and in places change strike by up to 30°, producing the variation in trend shown in Figure. 6b. In the field, these lineaments correspond to ductile shear zones containing mylonitised gneiss and pegmatite, 1m-35m wide. Shear criteria within the mylonites, and the regional drag of foliation into the shear zones, are indicative of normal movement. Given the dips of these shear zones and of the dykes which they displace, between 0.9km and 2km of vertical displacement would be required along these mylonite zones in order to give the observed horizontal offsets. Alternatively, later brittle strike-slip reactivation may have taken place; however, no such structures have been observed (but limited outcrop and erosion along these lineaments does not completely preclude this possibility).

Joints : The wide variation in the orientation of the joints measured in the gneisses and dolerite dykes of the Northampton Block is illustrated in Figure 7. Depending on area (Fig. 5), the gneisses are generally dominated by NW, N-S, NE and E-W trending joint sets. Dolerite dykes contain joint sets trending NW and NE, with the latter being similar in orientation and spread to the dykes themselves (cf. Fig. 6a). Unfortunately, the timing of formation of these joints in the gneisses and dolerite dykes is uncertain, but may reflect events associated with the development of the Perth Basin during the Palaeozoic and Mesozoic.

CROSS-CUTTING BRITTLE DUCTILE SHEAR ZONES IN THE LEEUWIN BLOCK

In addition to the high-grade ductile deformation referred to above, the Leeuwin Block is dissected by a network of brittle-ductile and brittle shear zones. Prominent ENE (dextral) and NW (sinistral) trending structures can be interpreted from contoured aeromagnetic maps and Landsat images. Gneisses are cross-cut by a network of joints and shear fractures showing little offset parallel to

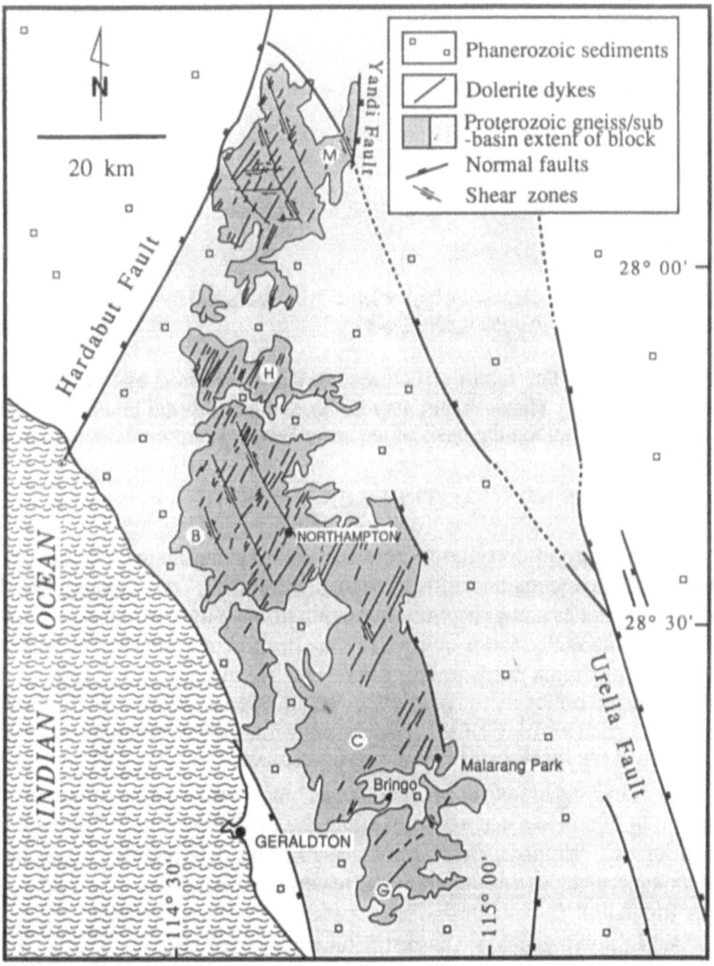

Figure 5. Map of the Northampton Block showing dolerite dykes, NNW and E-W shears, and sub-areas for Figure 8: M=Murchison River inlier, H=Hutt River inlier, B=Bowes River area, C=Chapman River area, G=Greenough River inlier.

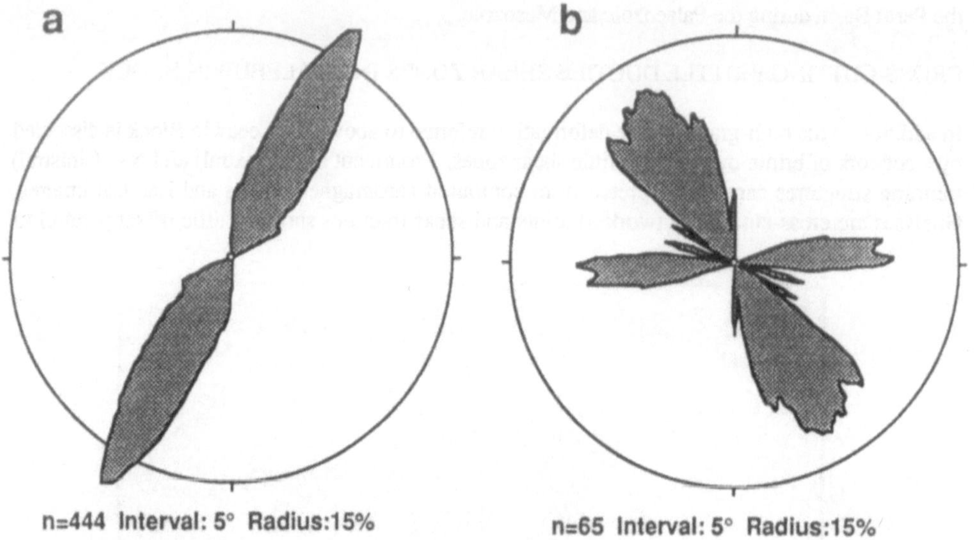

Figure 6. Moving-average rose diagrams of (a) dolerite dyke trends and (b) trends of the NNW "sinistral" and ENE "dextral" faults in the Northampton Block.

these regional features. (The major structures are rarely exposed as they have controlled the position of watercourses). These shears may be correlated with the reverse displacement event along the Proto-Darling Fault on the basis of the same WNW interpreted orientation of s_1.

CONTROLS OF BASEMENT STRUCTURES ON FAULTING

There appears to be a good correlation between the orientations of the basement structures described above and the orientations of faults in the Perth Basin. This is exemplified by the control of Precambrian shear zones along the western margin of the Yilgarn Craton on the localisation of the Darling Fault. This fault zone is oblique to the direction of plate separation at break-up, and thus to the regional maximum extension direction (Fig. 3). In a sinistral transtensional model, the Darling Fault originated obliquely to the implied NE direction of rifting. It remains however to be determined whether such a strong anisotropic zone has rotated the regional stress field at the time of rifting such that, at the local scale, s_3 was perpendicular to the basement shear zones. Detailed analysis of fault-plane striations (now commenced) may provide the answer to this problem. Minor reorientation of stresses across major structural discontinuities is a feature of the current stress distribution (C. Windsor, pers. comm. 1990), and it is therefore possible that such a phenomenon took place across this major crustal feature at earlier times.

NNW trending faults in the Perth Basin parallel similarly trending shear zones in the Northampton and Leeuwin Blocks. These structures may have been reactivated with dominantly normal displacements under NE extension, but would have been transtensional features during NW extension (with s_1 vertical). Most of the dolerite dykes parallel NE trending faults, such as the Hardabut Fault, involved in down-faulting the Tumblagooda Sandstone during the Palaeozoic. It is uncertain whether the fracturing event, involving intrusion of the dolerite dykes, or that

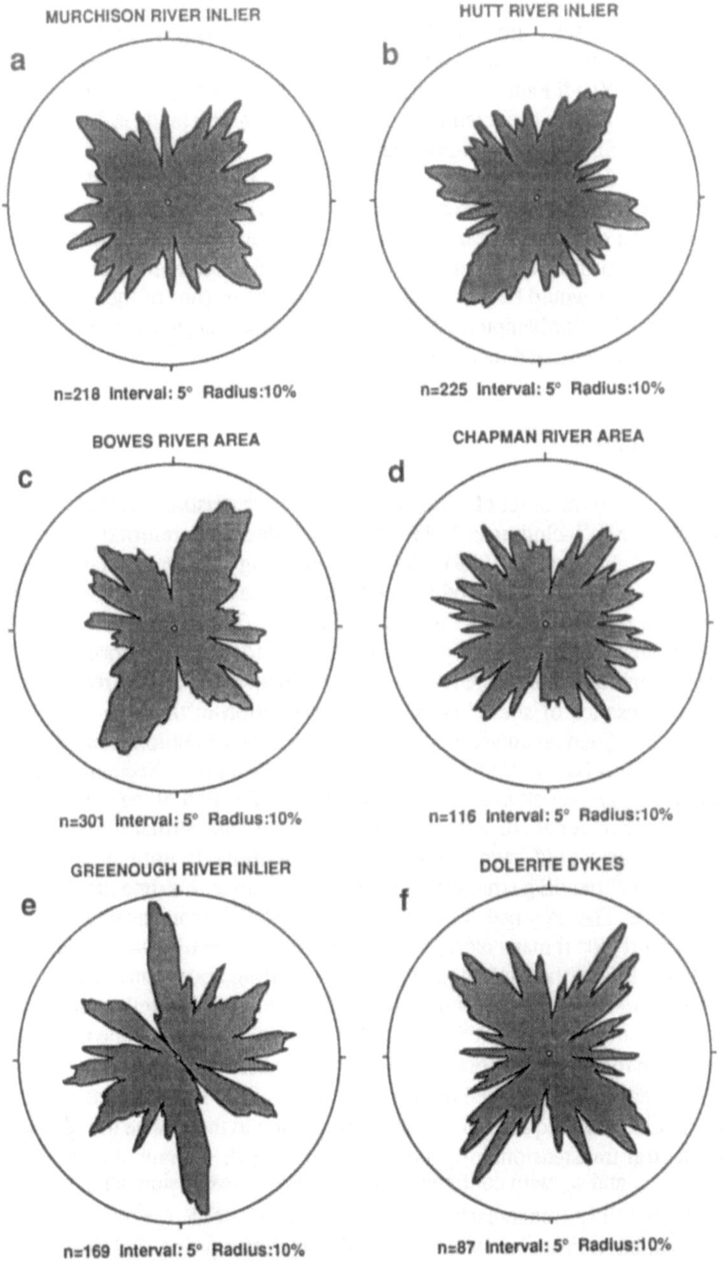

Figure 7. Moving-average rose diagrams of joints occurring in the gneisses (a-e) and dolerite dykes (f) of the Northampton Block. See Figure 5 for locations of areas.

involving the Hardabut Fault, was responsible for the NE trending joints found in the gneisses and dolerite dykes. The E-W striking faults southeast of the Northampton Block, which have been ignored in previous models, are parallel to the E-W trending dextral shear zones in the Northampton Block. Joints trending N-S in the gneisses may be associated with the formation of N-S faults, such as the Yandi Fault. Since the timing of formation of the joints in the gneisses and dolerite dykes is uncertain, the event which produced N-S trending faults is difficult to date, except that it is probably post-Silurian and pre-Permian.

At the south end of the basin ENE trending faults in the Leeuwin Block do not appear to have been reactivated, at least on a regional scale. Reactivation of these structures in addition to the NW ones would have been expected under orthogonal rifting or dextral transtension, and basement anisotropy alone is not sufficient to explain the asymmetrical pattern of faulting in the Perth Basin. No reactivation however would be expected of these structures (nor of those controlling dolerite dyke emplacement in the Northampton Block) during sinistral transtension as they almost parallel the implied maximum extension direction (ie. NE to ENE).

Tectonic Implications

Veevers (1984) portrayed the onset of rifting off northwestern Australia in the Cambro-Ordovician as a series of 120° arms including a NNW trending failed arm, referred to as the Westralian Aulacogen (Finkl & Fairbridge 1979). Whereas faulting in the Silurian suggests NW-SE extension, NE-SW extension as part of a failed rift may have controlled sedimentation in the northern Perth Basin from the Permian through the Jurassic. The NNW trend in the Perth Basin is reflected in the reconstructions of Falvey & Mutter (1981) for the Late Jurassic (Fig. 8) prior to a change in stress orientation before break-up. The development of NNW trending structures was facilitated by the presence of shear zones in this orientation in the basement (as seen in the Northampton Block). Such an aulacogen model is indicative of approximate NE-SW extension across the northern Perth Basin. Under this stress system, the strong Precambrian structural grain along the western Yilgarn Block margin controlled the localisation of the generally N-S striking Darling Fault. Under regional NE-SW extension, sinistral and normal movement components along the Darling Fault would have eventuated, with dominantly normal, and minor sinistral, movement on the NNW trending structures; ie. the Perth Basin at this time developed in a sinistral transtensional regime. The presence of N-S sinistral offsets of Devonian and Permian strata of the Carnarvon Basin on regional maps along the the Williambury and Lockwood Faults (eg. Myers and Hocking 1988), the northern prolongations of the Darling Fault Zone, adds further evidence for this interpretation. NE-SW extension in the Perth Basin may also reflect oblique extension at this time between Australia and Antarctica, when Greater India was still pinned to Antarctica.

A major change in the regional stress field to NW-SE extension is indicated during final break-up in the Early Cretaceous, as reflected in the pattern of sea floor magnetic anomalies in Figure 3a (although further studies are required to date the exact timing in the change in regional stress field). A change to a dextral transtensional regime can be envisaged. Permutation between principal-stress components, s_1 and s_2, with continued constant NW-SE extension has given rise to dextral transcurrent, and possibly transpressional, reactivation of NNW striking faults, and may be responsible for local basin inversion immediately prior to and/or during the Neocomian break-up as outlined by Stein (1989). Dextral NNW and NW striking faults cutting the Yilgarn Block and Albany Mobile Belt also developed at this time, in part reactivating Precambrian structures in this

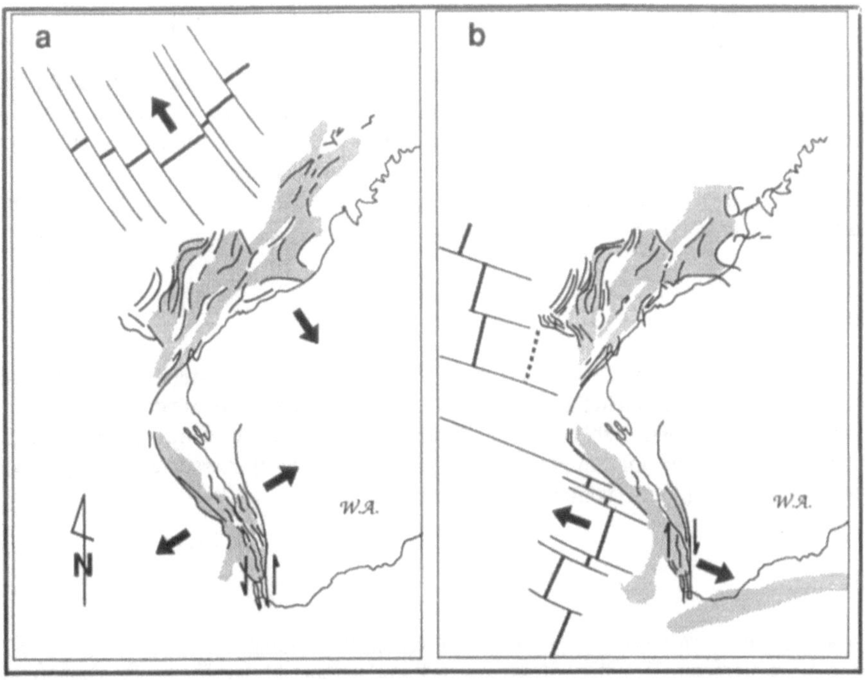

Figure 8. Major faults and thick rift-phase sedimentary accumulations in Western Australia (from Falvey & Mutter 1981) showing interpreted regional extension and transcurrent movement components, for (a) Late Jurassic and (b) Early Cretaceous. WA=Western Australia.

orientation. Evidence for post-Neocomian dextral movement along these structures comes from a quarry in the Bunbury Basalt, in the southern Perth Basin, which is cut by a NNW striking fault with horizontal striations where microstructural criteria indicate dextral displacement.

Conclusions

Experimental modelling and a review of areas where plate movements have been well documented indicates that distinct fault patterns are produced in rifting from which the regional tectonic regime may be interpreted. This study has shown that, by comparison with these fault geometries, previous tectonic models of orthogonal rifting or dextral transtension do not adequately explain the gross pattern of faulting within the Perth Basin. The overall geometry of the Perth Basin suggests that there was an initial phase of deposition in the Silurian, for which there remains uncertainty in the regional tectonic regime. Sinistral transtension due to oblique rifting along a failed arm from a triple point took place during late Palaeozoic-Early Mesozoic evolution. Basement reactivation provided a major control on the localisation and orientation of faults. A change in the regional tectonic regime towards the onset of "break-up" in the Lower Cretaceous is implied by the pattern of sea-floor magnetic anomalies indicating NW separation of Greater India and Australia. This has lead to a dextral transtensional regime, with local basin inversion. Permutation of s_1 and s_2 is thought to be responsible for the development of transcurrent faults in the basin and the Yilgarn Block and the Albany Mobile Belt.

Reactivation of basement structures in the above model can account for the various strikes of the basin faults, but does not fully explain the style of extension during the various stages of rifting, nor the precise orientation of, and changes in, the stress field. Further detailed examination of movements on faults in the Perth Basin and determination of the palaeo-stress tensors on the basis of fault-plane striation analysis (currently in progress) are now required to fully document the structural evolution evolution of this major crustal rift.

Acknowledgements

The study of the Northampton Block was funded by West Australian Metals NL. This project forms part of ongoing research in the Darling Mobile Belt and Perth Basin funded by the Australian Research Council. John Ashley from Southern Geoscience Consultants processed the aeromagnetic data of the Northampton Block which was flown by Aerodata Holdings Pty Ltd. John Beeson is thanked for his comments on an earlier draft.

References

Anderson, E.M. 1951. The Dynamics of Faulting. Oliver and Boyd, Edinburgh, 206p.

Angelier, J., & Bergerat, F. 1983. Stress systems and continental extension. In, Poppoff M., & Tiercelin, J. (ed.) Ancient Rifts and Troughs, Tectonics-Volcanism-Sedimentation Contribution of Actualism. Symposium of the French National Centre of Scientific Research (CNRS), Marseilles, Nov 30-Dec 2, 1982. 137-147.

Bergerat, F., Angelier, J., & Villemin, T. 1990. Fault systems and stress patterns on an emerged oceanic ridge: Iceland as a case example. Tectonophysics **179**, 183-197.

Blight, D.F., Compston, W., & Wilde, S.A. 1981. The Logue Brook Granite: age and significance of deformation zones along the Darling Scarp. Western Australian Geological Survey Annual Report 1980, 72-80.

Blockley, J.G. 1971. The lead, zinc and silver deposits of Western Australia. Western Australian Geological Survey Mineral Resources Bulletin **9**, 34-106.

Blockley, J.G. 1975. Lead and copper deposits of the Northampton Block. In, Knight C.L. (ed.) Economic Geology of Australia and Papua New Guinea Vol. 1, Metals. Monograph Series 5, Australasian Institute of Mining & Metallurgy, Melbourne. 409-410.

Bretan, P.G. 1985. Deformation processes within mylonite zones associated with some fundamental faults. PhD Thesis, Imperial College, (unpublished).

Chénet, P.Y., & Letouzey, J. 1983. Tectonics of the area between Abu Durba and Gebel Mezzazat (Sinai, Egypt) in the context of the evolution of the Suez rift. In, Poppoff, M., & Tiercelin, J. (eds) Ancient Rifts and Troughs, Tectonics-Volcanism-Sedimentation Contribution of Actualism. Symposium of the French National Centre of Scientific Research (CNRS), Marseilles, Nov 30-Dec 2, 1982, 201-215.

Chénet, P.Y., Colleta, B., Letouzey, J., Desforges, G., Ousset, E., & Zaghloul, E.A. 1987. Structures associated with extensional tectonics in the Suez rift. In, Coward, M.P, Dewey, J.F., & Hancock, P.L. (eds) Continental Extensional Tectonics. Geological Society of London, Special Publication **28**, 551-558

Compston, W., & Arriens, P.A. 1968. The Precambrian geochronology of Australia. Canadian Journal of Earth Sciences **5**, 561-583.

Cook, A., & Harris, L.B. 1987. Analogue modelling experiments of structuring during normal and oblique extension. Extended Abstracts - Applied Extension Tectonics. Bureau of Mineral Resources Record 1987/51, 116-124.

Daly, M.C., Chorowicz, J., & Fairhead, J.D. 1989. Rift basin evolution in Africa: the influence of reactivated steep basement structures. In, Cooper, M.A., & Williams, G.D. (eds) Inversion tectonics. Geological Society London, Special Publication **44**, 309-334.

Embleton, J.J., & Schmidt, P.W. 1985. Age and significance of magnetizations in dolerite dykes from the Northampton Block, Western Australia. Australian Journal of Earth Sciences **32**, 279-286.

Etheridge, M.A., Branson, J.C., & Stuart-Smith, P.G. 1985. Extensional basin forming structures in Bass Strait and their importance for hydrocarbon exploration. Australian Petroleum Exploration Association Journal **27**, 344-361.

Etheridge, M.A., Symonds, P.A., & Powell, T.G. 1988. Application of the detachment model for continental extension to hydrocarbon exploration in extensional basins. Australian Petroleum Exploration Association Journal **28**, 167-187.

Falvey, D.A , & Mutter, J.C. 1981. Regional plate tectonics and the evolution of Australia's passive continental margins. Bureau of Mineral Resources Journal of Australian Geology and Geophysics **6**, 1-29.

Finkl, C.W., & Fairbridge, R.W. 1979. Paleogeographic evolution of a rifted cratonic margin: S.W. Australia. Palaeogeography, Palaeoclimatology, Palaeoecology **26**, 221-252.

Fletcher, I.R., Wilde, S.A., & Rosman, K.J.R. 1985. Sm-Nd model ages across the margins of the Archaean Yilgarn block, Western Australia - lll. The western margin. Australian Journal of Earth Sciences **32**, 73-82.

Gibbs, A.D. 1984. Structural evolution of extensional basin margins. Journal of the Geological Society of London **141**, 609-620.

Gibson, A.A. 1974. Dolerite geochemical study, Northampton district, W.A. Western Australian Geological Survey Record 1974/1, 19.

Glickson, A.Y., & Lambert, I.B. 1973. Relations in space and time between major Pıecambrian shield units: An interpretation of Western Australian data. Earth and Planetary Science Letters **20**, 395-403.

Hall, P.B. 1989. The future prospectivity of the Perth Basin. Australian Petroleum Exploration Association Journal **28**, 440-449.

Harding, T.P. 1985. Seismic characteristics and identification of negative flower structures, positive flower structures, and positive structural inversion. American Association of Petroleum Geologists Bulletin **69**, 582-600.

Harris, L.B. 1987. A tectonic framework for the Western Australian Shield and its significance to gold mineralisation: a personal view. In, Ho, S.E., & Groves, D.I. (eds) Recent Advances in Understanding Precambrian Gold Deposits. Geology Department & University Extension, University of Western Australia, Perth, 1-27.

Harris, L.B., Delor, C.P., Beeson, J., & Standing, J.G. 1989. A major Palaeozoic event affecting Precambrian terrains of Australia and east Antarctica: implications for continental- scale crustal shortening. In, Proceedings of the Kangaroo Island Tectonics Conference 6-10 February 1989, Specialist Group in Tectonics & Structural Geology, Geological Society of Australia, Abstracts **24**, 65-66.

Hocking, R.M., van de Graaff, W.J.E., Blockley, J.G., & Butcher, B.P. 1982. 1:250000 Geological Series - Explanatory Notes for Ajana, Sheet SG/50-13. Western Australian Geological Survey.

Krantz, R.W. 1988. Multiple fault sets and three-dimensional strain: theory and application. Journal of Structural Geology **10**, 225-237.

Lowell, J.D. 1985. Structural Styles in Petroleum Exploration. O.G.C.I. Publications, Tulsa.

Marshall, J.F., Lee, C., Ramsay, D.C., & Moore, A.G. 1989. Tectonic controls on sedimentation and maturation in the offshore north Perth Basin. Australian Petroleum Exploration Association Journal **28**, 450-465.

Marston, R.J. 1979. Copper mineralisation in Western Australia. Western Australian Geological Survey Mineral Resources Bulletin **13**, 147-151.

Masse, P. 1983. The possible relations between grabens, wrench faults and uplifting in the intracontinental rifts. In, Poppoff, M., & Tiercelin, J. (eds) Ancient Rifts and Troughs, Tectonics-Volcanism-Sedimentation Contribution of Actualism. Symposium of the French National Centre of

Scientific Research (CNRS), Marseilles, Nov 30-Dec 2, 1982, 149-154.
Mauger, A.J. 1978. Computer analysis of the fracture pattern of the Northampton Block, Western Australia, with relation to Pb-Zn mineralisation. B.Sc.(Hons) Thesis, University of Western Australia (unpublished).
Middleton, M.F. 1990. Analysis of wrench tectonics in the Perth and Canning Basins, Western Australia. Western Australian Geological Survey Basin and Fossil Fuel Report No 3/1990 (unpublished).
Myers, J.S., & Hocking, R.M. 1988. 1: 1 000 000 geological map of Western Australia. Geological Survey of Western Australia.
Naylor, M.A., Mandl, G., & Sijpesteijn, C.H.K. 1986. Fault geometries in basement-induced wrench faulting under different initial stress states. Journal of Structural Geology **8**, 737-752.
Park, R.G. 1988. Geological Structures and Moving Plates. Blackie & Son, Glasgow.
Playford, P.E., Cockbain, A.E., & Low, G.H. 1976. Geology of the Perth Basin, Western Australia. Western Australian Geological Survey Bulletin **124**.
Playford, P.E., Horwitz, R.C, Peers, R., & Baxter, J.L. 1970. 1:250,000 Geological Series - Explanatory notes for Geraldton, Sheet SH/50-1. Western Australian Geological Survey.
Powell, C.M., Roots, S.R., & Veevers, J.J. 1988. Pre-breakup continental extension in East Gondwanaland and the early opening of the eastern Indian Ocean. Tectonophysics **155**, 261-283.
Prider, R.T. 1958. The granulites and associated rocks of Galena, Western Australia. Australasian Institute of Mining and Metallurgy Stillwell Anniversary Volume, 189-211.
Reches, Z. 1983. Faulting of rocks in three-dimensional strain fields II: theoretical analysis. Tectonophysics **95**, 133-156.
Reches, Z., & Dieterich, J.H. 1983. Faulting of rocks in three dimensional strain fields I: failure of rocks in polyaxial, servo-control experiments. Tectonophysics **95**, 111-132.
Richard, P. 1990. Champs de failles au dessus d'un décrochement de socle. Memoires et documents du Centre Armorican d'étude structurale des socles **34**, 342 p.
Richards, J.R., Blockley, J.G., & de Laeter, J.R. 1985. Rb-Sr and Pb isotope data from the Northampton Block, Western Australia. Australasian Institute of Mining and Metallurgy Proceedings **290**, 43-55.
Stein., A.M. 1989. Geometry of an oblique slip rifted margin: Perth Basin, Western Australia. Lecture program and abstracts, Tectonic studies group Christmas meeting 1989, Imperial College of Science, Technology and Medicine, University of London (unpublished), 58.
Tchalenko, J.S. 1968. The evolution of kink-bands and the development of compression textures in sheared clays. Tectonophysics **6**, 159-174.
Trendall, A.F., & Peers, R. 1975. Precambrian rocks beneath Phanerozoic basins. Western Australian Geological Survey Memoir **2**, 217-220.
Veevers, J.J. 1984. Phanerozoic Earth History of Australia. Oxford monographs on geology and geophysics **2**, Oxford Science Publications, Oxford.
Veevers, J.J., & Cotterill, D. 1978. Western margin of Australia: Evolution of a rifted arch system. Geological Society of America Bulletin **89**, 337-355.
Veevers, J.J., & Johnstone, M.H. 1974. Comparative stratigraphy and structure of the Western Australian margin and the adjacent deep ocean floor. Deep Sea Drilling Project Initial Report **27**, 571-586.
Veevers, J.J., Powell, C.McA., & Johnson, B.D. 1975. Greater India's place in Gondwanaland and in Asia. Earth and Planetary Science Letters **27**, 383-387.
Warren, H. 1973. The geology of the Northampton Area, Western Australia. B.Sc.(Hons) Thesis, University of Western Australia (unpublished).
Warris, B.J. 1988. The geology of the Mount Horner Oilfield, Perth Basin, Western Australia. Australian Petroleum Exploration Association Journal **28**, 88-99.
Wernicke, B., & Burchfiel, B.C. 1982. Modes of extension tectonics. Journal of Structural Geology **4**, 105-115.

THE PINE CREEK SHEAR ZONE, NORTH OF PINE CREEK (NORTHERN TERRITORY): STRUCTURAL EVOLUTION AND EXPERIMENTAL STUDIES

J. KROKOWSKI & S. OLISSOFF
9 Day Ave
Rostrevor S.A. 5073
Australia

ABSTRACT. A series of experiments on clay, plastic clay and bread dough have been performed in an attempt to explain the deformation in the Pine Creek Shear Zone (PCSZ) north of Pine Creek. The NNW-N trend and contrary plunging F1 fold system to the east and west of the PCSZ were the principal factors causing the deformation of the 15°-30° clockwise-inclined foliation in the zone. In the experiments these factors were simulated by complex deformation: simple coaxial compression together with rotational faulting in the basement.

A model for the deformation in the zone was derived, from which the genesis and geometry of the foliation, and subhorizontal attitude of the b1 lineation in the zone could be determined. It could also suggest a sinistral transpression during the deformation. The pattern, growth mechanism and evolution of structures (folds, cleavage, faults, thrusts) was examined in the experiment. The model corresponds to the D1 stage of the PCSZ evolution. Structures of the Darwin-Pine Creek-Katherine wrench system were subsequently superimposed on the already existing form of the foliation.

Introduction

The Early Proterozoic sediments and volcanic rocks of the Pine Creek Geosyncline in the region of Pine Creek (Fig. 1) were deformed, metamorphosed and intruded by granitic rocks during the Top End Orogeny at 1870-1780 Ma (Page et al. 1980). Regional metamorphism is generally of low-grade (greenschist) facies. The Pine Creek Shear Zone (PCSZ) originated as a granite embayment between two lobes of the Cullen Batholith (Stuart-Smith et al. 1987). Vertical and later strike-slip motions were probably the main components of the movement in the Lower Proterozoic metasediments between the lobes. A very characteristic feature is the NNW-N trending foliation which is different from the cleavage in the adjoining region to the west and east of the zone. The PCSZ belongs to the wrench regime of the Darwin-Pine Creek-Katherine Fault System (Stuart-Smith et al. 1986, 1987). Summaries of the distribution of Early Proterozoic metasediments and the geology of the Pine Creek Inlier have recently been described by Fergusson (1980), Johnston (1984), Needham et al. (1980, 1988), Stuart-Smith et al. (1980, 1985, 1986, 1987), Nicholson and Eupene (1984), and others.

The geology of the Pine Creek Shear Zone

Lower Proterozoic metasediments of the Pine Creek Shear Zone between Boomleera Siding and Mt Wells (BS-MtW) and adjoining regions (Figs 1 & 2) are affected by two phases of regional folding (Johnston 1984, Stuart-Smith et al. 1986, 1987). The major folds (F1) are moderately tight to

M. J. Rickard et al. (eds.), Basement Tectonics 9, 43–53.

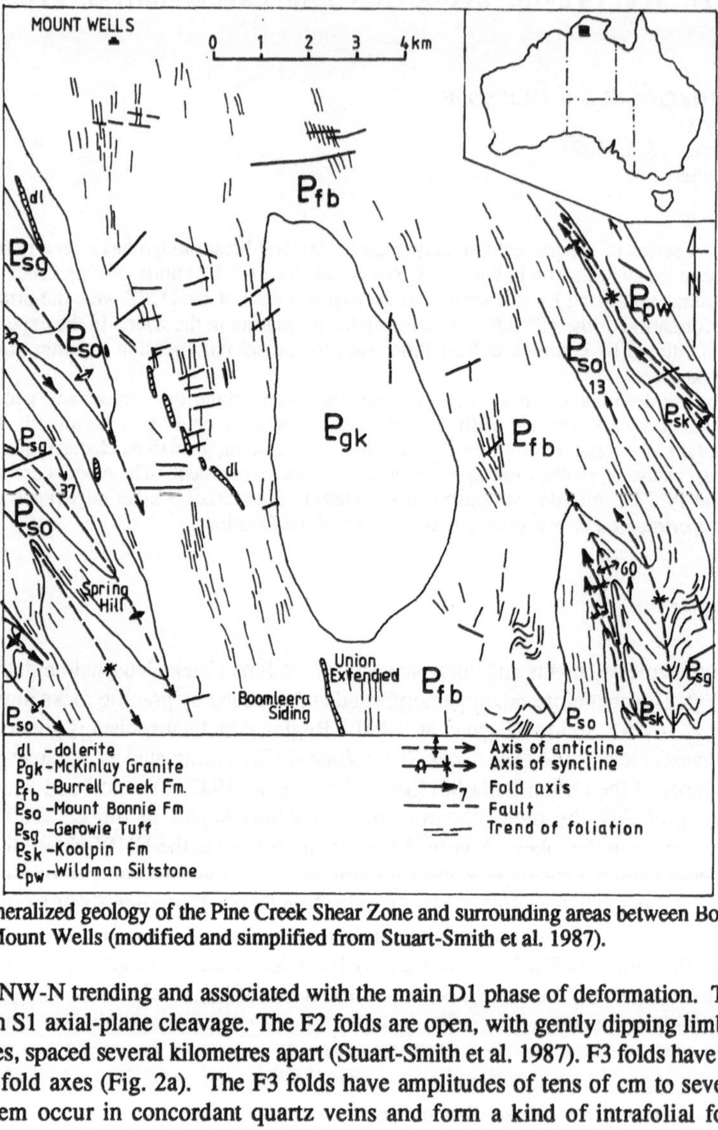

Figure 1. Generalized geology of the Pine Creek Shear Zone and surrounding areas between Boomleera Siding and Mount Wells (modified and simplified from Stuart-Smith et al. 1987).

isoclinal, NNW-N trending and associated with the main D1 phase of deformation. This folding produced an S1 axial-plane cleavage. The F2 folds are open, with gently dipping limbs and E-W trending axes, spaced several kilometres apart (Stuart-Smith et al. 1987). F3 folds have dominantly subvertical fold axes (Fig. 2a). The F3 folds have amplitudes of tens of cm to several metres; some of them occur in concordant quartz veins and form a kind of intrafolial folding with amplitudes of tens of centimetres. The F3 folds are associated with the NNW-N trending flexural-slip movements or faults. The movement was parallel to foliation and mostly dextral; the folds are probably drag structures related to the shearing of the right-lateral strike-slip component of the PCSZ.

In the BS-MtW area there is a strong dominance of a subvertical and vertical foliation in the NNW-N direction (Fig. 2). The foliation cross-cuts the bedding, and in: the Spring Hill anticline and structures immediately to the east the S1 foliation in the zone is inclined clockwise about 15°-30°. East of the PCSZ and the McKinlay Granite, within rocks older than the Burrell Creek

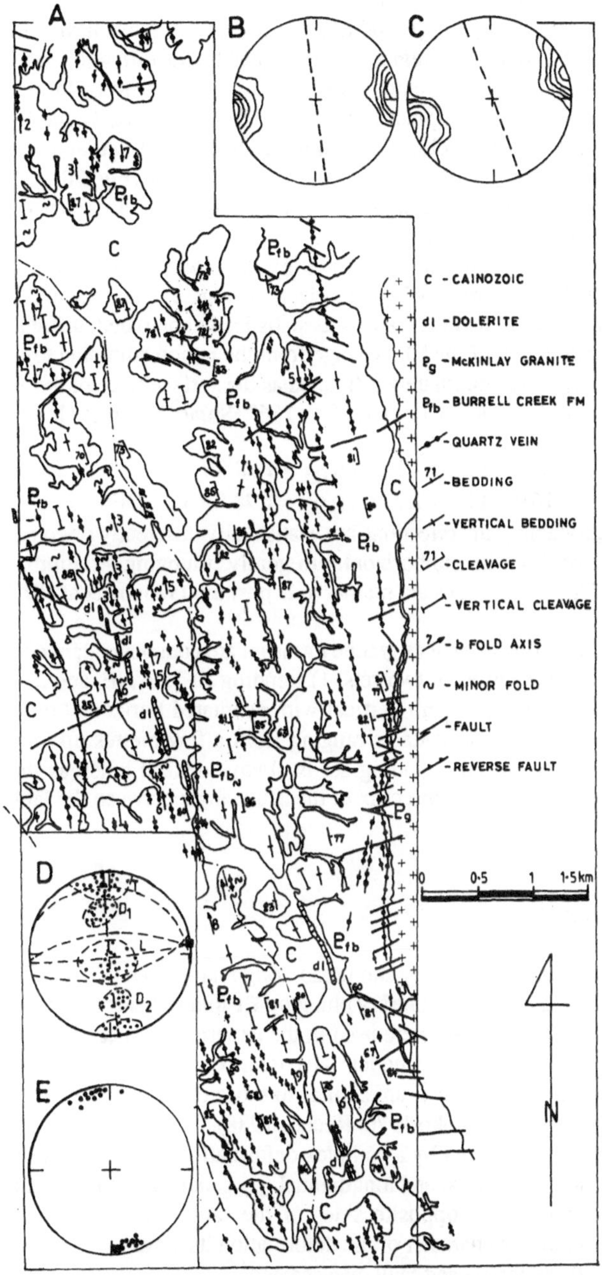

Figure 2A. Geological map of the Pine Creek Shear Zone between Boomleera Siding and Mount Wells; diagrams: B, C. Contoured foliation poles in the north and south part of the area respectively; D. Poles to joints (square = pole to average bedding); E.b lineations in the area.

Formation, the F1 folds plunge to the N and NNW (Fig. 1). Whereas to the west of the PCSZ the F1 folds trend NW-NNW and plunge to SE and SSE. These two regions have similar trending, although contrary plunging F1 folds. The regions are separated by the differently oriented foliation of the PCSZ. Well-developed joints oriented approximately normal to bedding are probably temporally related to the F1 folds. In places, the joints form two sets: L and T as extension fractures in bc and ac planes respectively. Commonly, an additional two sets, D1 and D2 are present as conjugate hkO fractures enclosing an acute angle about a_1 (Fig. 2d). The transverse sets can in places be subdivided into hkO conjugate hybrid fractures T_1+T_2 (Hancock 1985) with a small dihedral angle (Muehlberger 1961).

Within the BS-MtW area the b1 lineations formed by the intersection of the bedding or S1 cleavage and L longitudinal joints are mostly subhorizontal (Fig.2e). The mean direction of the foliation from the BS-MtW part of the PCSZ area was statistically compared, using the Student T-test (Gren 1970), with the direction of the S1 cleavage from the region adjoining to the west of the zone. The F1 fold axes may be approximately estimated as the surface trace of the S1 cleavage in the region. The direction of axes ranges from 138° to 156° (Stuart-Smith et al. 1987). The test results are as follows:
BS-MtW north part: $|S| = 175°$ $|t| = 17.7 > t_{0.05} = 1.98$ for 113 DF
BS-MtW south part: $|S| = 161°$ $|t| = 7.7 > t_{0.05} = 1.98$ for 118 DF
and show that with 0.95 probability the directions are statistically differently oriented.

Many strike-slip slickensides with subhorizontal to gently plunging striations were formed along foliation surfaces. The slickensides are mainly in D position (Wilcox et al. 1973) in relation to the PCSZ.

Post-dating the previous faulting as well as granite intrusion are complementary conjugate strike-slip or oblique-slip faults and fractures of two sets: 1)-trending ENE to NE, and 2)-trending ESE to SE (Fig. 2). The faults are steep or sub-vertical with estimated horizontal displacements of quartz veins or granite-metasediment contacts ranging from a few metres up to tens or rarely hundreds of metres. The ENE to NE trending set is always dextral whereas the ESE to SE trending set is dominantly sinistral. The bisectrix of the acute angle between the fault sets is in an easterly direction.

Description of experiments

A series of experiments on clay, plastic clay and bread dough have been performed to model the deformation in the PCSZ and explain the contrary plunging F1 fold systems to the east and west of the PCSZ. Although, different materials were used in the experiment, special attention has been paid to results of the clay experiment. Clay models have long been used for analysis of faulting as the fault patterns produced in a clay model are similar to those observed in the field (e.g. Cloos 1928, Riedel 1929, Cloos 1955, Tchalenko 1970, Reches 1988).

The observed fold pattern was caused by a complex deformation comprising a simple coaxial compression and folding, and a rotational and longitudinal (perpendicular to the compression) fault in the basement. This fault caused opposite sides of the fault walls to plunge in contrary directions. Because of the complex character of the deformation its magnitude is always shown as an average sum of both fold shortening and fault displacement.

The deformation apparatus was built for samples up to 25 cm wide and 37 cm long (Fig. 3). In a typical experiment the material is molded into the apparatus space up to 3 cm high. The bottom

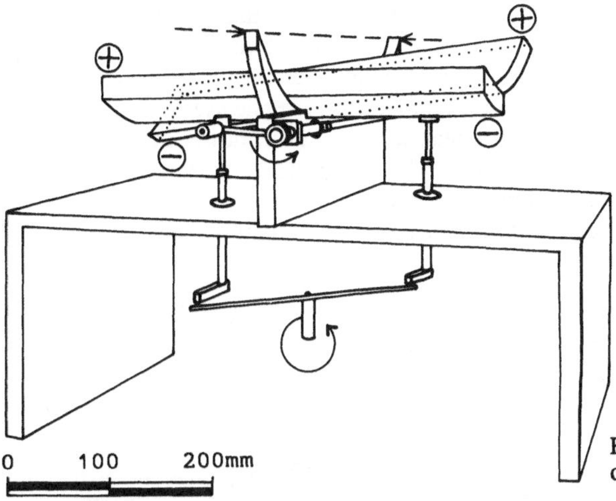

Figure 3. The deformation frame of the experimental apparatus.

side of the apparatus was a rubber sheet with longitudinal metal ribs and edges. The sheet was fixed to the frame in its centre to allow complex folding to occur. Bending of the major fold was up to 45 causing the shortening up to 35%, while the variation in a plunge of limbs in opposite directions caused by the fault was up to 25°. The experiments were analysed in situ , and some of them were videotaped for later analysis. The experiment was performed with a 4 x 10^{-4} sec^{-1} strain rate produced by hand operation of special screws (Fig. 3). However, some experiments were slowed down and generated with a 2 - 4 x $10^{-5}sec^{-1}$strain rate. Only the central 1/2-1/3 part of the layer (Part B, Fig. 4c) has been analysed, as in marginal parts (A and C) the basement rotational fault caused the layer continuity to be broken.

Results of experiments

A geometric description of the evolution of fold and fault pattern observedin the PCSZ and margins is documented here; there is no intention to derive the quantitative parameters. It was found that the growth mechanisms and pattern of evolution were similar in all experiments, even though the patterns differed in detail from one experiment to the other. There were differencies in the deformation between the clay, plastic clay and dough. The experiment on the clay and plastic clay better reflects faulting while those with dough give more expression to folding.

During the experiment the deformation was inhomogenously distributed in thelayer (part B). The effect of compression and fault rotation was the strongest in a narrow central zone of the model (Fig. 4b). The zone is 5-7 cm wide in the clay and 6-9 cm wide in dough. The zone is inclined from the longitudinal direction (the direction perpendicular to the maximum compression). The inclination is 15-40° (Figs. 4, 5 & Pl. 1) and depends mainly on the material. However, an experiment with only basement rotational faulting without compression produced the largest inclination in the range 45-50°.

The strike-slip component of displacement in the deformed layer was shown by the movement of marked lines (Fig. 4c). As a first approximation, the magnitude of the strike-slip component depends on the angle of plunge caused by contrary rotation on the main fault in the basement. In the experiment, sinistral transpression took place mainly in the deformation zone, and decreased on the limbs of the primary syncline. Initial sinistral displacement of marked lines was continuous (ductile). As strain increased, however, the displacement was mainly on faults. During the experiment folds were formed after 6-9 % of the deformation. The fold axes were 20°-40° inclined from the Y-Y line (Fig. 4), and were both linear and curviinear, however, the curvilinearity increased only in some stages of the deformation. The folds were from a few mm to several tens of mm long. Some folds grew longer during the experiment, and some divided into new ones. The growing folds were the active, dominant structures and account for a large amount of the deformation, especially in later stages.

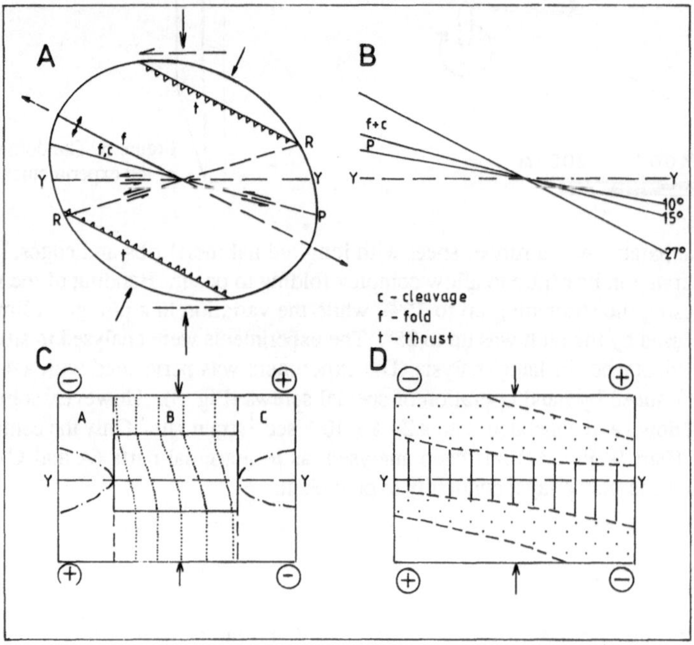

Figure 4A. Diagram of structure orientations in the deformed material; B. average angles between structural trends; C. Deformation layer; D. Distribution of strain in analysed part from maximum (lines) to minimum (blank).

Fractures of cleavage density were formed after and during the fold formation. Some of the fractures developed into faults. Usually one set of faults was formed, however, in some cases fractures of different orientatio were observed (Fig. 4). The main set of faults was dominantly reverse and parallel to the axial plane of the folds. Some are also strike-slip or oblique reverse. The faultswere from 1 mm to a few or several tens mm long, however, some were longer. The initial length of new faults is unknown, as faults shorter than 1 mm were very difficult to see (Reches, 1988). The strike-slip displacements were greatest in the central part of the zone and ranged from 1 mm or smaller to a few mm Dip-slip displacements were harder to measure, but they were probably also of a similar range. The faults were more commonly associated with synclines. In the clay and plastic-clay experiments, the faults were formed earlier than in the dough.

In some cases other sets of fractures occured. One of them (R) seems to be symmetrical to a previous one (Fig. 4a). The fractures were very weak however, sometimes they had a cleavage. Very rarely, especially in the end parts of the deformation zone, fractures parallel to the Y-Y main direction were observed. The fractures are second-order to the primary faulting in the basement. In regard to the strike-slip component of the primary deformation, the faults are P, R and D respectively, and their geometry is similar to those in strike-slip experiments (Tchalenko 1968, 1970, Wilcox et al. 1973, Bartlett et al. 1981). The transpression produced in the zone corresponds to the strike-slip compressive model (Sanderson & Marchini 1984, Fig. 5a). In the experiment, the P shears developed concurrently (Bartlett et al. 1981) or even initially then subsequently to R fractures. The latter sequence was described by Morgenstern & Tchalenko (1967), Tchalenko (1970) and Wilcox et al. (1973). In experiments with only rotational faulting in the basement, without compression a similar fault pattern with only slightly different geometry has been observed. The transpression in the experiments corresponds rather to the classical wrench tectonics model. Some of the fractures may be caused by the folding component of the primary deformation. They are hkO fractures with an acute angle about the fold axis b, possibly formed by the folding (Price 1966, Hancock 1985) and later developing into faults.

In late stages of the experiment, after 25% shortening and more than 15° fault rotation in opposite directions, a phenomenon of refolding, thrusting and reverse faulting took place (Pl. 1b). The vergence is outside ofthe central part of the zone in the direction of the primary syncline limbs. However, in some cases the structures refolded on the limbs were verging to the centre of the zone. The thrusting is parallel to fold structures. The fold and thrust direction was close to that suggested by compressive models of wrench tectonics (Tchalenko 1968, 1970, Wilcox et al. 1973, Harding 1974 and Sanderson & Marchini 1984, Fig. 5).

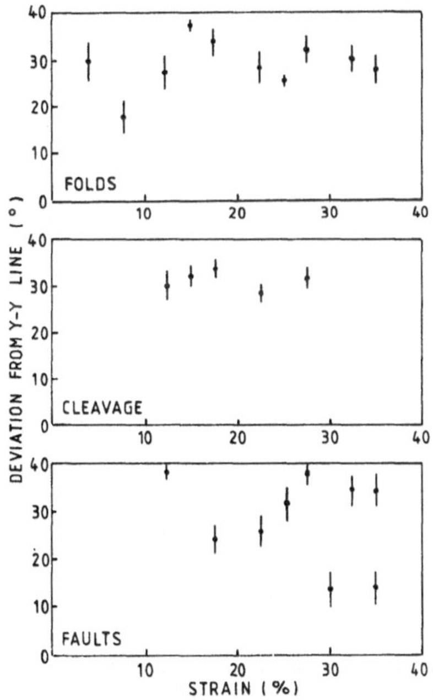

Figure 5. Orientation of new structures observed during experiment (angles from Y-Y line). Vertical bar indicates two standard deviations.

Two mechanisms have been recognized for growth of folds and faults: propagation and coalecsence. Both mechanisms occured during all stages of the experiment. Folds and faults start growing by in-plane propagation. At later stages however, the propagation is often out-of-plane. The out-of-plane propagation always developed into coalescence. The coalescence comprises several fold or fault segments always in en echelon arrangement. Longitudinal folds and faults are mainly homothetically organised. During early stages of the deformation the coalescence and propagation occured mainly among folds. This phenomenon took place when the folding and faulting on the primary structure were developed proportionally. However, when the early strain was expressed by folding, the newly formed folds were rebuilt and reorganised when faulting was superimposed. Similarly, when the first structures developed were faults their role decreased when the folding was superimposed. In some experiments, in the beginning stages, pull-apart and push-up structures were observed. Fault coalescence was achieved mainly by antidilational jogs (Sibson 1985), although, dilational jogs have also been observed. Structures associated with the jogs (Reches 1988) have been recognized.

The occurrence of fault jogs probably reflects the initial en echelon arrangement of the structures. The interaction between structures was recognized by the deviation of a fold axis or fault trace from its linear shape in the proximity of other folds or faults. The en echelon systems were caused by the sinistral strike-slip component of the deformation. The interaction between structures was relatively common in areas where the structures were closely spaced and overlaped each other. In places new structures were independent and not influenced by the pre-existing ones. In late stages some newly formed faults cut older folds and faults. However, it was also observed

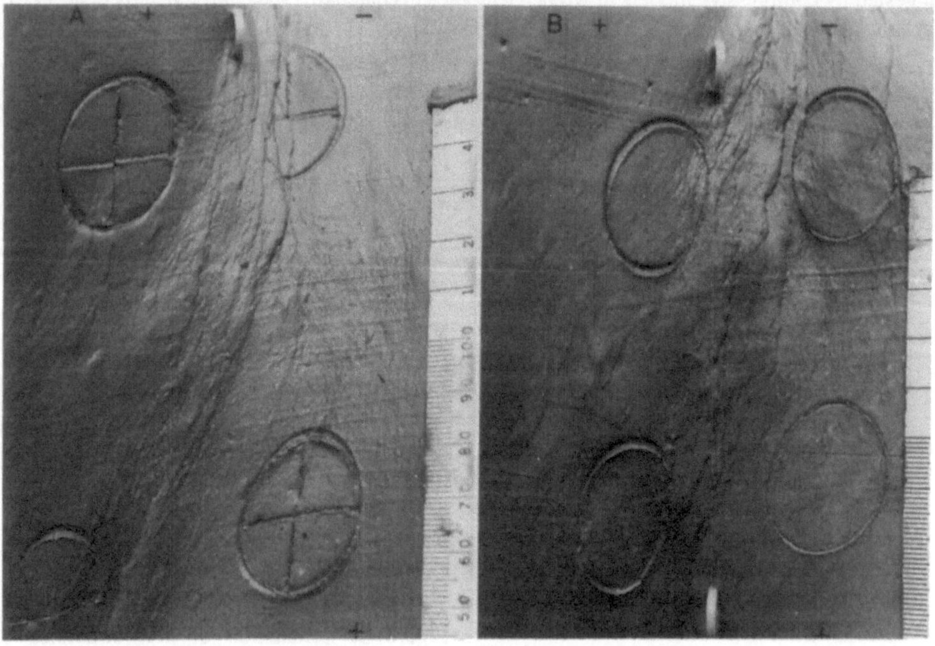

Plate 1. A top view of the clay experiment. The primary rotational fault is marked on the top, (+) and (-) show a plunge of the fault walls. Axis of the maximum compression is perpendicular to the longer side of photographs. A - 24 % deformation, B - 35 % deformation.

that some folds and faults maintained their original length and shape for a long time during the experiment, almost until its end.

Discussion

Results of the experiment show that folds produced by simple compression were reoriented and rebuilt when rotational movement caused by contrary plunging fold systems was superimposed. The experiment can explain the different orientation of the foliation in the BS-MtW area from the adjoining regions as well as the horizontal attitude of the b1 lineation within the area. It also can suggest a sinistral transpression during the deformation. Some of the strike-slip and oblique-slip slickensides parallel to the foliation may belong to this deformation.

The different plunge of the F1 folds in the east and west of the PCSZ has taken place during D_1 deformation. The PCSZ represents a less competent crustal zone developed between two more rigid lobes of the Cullen Batholith. The zone is a part of the Darwin-Pine Creek-Katherine wrench system (Stuart-Smith et al. 1986, 1987), Needham et al. (1988). The structures of the system were probably superimposed on the already existing foliation in the area. The doleritic dyke which occurs in the BS-MtW part of the PCSZ is probably en echelon and dextrally disrupted. The dyke was intruded parallel to the foliation surface. The PCSZ is probably associated with deep crustal fracture.

Conclusions

The present study provides results on the evolution of the PCSZ structure pattern which can be summarized as follows:

1) The different attitude of the foliation in the PCSZ may have originated from the contrary plunge of the F1 fold systems (Fig. 1).
2) Subhorizontal attitude of the b1 lineation probably reflects the same position of fold axes.
3) Structures of the wrench system were superimposed and used an already existing structure of the PCSZ with NNW-N trending foliation.

Results of the experiment can be summarized as follows:

4) The deformation contributed to a narrow zone which is 20-35° inclined from the direction of the primary fold system. Folds and three sets of differently oriented fractures of cleavage and fault type form structures in the zone. In late stages of the deformation refolding and reverse faulting took place
5) The deformation produced a sinistral transpression in the zone.
6) The structures of the zone grew by in-plane and out-of-plane propagation and coalescence.
7) Interaction between structures was relatively common in areas where closely spaced subparallel overlap each other. The folds and parallel faults are mainly en echelon. The main set of faults are homothetical in regard to the pair of forces (sinistral transpression).
8) The faults are antidilational and dilational jogs. Structures associated with the jogs have been observed.
9) A major structure may develop from a pre-existing one and may dominate deformation in late stages of the experiment.

Acknowledgments

We wish to acknowledge the directors of Zapopan N.L. for giving their permission and the Flinders University of South Australia for material support. In particular, we are very grateful to Dr Alex Grady for support and Trevor McGrath for building the apparatus.

References

Bartlett, W.L., Friedman, M. & Logan, J.M. 1981. Experimental folding and faulting of rocks under confining pressure: Part IX: wrench faults in limestone layers. Tectonophysics **79**, 255-277.

Cloos, H. 1928. Experimente zur inneren Tektoniko Zentralblatt für Mineralogie Geologie und Palaeontologie p. 609-621

Cloos, E. 1955. Experimental analysis of fracture patterns. Bulletin of Geological Society of America. **66**, pp. 241-256.

Ferguson, J. 1980. Metamorphism in the Pine Creek Geosyncline and its bearing on stratigraphic correlations. In, Ferguson, J. & Goleby, A. (eds) Uranium in the Pine Creek Geosyncline International Atomic Energy Agency, Vienna, p. 91-100.

Hancock, P.L. 1985. Brittle microtectonics: principles and practice. Journal of Structural Geology **7**, 437-457.

Harding, T.P. 1974. Petroleum traps associated with wrench faults. American Association of Petroleum Geologists, Bulletin **60**, 365-378

Gren, J. 1970. Modele i zadania statystyki matematycznej (Models and problems of the mathematical statistics). PWN, 324 p, Warsaw.

Johnston, D.J. 1984. Structural evolution of the Pine Creek Inlier and mineralisation therein, Northern Territory, Australia. Ph.D thesis, Monash University, Clayton, Victoria (unpublished.).

Morgenstern, N.R. & Tchalenko, J.S. 1967. Microscopic structures in kaolin subjected to direct shear. Geotechnique **17**, 309-328.

Muehlberger, W.R. 1961. Conjugate joints sets of small dihedral angle. Journal of Geology **69**, 211-219.

Needham, R.S., Crick, I.H., & Stuart-Smith, P.G. 1980. Regional geology of the Pine Creek Geosyncline. In, Ferguson J & Goleby, A. (eds) Uranium in the Pine Creek Geosyncline. International Atomic Energy Agency, Vienna, p. 1-22.

Needham, R.S., Stuart-Smith, P.G. & Page, R.W. 1988. Tectonic evolution of the Pine Creek Inlier, N.T. Precambrian Research **40/41**, 543-564.

Nicholson, P.M & Eupene, G.S. 1984. Controls on gold mineralisation in the Pine Creek Geosyncline. The Australasian Institute of Mining & Metallurgy Conference, Darwin, N.T., p.377-396.

Page, R.W., Compston, W., & Needham, R.S. 1980. Geochronology and evolution of basement and Proterozoic rocks in the Alligator Rivers Uranium field, Northern Territory, Australia. In: Ferguson, J. & Goleby, A. (eds) Uranium in the Pine Creek Geosyncline. International Atomic Energy Agency, Vienna, p. 39-68.

Price, N.J. 1966. Fault and joint development in brittle and semibrittle rock. Pergamon Press, New York.

Reches, Z. 1988. Evolution of fault patterns in clay experiments. Tectonophysics **145**, 141-156.

Riedel, W. 1929. Zur Mechanik geologischer Brucherscheinungen. Zentralblatt für Mineralogie Geologie und Palaeontologie, **1929B**, 354-368.

Sanderson, D.J., & Marchini, W.R.D. 1984. Transpression. Journal of Structural Geology **6**, 449-458.

Sibson, R.H. 1985. Stopping of earthquake ruptures at dilational fault jogs. Nature **316**, 248-251.

Stuart-Smith, P.G., Wills, K., Crick, I.H., & Needham, R.S. 1980. Evolution of the Pine Creek Geosyncline. In, Ferguson, J. & Goleby, A. (eds) Uranium in the Pine Creek Geosyncline. International Atomic Energy Agency, Vienna, p.23-38.

Stuart-Smith, P.G., Needham, R.S., Wallace, D.A., & Roarty, M.J. 1986. McKinlay River, Northern Territory. Bureau of Mineral Resources, Australia. 1: 100 000 Geological Map and Commentary.

Stuart-Smith, P.G., Needham, R.S., & Wyborn, L.A. 1985. Geochemistry and mineralisation of the Pine Creek Geosyncline, Northern Territory. In, Tectonics and Geochemistry of Early to Middle Proterozoic Fold Belts, Darwin, NT, Bureau of Mineral Resources Record **28**.

Stuart-Smith, P.G., Needham, R.S., Bagas, L., & Wallace, D.A. 1987. Pine Creek, Northern Territory (Sheet 5270). Bureau of Mineral Resources, Australia. 1: 100 000 Geological Map and Commentary.

Tchalenko, J. 1968. The evolution of kink-bands and the development of compression textures in sheared clays. Tectonophysics **6**, 159-174.

Tchalenko, J.S. 1970. Similarities between shear zones of different magnitudes. Bulletin of Geological Society of America **81**, 1625-1640.

Wilcox, R., Harding, T., & Seely, D. 1973. Basic wrench tectonics. Bulletin American Association of Petroleum Geologists **57**, 74-97.

BASEMENT AND COVER THRUST TECTONICS IN CENTRAL AUSTRALIA BASED ON THE ARUNTA-AMADEUS SEISMIC-REFLECTION PROFILE

R.D. SHAW, B.R. GOLEBY, R.J. KORSCH AND C. WRIGHT
Bureau of Mineral Resources
GPO Box 378
Canberra ACT 2601
Australia

ABSTRACT. The integrated results of surface geology and deep seismic-reflection profiling experiments across the Early Proterozoic Arunta basement and the Late Proterozoic to Middle Palaeozoic Amadeus Basin in central Australia has led to a new understanding of thrust tectonics in central Australia. Two convergent thrust systems with both 'thick-skinned' and 'thin-skinned' elements are recognised in this unique intracratonic setting. The seismic reflection results show the thrust belt at the northern margin of the basin to be dominated by a thick-skinned, south directed, basement-thrust feature, the Redbank Thrust Zone, which has been imaged in two parallel profiles as a planar feature to mantle depths. Other, more southerly, basement-cored thrusts appear to splay southwards from the Redbank Thrust Zone toward the basin and form a basement wedge. Based on various structural models, crustal shortening across the thrust belt has been estimated from as low as 25-35 km (15-25%) to as high as 66 km (50%) by others. Here we suggest a crustal shortening in the order of 30 km in the Redbank Thrust Zone region and a minimum of 30 km of shortening in the central Amadeus Basin succession. For the most part, the structures imaged by the seismic reflection profiles date from the Late Devonian-Carboniferous (Alice Springs Orogeny). As thrusting associated with the Alice Springs Orogeny progressed, thick conglomeratic sediments accumulated in a narrow (foreland-like) footwall trough that forms the northern, thickest part of the Amadeus Basin. In the central and southern parts of the basin, the seismic reflection results indicate north-directed, thin-skinned overthrusting on shallow detachments that were accompanied by Jura-style folding, also associated with the Alice Springs Orogeny. These detachments sole out in a salt horizon near the base of the succession. Some of these thrusts were also active in the Late Proterozoic. Duplication of the sedimentary section over a horizontal distance of at least 30 km has occurred at the leading northern edge of the thrust complex. A crust-cutting thrust is postulated at the southern margin of the basin to accommodate shortening on a crustal scale. Although both northern and southern thrust complexes were active at about the same time, the southern thrust complex continued longer and over-rode the footwall trough of the northern complex. It is shown that thick-skinned and thin-skinned styles of thrusting are not mutually exclusive. Whether one style or the other tends to develop depends on the interplay between a variety of factors including the orientation of pre-existing structures and on the differing mechanical properties of crystalline basement and sedimentary cover rocks.

Introduction

The Alice Springs region of Central Australia contains some unusual geological and geophysical features in a unique intracratonic setting. The region consists of metamorphosed Early Proterozoic basement rocks of the Arunta and Musgrave Blocks which form uplifted basement blocks, separated by the Late Proterozoic to Middle Palaeozoic Amadeus and Ngalia Basins (Fig.1). The stratigraphic sequence for the Amadeus Basin is set out in Table 1. Some of the largest gravity anomalies associated with the earth's continental crust occur in this region. These are near-linear,

M. J. Rickard et al. (eds.), Basement Tectonics 9, 55–84.

E-W trending anomalies where, at first sight, the strong negative Bouguer anomalies appear to correspond to sedimentary basins and the positive anomalies appear to correspond to intervening regions of exposed basement. In fact, the strongest gravity gradient between negative and positive anomalies occurs in the basement north of the Amadeus Basin across the Redbank Thrust Zone (Goleby et al. 1989). Similarly, in the south of the Amadeus Basin, some of the thickest sections of Proterozoic sediment correspond to a regional gravity ridge (Edgoose et al. 1990).

Several tectonic models for the development of the Amadeus Basin and surrounding basement have been proposed. Using a mechanical model, Lambeck (1984) invoked compression in the crust for long periods of time to explain the progressive development of the Amadeus Basin. He suggested that a crust with an initial perturbation in its structure was subjected to compression. The initial deformation is magnified and grows with time. Forman & Shaw (1973) & Shaw (1987) proposed a modified 'thick-skinned' deformational model in which the Redbank Thrust Zone and other major faults in the Arunta Block cut deep into the crust and appear to displace the crust-mantle boundary. A similar model, invoking basement faulting in the northern Amadeus Basin, was also proposed by Schroder & Gorter (1984). This contrasts with the model of Teyssier (1985) who proposed a modified 'thin-skinned' model (Coward 1983) for the deformation of the southern Arunta Block and the northern Amadeus Basin in which steep thrusts fan upwards from a shallow-dipping, master sole thrust which becomes horizontal at mid crustal

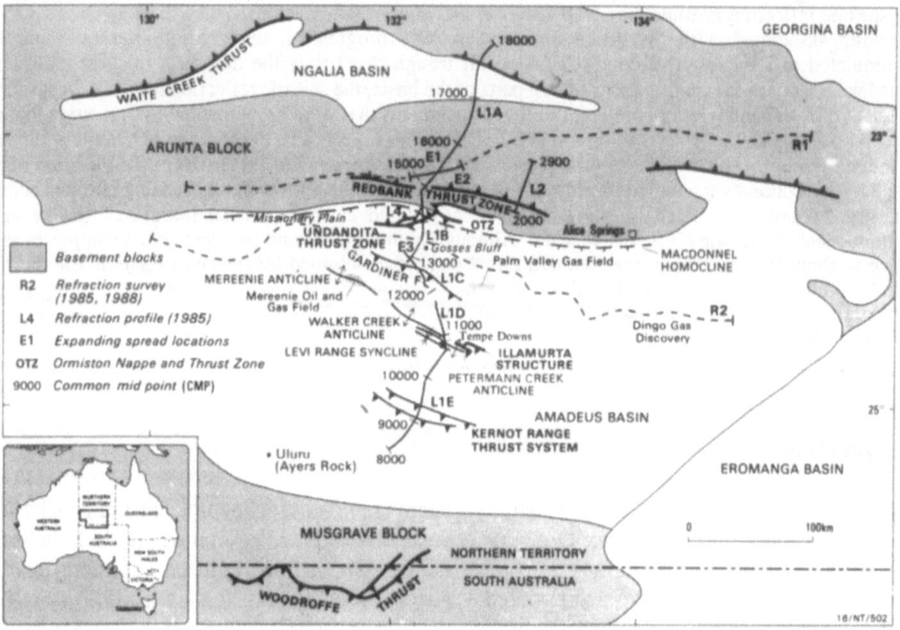

Figure 1. Locality map of the Amadeus Basin and surrounding regions showing the location of the deep seismic reflection profiles (L1A-L1E, L2, L4), the location of expanding spreads (E1, E2 and E3) and refraction profiles (R1 and R2), and the location of the main structural features. CMP numbers are marked for the deep seismic-reflection profiles. (after Shaw et al. 1991a).

depths. Bradshaw and Evans (1988) also invoked a modified thin-skinned model for deformation in the Amadeus Basin. More complex tectonic models for the evolution of the Amadeus Basin involving extensional and compressional episodes have been proposed by Lindsay and Korsch (1991), and Shaw at al. (1991a).

In 1985, the Bureau of Mineral Resources (BMR) conducted deep seismic-reflection experiments in central Australia to examine the geometry of the basins and to test models for the tectonic development of the region (Goleby et al. 1988). Key results pertaining mainly to basement structures have been published elsewhere (Goleby et al. 1988; Wright et al. 1991a), as has a model for the structure of the Arunta Block. This paper follows papers by Korsch et al. (1990) and Shaw et al. (1991a) which were concerned with that portion of the BMR seismic line which crosses the Amadeus Basin (Fig. 1), and by Goleby et al. (1989) for the Arunta Block. Here, we concentrate on the upper 5 seconds two-way time (TWT) of the seismic-reflection section in an attempt to examine aspects of the structure of the Amadeus Basin and of the adjoining basement of the Arunta Block.

The purpose of this paper is to contribute to the current debate on styles of thrusting and the extent to which thrusting is controlled by detachment zones within the crust (thin-skinned) or by fault structures that involve the entire crust as well as the upper mantle (thick-skinned). This is achieved by discussing the implication of the 1985 BMR deep seismic-reflection results for the tectonic models that have been proposed for the central Australian region. At the same time we further clarify the results of the BMR deep seismic-reflection traverse in central Australia (Fig. 1),

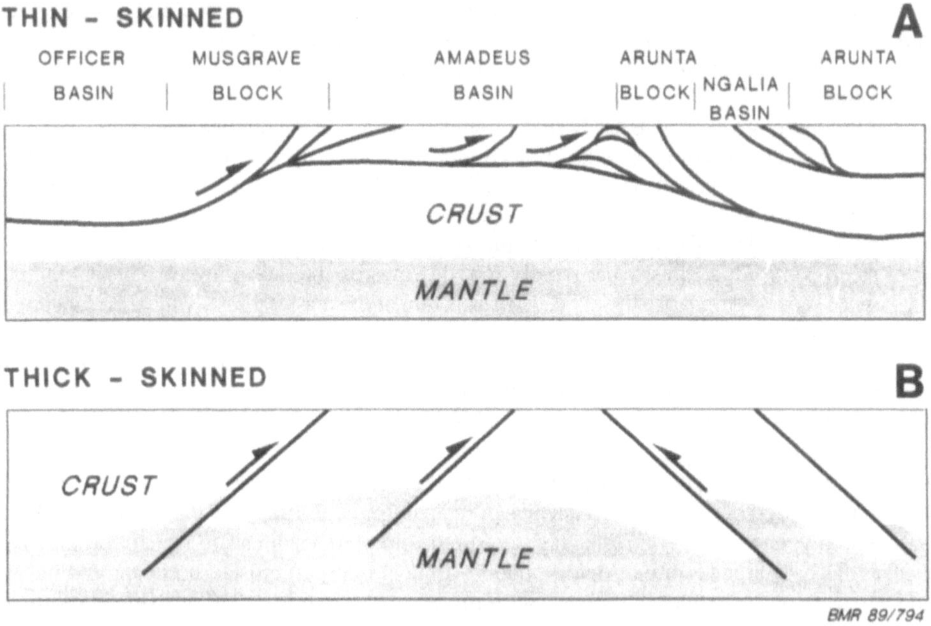

Figure 2. Plausible tectonic models previously proposed for the central Australian region. A) Model emphasising 'thin-skinned' thrusting features (after Teyssier 1985), and B) Model emphasising 'thick-skinned' crustal fault zones (after Mathur 1976; Lambeck 1984; Shaw 1987).

carried out in the region between the Ngalia and southern Amadeus Basins. Two plausible tectonic models have been proposed recently for the central Australian region (Fig. 2). The first model (Fig. 2A), examined and discussed by Teyssier (1985), is a hybrid model in which basement thrusting is dominated by a thin-skinned style of thrust system, but moderately steep dips. The second model (Fig. 2B) is a thick-skinned model for the region in the gravity models of Mathur (1976) and adopted as the final tectonic configuration in the mechanical models of Lambeck (1984).

THICK-SKINNED AND THIN-SKINNED STYLES OF THRUSTING

When combined with our knowledge of the surface geology, the seismic-reflection results reveal two distinctly different styles of thrusting developed entirely in an intracratonic setting. One thrusting style, which is thick-skinned in character, is related to basement uplift. The other style is confined to the Amadeus Basin and is thin-skinned. At first glance, the development of these two contrasting styles in the same region appears to be contradictory. We examine the nature of thrusting in each of these regions and provide an explanation why these two different styles of crustal thrusting have developed in the same region.

Thrust development in central Australia is unusual in that it involved both nappe formation and imbrication of the cover succession, but it took place entirely in the centre of the Australian continent during a Devonian-Carboniferous event, referred to as the Alice Springs Orogeny (Shaw et al. 1984). It is important to realise that an intracratonic setting such as central Australia differs significantly from convergent orogens at plate boundaries such as those in the European Pyrenees and Alps where classical models of thrusting have been developed. In these latter regions there is evidence of prior extension or passive-margin development, a situation very different to that in central Australia.

In the European orogens there has been considerable debate on the relative role of thin-skinned and thick-skinned modes of crustal thrusting (Butler 1983; McCaig 1988). The same debate has taken place in central Australia (Teyssier 1985; Shaw 1987; Goleby et al. 1989). Thick-skinned and thin-skinned models of thrusting are extreme end-members of a broad spectrum of tectonic styles and many regions show elements of both systems. Thick-skinned models emphasise the role of steep controlling thrusts that rapidly root downwards into steep to near-vertical faults. Thin-skinned models emphasise the allochthonous nature of the master thrusts which are suggested to flatten out at depth.

In a thrust system dominated by thin-skinned thrusting, the overall movement is accommodated along one or more sub-horizontal detachment zones or sole thrusts which may develop at one or more levels in the crust. Imbricate thrusts or duplexes within these thrust systems typically take up the movement at the leading edge of the master thrust and commonly show a sequential younging towards the foreland. A good example of the thin-skinned style of thrusting is provided by the allochthonous region of northeastern America (Allmendinger et al. 1983). In thin-skinned models, movement on younger master thrusts may steepen the dip of older thrusts to which they may be linked, as has been argued for the Alpine model by Butler (1983). If crustal shortening continues, then late-stage, out-of-sequence thrusts may form behind an imbricate fan or rising antiformal stack, or as back-thrusts in front of the stack (e.g. Tabasco Fault of the MacKenzie Mountains, Vann et al. 1986). Two important features of thin-skinned models are that (1) uplift is minimised relative to the amount of crustal shortening so energy requirements are reduced and (2) the stress levels implied within the crust are also relatively low. Significant crustal shortening inevitably leads to thickening of the crust, as argued for the Himalayas in Pakistan by Coward et al. (1987).

Where plastic deformation becomes penetrative over a wide zone, other complications may develop, in which case continued crustal shortening may be accommodated by heterogeneous upward and downward extension in a crustal lozenge (e.g. 'Hedgeback' model; Coward et al. 1987). In the Hedgeback model, a progression occurs to a tectonism more typical of thick-skinned style of thrust zones.

In thick-skinned thrust systems, most of the movement takes place on moderately steep faults which cut the crust. Consequently, uplift is of similar magnitude to the horizontal displacement. Thickening of the crust is thereby minimised. Energy requirements are presumed to be high, but may be considerably reduced if thrusting takes place by small incremental movements, and exhumation keeps pace with uplift. Such structures are expected to be accompanied by relatively high states of crustal stress and local isostatic imbalance. The best documented example of the thick-skinned style is provided by the Laramide uplifts in the Wyoming foreland of northwestern America (Brown 1989). Thick-skinned imbrication of crust and lithosphere, resulting from crustal shortening during continental collision, is also illustrated by seismic profiles from the Himalayas (Allegre et al. 1984).

Acquisition and Processing of the Seismic Data

The recording parameters used in the routine reflection profiling have been summarised by Goleby et al. (1988). Briefly, the data were recorded using explosive sources, 6-fold CMP (common mid-point), a split-spread shot-receiver configuration, 48 channels of recording, and a geophone group spacing of 83 m. These recording parameters were chosen to be optimal for imaging deep basement reflectors. While the low fold of recording and large group interval are generally good for deep crustal work, they are less satisfactory for studying shallow sedimentary structures. Nevertheless, the data provide adequate resolution of many shallow structural features.

Irregularities in low-velocity overburden (P-wave velocities less than 2 km/s) and weathered bedrock resulted in severe static correction problems that were not adequately solved with available industry refraction-statics computer software. It was also difficult to use autostatic computing routines, because of the common absence of shallow reflections that were continuous over an appreciable distance. These difficulties resulted in the development in BMR of an alternative approach to calculating refraction static corrections based on a reciprocal method which resulted in significant improvement of shallow seismic sections (Wright et al. 1991b).

Structure of the Amadeus Basin and Bordering Basement

The main structural features of the Amadeus and Ngalia Basins and the surrounding basement crossed by the seismic traverse are shown in Figure 1. Foreland-like thrust belts mark the northern and southern margins of the Amadeus Basin (Wells et al. 1970). Major mylonite zones on the hinterland side of the thrust belts mark a jump in the regional metamorphic grade and a steep Bouguer anomaly gradient. Forman and Shaw (1973) proposed that these mylonite zones, namely the Redbank Thrust Zone in the north and the Woodroffe Thrust Zone in the south, continue through the crust to mantle depths. The deep structure of the northern margin has been inferred from structural mapping (Shaw, 1987), modelling of long-range seismic travel-time residual anomalies (Lambeck et al. 1988), and deep reflection-seismic imaging (Goleby et al. 1989). These studies show that the Redbank Thrust Zone continues at a dip of about 45° to mantle depths and may displace the crust by possibly as much as 10-15 km. Preliminary teleseismic studies of the Woodroffe-Mann Fault system (Lambeck 1991) suggest that a similar, mirror-image, crustal structure exists there. These studies show that a thick-skinned tectonic mode dominates in at least these regions, rather than the thin-skinned style of imbricate thrust complexes proposed across the Redbank Thrust Zone by Teyssier (1985). As will be shown below, however, a thick-skinned tectonic style does not dominate in all areas.

In the south, the Petermann Ranges Nappe (Wells et al. 1970) was interpreted to lie between the basin and the Woodroffe Thrust. Recent work, however, raises doubts concerning the nappe interpretation, and suggests that the magnitude of shortening due to the Petermann Ranges Nappe and fold and thrust structures in the basin have been over-estimated. Indeed, they may have formed during several tectonic events (Shaw at al. 1991a). At the northern basin margin, the thrust belt south of the Redbank Thrust Zone is dominated by the Ormiston Thrust Zone and a monoclinal upturn, the MacDonnell Homocline (Korsch et al. 1990; Shaw et al. 1991a). Individual, basement-cored, thrust nappe structures in the northern thrust belt have been described in detail (see Shaw et al. 1984).

The locally developed upthrust geometry and a general similarity to the Wyoming style of tectonics, however, led Schroder and Gorter (1984) to postulate basement involvement within the basin and, in particular, for the Gardiner Thrust. Fold and thrust structures within the basin are described by Ranford et al. (1965) and by Wells et al. (1970), and analysed further by Bradshaw and Evans (1988).

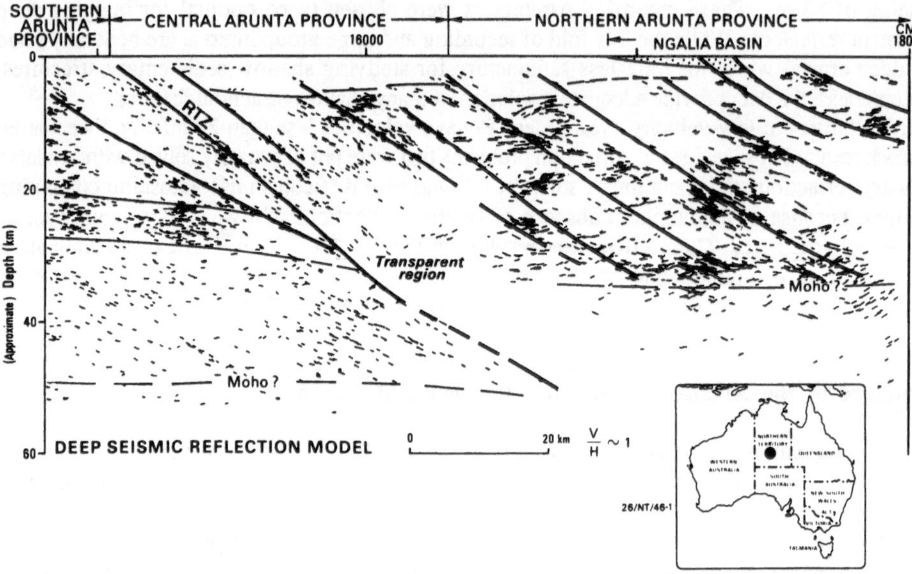

Figure 3. Interpreted crustal structure in the region of the Ormiston Thrust Zone (OTZE), the Redbank Thrust Zone (RTZ) and northwards to the Ngalia Basin based on the deep seismic reflection model of Goleby et al. (1989). (after Shaw et al. 1991a).

THE REDBANK THRUST ZONE, (AND MINOR SUB-PARALLEL STRUCTURES TO THE NORTH)

The reflection-seismic image of the Redbank Thrust Zone (RTZ) and related structures has been described previously by Goleby et al. (1989) and its along-strike continuity confirmed by Wright et al. (1991a), (Fig. 3). Prominent surface faults marked by mylonite zones outline the RTZ and correspond to a strong band of reflectivity which can be traced to depths of at least 30-35 km and possibly down to 50 km (Fig. 3; Goleby et al. 1989). This zone of strong, steep reflection events dips at an angle of around 45° and is remarkably planar. Supporting evidence that the RTZ displaces the Moho comes from a study of teleseismic travel-time anomalies (residuals) recorded across the RTZ region (Lambeck et al. 1988). Shaw & Black (1991) have demonstrated that the RTZ formed as a result of reactivation of a pre-existing Proterozoic province boundary due to the imposition of exceptionally high compressive stresses during the Alice Springs Orogeny.

North of the RTZ, a series of steeply dipping seismic events (Fig. 3), which truncate sub-horizontal events, are considered to be north-dipping lag faults that moved parallel to the RTZ in the final stages of overthrusting during the Alice Springs Orogeny. The impression gained from the seismic image is that these faults moved like a stack of dominoes with little or no rotation of the crustal blocks.

MACDONNELL HOMOCLINE, ORMISTON THRUST ZONE AND RELATED STRUCTURES

The regional monoclinal flexure at the basin margin is referred to as the MacDonnell Homocline, and is marked in the seismic section (Fig. 4 &5) by an abrupt change from clear sub-horizontal reflections to absence of reflectivity which accompanies the transition from about 10 km thickness of sedimentary rocks (i.e. down to reflector 15; Table 2) to basement. At the point where the seismic line crosses the homocline, the structure is complicated by oblique intersection with the northeasterly continuation of the Undandita Thrust Zone (Figure 1; Schroder & Gorter 1984).

The basis for important interpretative features in the seismic section (Fig. 4) are given in Shaw et al. (1991a). The MacDonnell Homocline and the Undandita Thrust Zone have been confirmed in a second, short branching seismic traverse (Fig. 5). Immediately north of the monocline, the along-strike equivalent of the Ormiston Thrust Zone (OTZE) is imaged as a discontinuity between a region of poor reflectors above from a region of shallow-dipping reflectors and diffractions below. The Undandita Thrust Zone (UTZ) is traced from the alignment of migrated diffractions and the termination of reflections.

A mid-crustal reflectivity boundary appears to be continuous across the region of the homocline and the OTZE (Fig. 5A; Wright et al. 1991b), and eliminates the possibility that the homocline is controlled by a steep-dipping crustal fault. However, diffractions below 6 s at CMP 14300 disrupt the most prominent zone of high reflectivity to the north.

At the base of the upturned sedimentary succession at the homocline edge, multiple detachment zones (Fig. 4) are exposed at the surface in basal carbonate succession. These detachment zones are thought to pre-date the faults of the Undandita Thrust Zone and to have formed as the basement to the north was uplifted. West of the seismic profile, the thrust zone is complicated by salt migration from near the base of the succession which may have begun in the early Cambrian (Schroder & Gorter 1984). Beneath and north of the homocline (see deep seismic-reflection image in Goleby et al. 1989, 1990), a thrust wedge or triangular zone is apparent, rather than a simple antiformal stack as suggested by Vann et al. (1986) for similar upturns in other regions. This thrust wedge (Shaw et al. 1991a) resulted from progressive movements on a series of thrusts (including the Razorback Klippe, Mount Sonder Thrust and Ormiston Thrust Zone). The succession was progressively overturned ahead of the developing thrust wedge.

The MacDonnell Homocline is oblique to fold and thrust trends in the western part of the basin, suggesting some sinistral strike-slip faulting along the line of the homocline in that region. Faults corresponding to the Undandita Thrust Zone and other minor faults may have undergone oblique or strike-slip movements late in their history.

The thrusts, including the OTZE, do not appear to shallow progressively towards the foreland as is the case in a typical thrust duplex forming an antiformal stack. For example, underlying basement-thrust faults, which have been mapped in outcrop and structurally underlie the northeast-dipping Ormiston Thrust Zone, dip at about 70° north (ONTZ in Fig. 1).

A second thrust wedge is interpreted to occur between the Ormiston and Redbank Thrust Zones. In this region, several discontinuous, branching and anastomosing thrusts occur between the main thrust zones (Shaw & Black 1991). In outcrop, these thrusts within the wedge: (1) are arranged

Table 1. Simplified Stratigraphy of Amadeus Basin

AGE		FORMATION	
		Brewer Conglomerate*	
Late Devonian		Hermannsburg Sandstone*	
		Parke Siltstone*	
Silurian to Early Devonian		Mereenie Sandstone	
		Carmichael Sandstone#	
Late Ordovician		Stokes Siltstone#	
		Stairway Sandstone#	
Early Ordovician		Horn Valley Siltstone#	
----------	----------	Pacoota Sandstone#	----------
		Goyder Formation+	
		Petermann Sandstone+	
Cambrian		Deception Formation+	
		Illara Sandstone+	
		Tempe Formation+	
		Chandler Formation+	
----------	----------	Arumbera Sandstone+	----------
	SOUTH	NORTH	
	Winnall Beds		Julie Formation
		Pertatataka Formation	
	Inindia Beds		Areyonga Formation
Late Proterozoic		Bitter Springs Formation	
		(Gillen Member)	
		(Loves Creek Member)	
		Heavitree Quartzite	

* - Pertnjara Group
\# - Larapinta Group
\+ - Pertaoorrta Group

Dashed lines indicate that a formation straddles period boundaries.

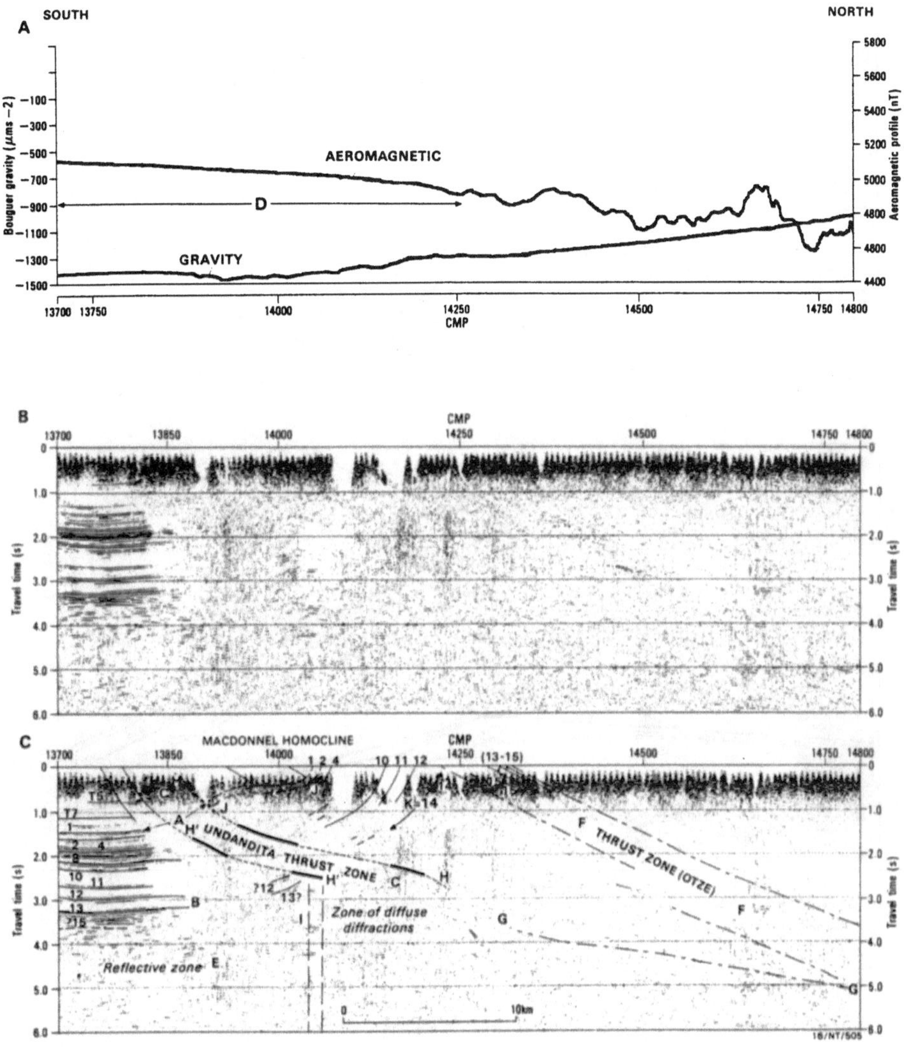

Figure 4. The northern part of the deep seismic reflection section L1B across the upturned northern margin of the basin (MacDonnell Homocline) and the along-strike equivalent of the Ormiston Thrust Zone (OTZE) showing the relationship to total magnetic intensity and Bouguer gravity profiles. A) Aeromagnetic and Bouguer gravity profiles. D shows the location of the Amadeus Basin. B) Unmigrated seismic section, and C) Interpreted seismic section. (Formation picks for numbered reflections are given in Table 2). The letters A through I refer to identified basement reflectors and zones of migrated diffractions (after Shaw et al. 1991a).

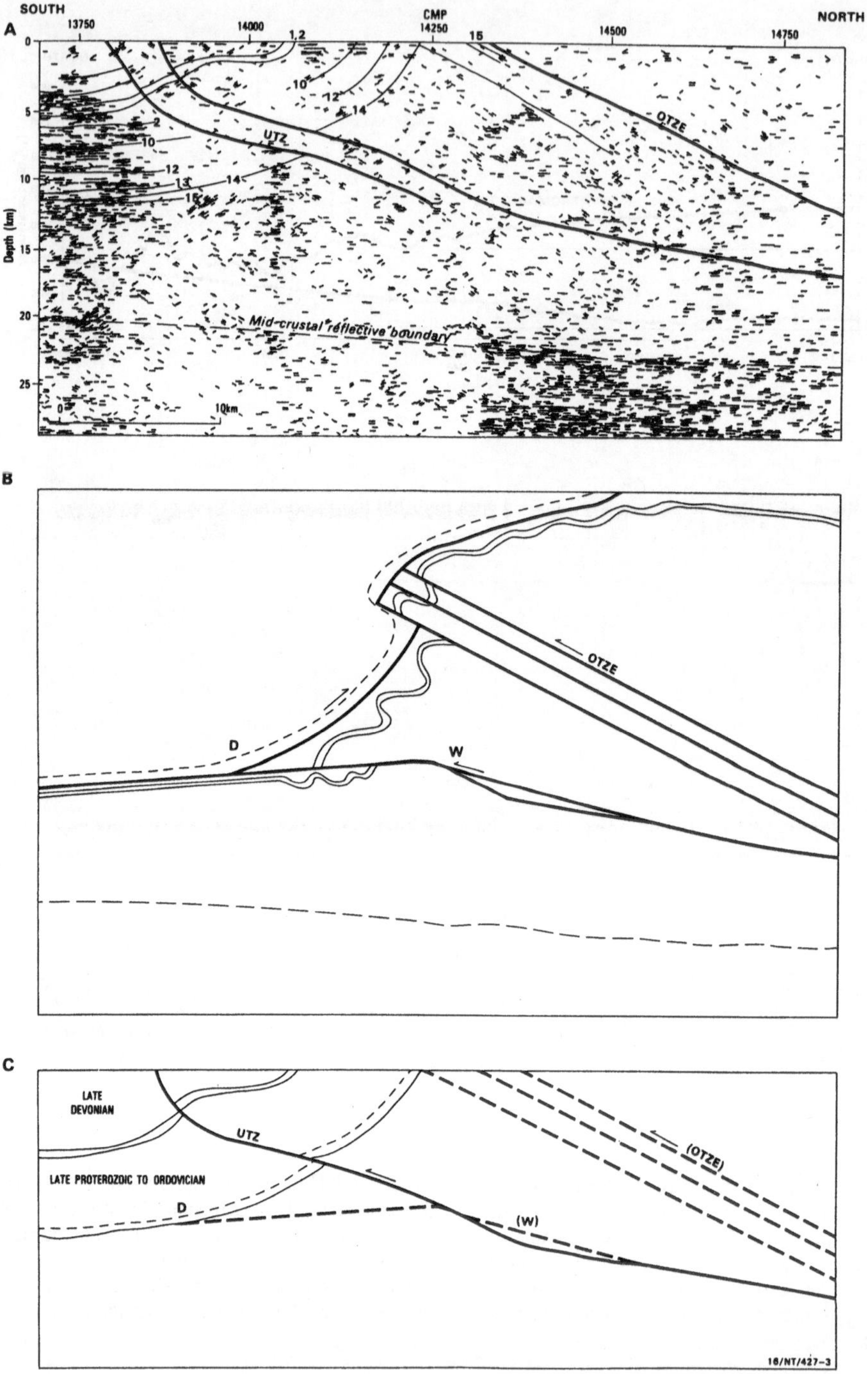
A
SOUTH
NORTH
CMP
13750
14000
14250
14500
14750
1,2
15
10
12
14
2
13
UTZ
OTZE
Depth (km)
0
5
10
15
20
25
Mid-crustal reflective boundary
0
10km
B
OTZE
D
W
C
LATE DEVONIAN
LATE PROTEROZOIC TO ORDOVICIAN
UTZ
(OTZE)
D
(W)
16/NT/427-3

Table 2. Identification of primary reflectors from the northern part of the Missionary Plain and their TWT's at CMP 13142 (after Shaw et al. 1991a; Figure 10, and J.D. Gorter, personal communication, 1990).
* approximate only

Reflector	Stratigraphic Unit	Two-Way Time (TWT) seconds
T5	Within Brewer Conglomerate 1	0.54 *
T6	Within Brewer Conglomerate 2	0.61 *
T7	Top Hermannsburg Sandstone	0.77 *
1	Top Parke Siltstone	1.05
2	Top Mereenie Sandstone	1.19
3	Within Mereenie Sandstone	1.28
4	Base Carmichael Sandstone (thin unit)	1.47
5	Near top Stokes Siltstone	1.51
6	Near base Stokes Siltstone	1.56
7	Top Horn Valley Siltstone	1.70
8	Near top Pacoota Sandstone	1.77
9	Within Pacoota Sandstone	1.84 *
10	Top Goyder Formation	1.88
10A	Base Goyder Formation	2.01
11	Near base Tempe Formation (base Chandler Fm Equivalent)	2.20
12	Top Julie Formation (near base Arumbera Sandstone)	2.36
12A	Within Late Proterozoic (?near base Julie Formation)	2.40
12B	Within Late Proterozoic (?near top Areyonga Formation)	2.65
13	Top Loves Creek Member Bitter Springs Formation	2.77
13A	Near top Gillen Member Bitter Springs Formation	2.80
14	Within Gillen Member Bitter Springs Formation (possibly base of salt unit)	2.87
15	Near base of Heavitree Quartzite	2.96 *

Figure 5. Interpretation of seismic section line L1B across the northern margin of the Amadeus Basin. A) Interpreted line drawing of section line L1B. UTZ = Undandita Thrust Zone, OTZE = the along-strike equivalent of the Ormiston Thrust Zone. (Numbered formation picks are given in Table 2). B) First stage tectonic model of the MacDonnell Homocline: a monocline developed due to localised back-thrusting of the succession on a basal detachment zone (D) over a rising basement wedge (W), followed by progressive overthrusting, and C) Second stage tectonic model of the MacDonnell Homocline: the MacDonnell Homocline was cut by the Undandita Thrust Zone (UTZ), which lies slightly oblique to the homocline. (after Shaw et al. 1991a).

en-echelon, (2) do not appear to be linked in surface outcrop, (3) are oblique to the homocline, and (4) splay at depth southwards away from the Redbank Thrust Zone at a shallow angle. In the deep seismic image of Goleby et al. (1989, 1990), the Ormiston Thrust Zone dips at 30° northwards and converges at depth with the Redbank Thrust Zone.

There is an inferred two-stage development of the MacDonnell Homocline (Fig. 5B). In the first stage, the sedimentary succession is rotated to form the Homocline as a result of tilting of a basement wedge (W in Fig. 5B) formed at the front of an imbricate thrust fan or wedge. Basement wedges (triangular zones) of this type have been described from the Wyoming region of the North American Rocky Mountains (e.g. the Rattlesnake Mountain structure, Lowell 1985) and the Ellesmerian mountain front in Northern Greenland, Soper & Higgins 1990). Some internal thrust imbrication and/or thrust duplexing might be expected in the basement wedge in order to maintain strain compatibility with the surrounding structural blocks. Such structural complexities may explain the incoherent seismic image in the region of the inferred basement wedge.

To accommodate local shortening of the section, the sedimentary succession is interpreted to have been back-thrust on a basal detachment zone (D in Fig. 5B) over the rising basement wedge (W). This complex detachment zone is visible in outcrop. As thrusting progressed, the homocline was overthrust by the sub-parallel Ormiston Thrust Zone Equivalent (OTZE in Fig. 5B; taken to approximate the Mount Sonder Thrust (ST) in Fig. 1). In the second stage, the monoclinal flexure, which became preserved as the MacDonnell Homocline, was cut by the Undandita Thrust Zone (UTZ in Fig. 5C).

LACK OF BASEMENT INVOLVEMENT IN THE MISSIONARY PLAIN SECTION

The section across the Missionary Plain (Line L1B, Fig. 6) is noteworthy for its lack of structure. The curvilinear and discontinuous nature of the reflectors at the base of the succession may be due to irregular basement topography of the type modelled synthetically by Cao et al. (1991). There is also a suggestion of semi-continuous basement reflectors at about 3.7 s TWT. Tracing of reflector 15 (basal unit - Heavitree Quartzite; Table 2) across the Missionary Plain shows that it has an irregular dip and appears to be locally disrupted by gently-dipping thrust faults of small displacement. It is significant that north of the Gardiner Range, across the Missionary Plain, there is no evidence of major thrusting in industry seismic data (e.g. Schroder & Gorter 1984; Lindsay & Korsch 1991).

Reflector 14, near the base of the succession, is interpreted as the base of a salt horizon, because it marks a rapid velocity increase below a region of low seismic velocity (4.8 km/s) (Wright et al. 1991a). It can be traced as a relatively continuous horizon across most of the section shown in Figure 6. The continuity of reflector 14 demonstrates that there is no significant salt pillowing in this section.

The lack of substantial deformation suggests that the Missionary Plain region can be considered para-autochthonous relative to the region south of the Gardiner Thrust. Minor detachment at the base of the sedimentary succession is suggested by lack of continuity shown by reflector 15 (Fig. 6). This contrasts with the model of Teyssier (1985) who proposed a major sub-horizontal thrust beneath the Missionary Plain, with back-thrusting on the Gardiner Thrust and at the MacDonnell Homocline, and some repetition of the Late Proterozoic succession with northward-directed overthrusting within the basin at the northern margin.

GARDINER RANGE

In the vicinity of the Gardiner Range, the distinctive reflectors of the Bitter Springs Formation are imaged in the seismic profile and allow the Gardiner Thrust, lying at the base of these reflectors, to be traced at depth. The Gardiner Thrust is shown to be a south dipping, listric feature (Fig. 7). The numbered reflections (Table 2) are determined by extrapolation to surface outcrops and by comparison with the character of the reflections under the Missionary Plain immediately north of the Gardiner Range. The interpreted reflections down to 2.65 TWT shown at the northern end of the section in Figure 7 include those reflectors identified at the southern end of the Missionary Plains segment of the line (Fig. 6; see Shaw et al. 1991a for further discussion).

The distinctive Bitter Springs Formation (reflectors 13, 13A and 14) and overlying succession of alternating weak and strong reflectors of the sedimentary succession can be traced in the seismic section, at approximately the same depth (6.9 - 8.8 km or 2.6 - 3.3 s TWT), for at least 30 km to the south of the exposed Gardiner Thrust (CMP 12350 to 11550; Figs. 7 & 8) without significant disruption. At the surface, however, the Bitter Springs Formation occurs immediately to the south of the Gardiner Thrust. Thus, in this seismic section, the Bitter Springs Formation is repeated due to a large amount of shortening on the Gardiner Thrust.

Because the Hermannsburg Sandstone of the Devonian Pertnjara Group (Table 1) is juxtaposed against the Late Proterozoic Bitter Springs Formation, at least some of the movement on this fault must have occurred late in the Alice Springs Orogeny. Using seismic root-mean-square (r.m.s.) velocities, the thickness for the section from the top of the Bitter Springs Formation to the base of the Hermannsburg Sandstone in the southern Missionary Plain is 5.13 km (Wright et al. 1991b). Taking subsequent erosion into account, the vertical displacement on the ramp is likely to be close to 6 km.

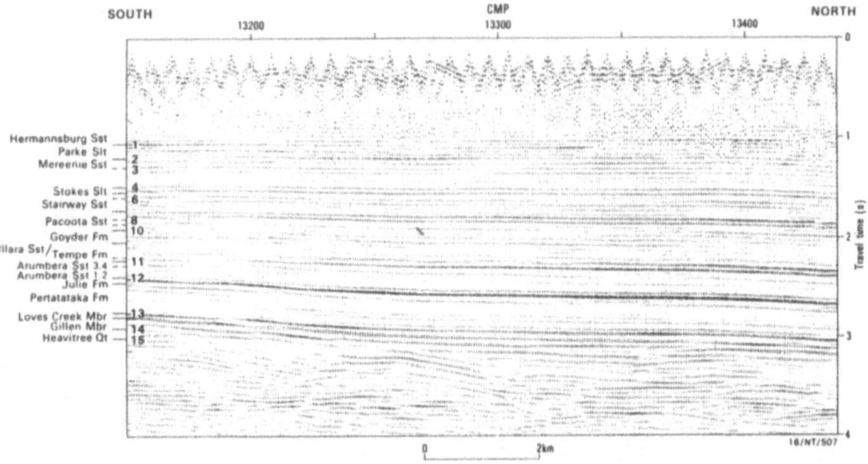

Figure 6. Section of the southern part of the deep seismic reflection line L1B across the Missionary Plain region. Accurate formation picks for numbered reflectors are given in Table 2. (after Shaw et al. 1991a).

The Gardiner Thrust has a ramp and flat geometry. Several leading imbricate thrust splays lying north of the Gardiner Thrust cannot be discounted; such splays would explain a space problem created by the presence of the overturned dips in the Larapinta Group (below reflector 4) and Mereenie Sandstone (dashed position of reflectors 2 and 4 in Fig. 7) observed in outcrops south of Missionary Plain (Wells et al. 1970-Fig. 52D). The Mereenie Sandstone has been identified at 1.2 s TWT on Missionary Plain north of Gosse Bluff (Fig. 1) in an earlier BMR seismic-reflection survey (immediately north of Fig. 6; Shaw et al. 1991a), implying that a steep to overturned Mereenie Sandstone section continues to a depth of about 3 km in order to maintain section continuity (dashed lines in Fig. 7). Beneath the flat is a section from the Bitter Springs Formation (and probably the Heavitree Quartzite, reflector 15) to possibly as high as the Larapinta Group (reflector 6). Hence there is a duplication of the Late Proterozoic to early Cambrian succession above the flat. The section beneath the flat appears to be complicated by faulting. The interpretation here of the Gardiner Thrust as a ramp and flat with at least 30 km of horizontal displacement differs from the interpretation of Schroder & Gorter (1984) who show the Gardiner Thrust cutting the Bitter Springs Formation at depth, and hence in their model the amount of horizontal displacement is limited to about 1 km.

From the geological maps of the Gardiner Range Anticline (Ranford et al. 1965; Wells et al. 1970) it can be seen that the trace of the Gardiner Thrust changes from being emergent to blind within the Cambrian formations at both the northwestern and southeastern ends of the anticline. At these localities, the surface expression of the shortening in the cover is converted from thrusting to folding above a blind-thrust ramp. Early movement on the blind-thrust during the Late Proterozoic Petermann Ranges Orogeny is implied because a section of Pertatataka Formation has been removed by erosion from the anticlinal core, before the deposition of the latest Proterozoic - early Cambrian Arumbera Sandstone.

Large horizontal transport of the cover (upper plate) above the ramp of the Gardiner Thrust implies that the present-day distribution of the sedimentary succession is not palinspastic. Thus, the original southern depositional limit of the Arumbera Sandstone, for example, would have been well to the south of its present outcrop limit in the Gardiner Range (Ranford et al. 1965). Arumbera Sandstone can be traced in the footwall section of the Gardiner Thrust as far south as CMP 11600 (i.e the unit between reflectors 11 and 12 in Fig. 8).

TEMPE DOWNS AREA

South of the Gardiner Range, the seismic line crosses outcrops of the Hermannsburg Sandstone, until the eastern projection of the Walker Creek Anticline is crossed near Tempe Downs (Fig. 1). Here, the Late Proterozoic Areyonga Formation (Table 1) is exposed, and the Walker Creek Anticline is overturned on its northern limb.

Some 6-7 km to the north of the Walker Creek Anticline, the seismic section shows another gentle, subsurface anticlinal structure (at about CMP 11550, Fig. 8). Field mapping (Ranford et al. 1965) has not detected any surface expression of this structure, which is predicted to occur within the outcropping Mereenie Sandstone. In the subsurface, there is a rollover structure outlined by the Pacoota Sandstone (the reservoir at the Mereenie Field) at a depth of 1.4 km (0.6 s TWT on seismic section, reflectors 8-10 in Fig. 8) and by the Tempe Formation (reflector 11, Fig. 8). We speculate that this structure is related to minor backthrusting.

Slightly farther south, partial repetition of the Late Proterozoic to Early Cambrian succession above the thrust is implied in outcrop in the eastern section of the Walker Creek Anticline north of Tempe Downs (Ranford et al. 1965), and continues as a paired anticline where the Illamurta

Diapiric Complex is developed. Hence there will be duplication of the Chandler to Bitter Springs succession beneath the Areyonga and Bitter Springs Formations exposed in the anticlinal core at the surface. For the anticline to be cored by the Bitter Springs Formation so close to the surface, some underlying complex structure, such as a thrust-duplex fault zone, must occur under the Walker Creek and Petermann Creek Anticlines where northerly directed thrusts ramp towards the surface. This also appears to be the case in the Illamurta Structure exposed along strike (Ranford et al. 1965).

SOUTHERN END OF THE SEISMIC LINE

Between the Gardiner Thrust Zone and the southern limit of the BMR seismic line, there are several surface expressions of tight macroscopic anticlines, some with overturned northern limbs. Although the seismic images of these structures are generally poor, principally due to steep dips as well as rapid variations in overburden and weathering thicknesses, we infer that most of the structures are related to a north-directed overthrust system. The structural pattern appears to be one where several blind to emergent thrusts branch off the main sub-horizontal thrust.

The thickness of sediment estimated from the seismic section in synclinal regions is commonly thicker than that measured in outcrop from anticlinal cores (e.g. at about CMP 11000, Fig. 1, where the thickness of the post-Bitter Springs succession is about 30% greater than that estimated from surface exposures). This discrepancy appears to be due to growth of the anticlines during the Petermann Ranges Orogeny (and perhaps precursor orogenies). In the Petermann Creek and Walker Creek Anticlines, Cambrian sandstones rest directly on the Areyonga Formation, due to erosion of the Pertatataka Formation during the Petermann Ranges Orogeny. These relationships indicate that these thrust-cored anticlines were active during the Late Proterozoic Petermann Ranges Orogeny and were subsequently reactivated during the Alice Springs Orogeny (Ranford et al. 1965).

Another thrust zone, the Kernot Range Thrust. System, is postulated to occur immediately north of the Kernot Range where, on the surface, Late Proterozoic Winnall and Inindia beds dip consistently at 25° to the SSW and appear to overlie exposed Bitter Springs Formation. We interpret these beds as a uniformly-dipping sedimentary package sitting above a thrust ramp that eventually flattens to the south (Fig. 9). Repetition of the same outcrop relationships and repetition of the distinctive Bitter Springs Formation reflectors suggest that several imbricate thrusts fan northwards from the ramp-thrust structure. We note that this thrust zone occurs about 2 km north of a planar feature (E in Fig. 9) observed dipping in basement to depths of at least 18 km (Wright et al. 1991b, Fig. 8). This planar feature is interpreted as a Proterozoic thrust and corresponds to an aeromagnetic domain boundary between the Central and Southwest Magnetic Domains (Fig. 10). The extent to which the thin-skinned structural deformation in the vicinity of the Kernot Range has been influenced by the presence of the older basement feature, however, is not clear.

In summary, the tight to overturned surface anticlines occur behind major thrust ramps in the Gardiner and Kernot Ranges, and possibly in the region of the Walker Creek Anticline. These anticlines appear to represent deformation of the cover above a blind thrust network formed by a combination of coupling and forethrusting.

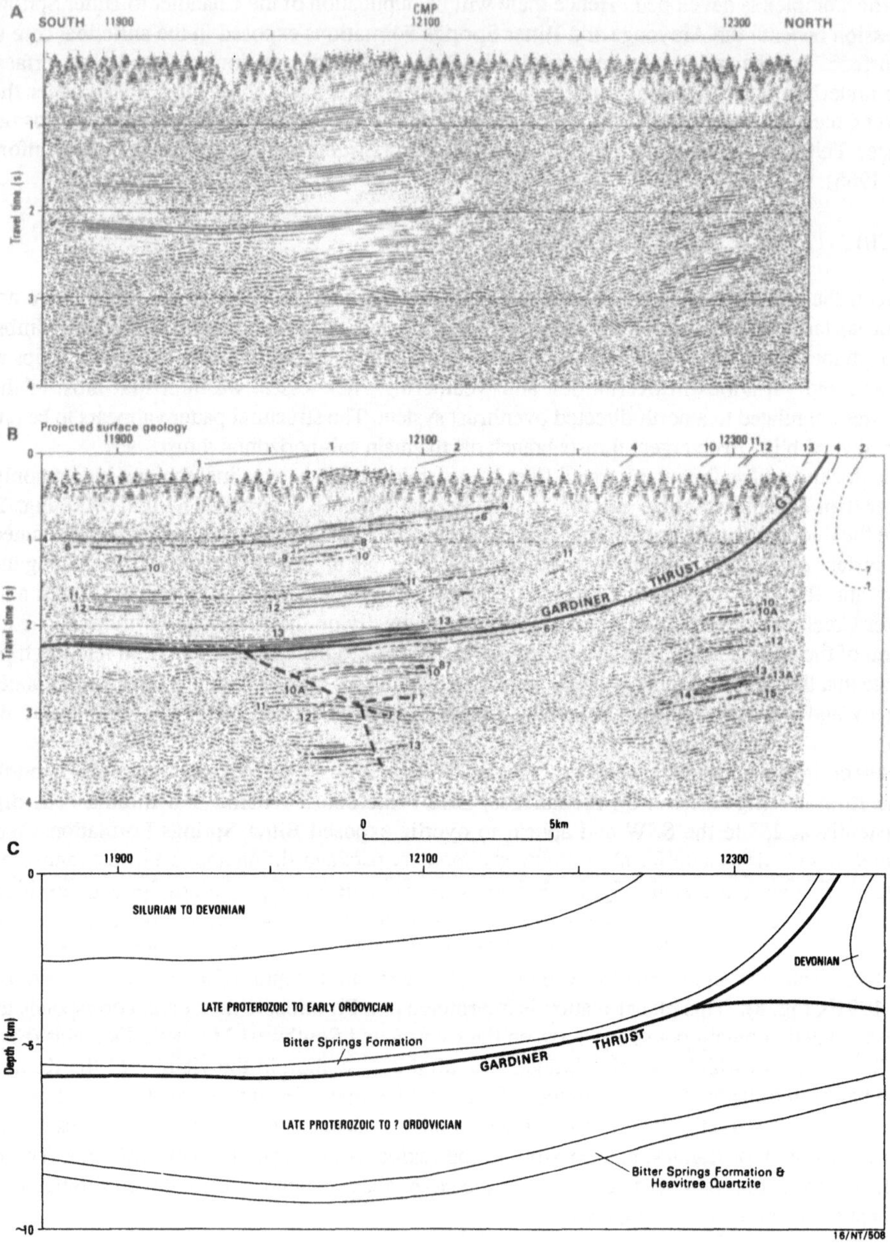

Figure 7. Northern part of the deep seismic reflection line L1D from north of the Gardiner Thrust (GT) to the flat of the thrust. A) Unmigrated seismic section. B) Interpreted seismic section. Lines above seismic section are based on the downward projection of the surface geology (see Table 2 for code to reflectors; F = fault, inferred or mapped), and C) Simplified geological interpretation. (after Shaw et al. 1991a).

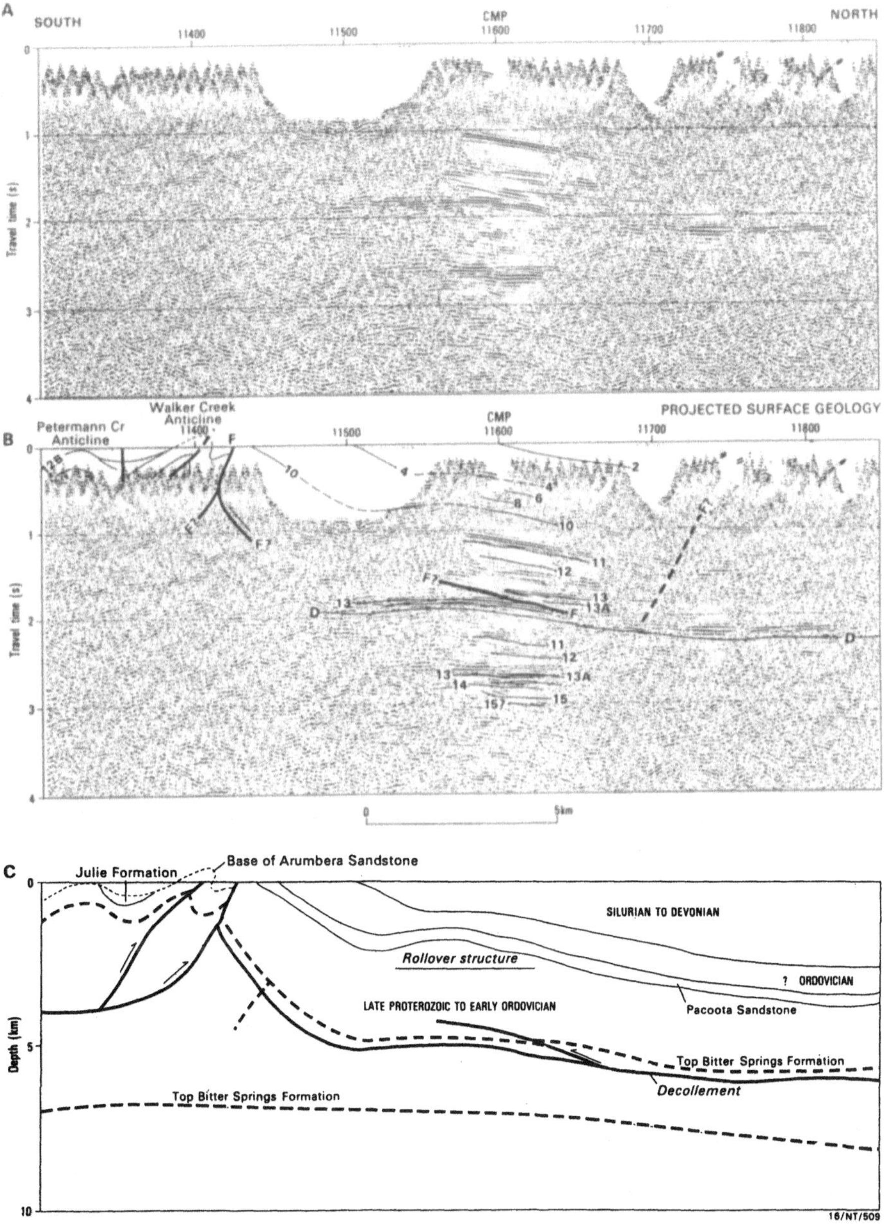

Figure 8. Section of the deep seismic reflection line L1D from north of Tempe Downs northwards towards the flat of the thrust running south from the Gardiner Range. A) Unmigrated section. B) Interpreted section. Lines above seismic section are based on the downward projection of the surface geology (see Table 2 for code to reflectors; F = fault, inferred or mapped). Note that the decollement (D; flat) is linked to the Gardiner Thrust Zone, and C) Simplified geological interpretation. (after Shaw et al. 1991a).

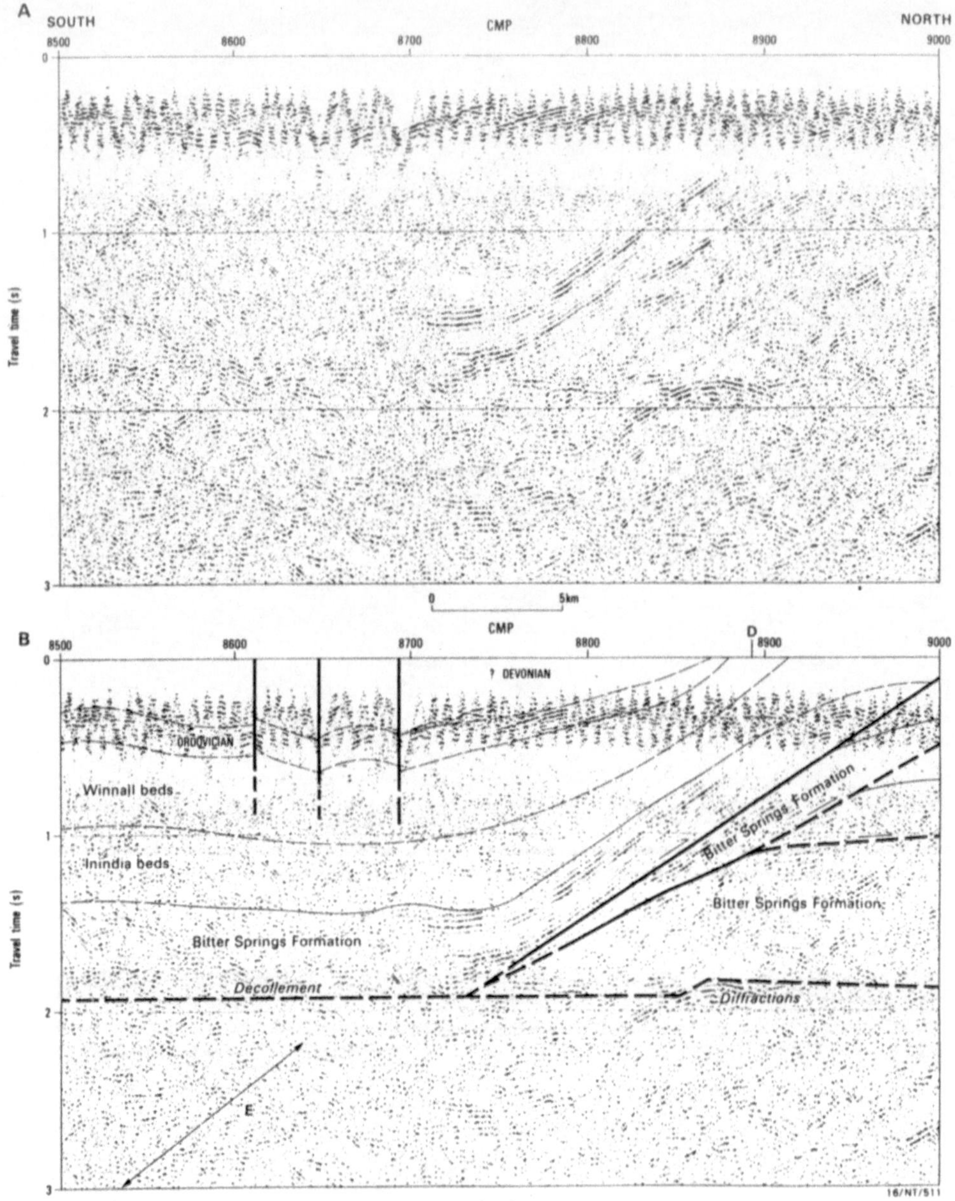

Figure 9. Seismic section in the vicinity of the Kernot Range showing the ramp and flat structure and several linked imbricate thrusts making up a thrust zone. A) Unmigrated section, and B) Interpreted section. Also shown is the extrapolated trend of strong bands of reflected energy (E) in the basement (see Wright et al. 1991b, for more details). (Figure after Shaw et al. 1991a).

Timing of Thrust Movements

Deformation in the southern two-thirds of the Amadeus Basin is related to north-directed overthrusting, whereas deformation of the northern margin of the basin (MacDonnell Homocline) is related to south-directed overthrusting associated with thrusting also directed southwards within the Arunta Block (Shaw et al. 1984; Shaw 1987).

In the region due south of the Gardiner Range to as far south as the Kernot Range, the tight folding is related to the linked thrust network that affected the Silurian-Early Devonian Mereenie Sandstone (Edgoose et al. 1990); that is, the effects of the Alice Springs Orogeny extent almost to the southern limit of the basin. The recognition that the effects of the Late Devonian to Carboniferous Alice Springs Orogeny can be traced to the southern margin of the basin raises the question of the role of the latest Proterozoic Petermann Ranges Orogeny in the basin. If the Petermann Ranges Orogeny produced folds, then the Alice Springs Orogeny tightened and flattened those folds.

Structural growth leading to stratigraphic thinning in the Walker Creek and Petermann Creek Anticlines (Fig. 1) suggests that these are thrust-cored anticlines in which movement occurred during the Late Proterozoic Petermann Ranges and precursor orogenies (Ranford et al. 1965; Shaw et al. 1991a). The en-echelon pattern displayed by anticlines northwest of Tempe Downs, ascribed by Bradshaw & Evans (1988) to right-lateral shear, may be the result of interference between Late Proterozoic and Devono-Carboniferous fold and thrust systems. These early thrust movements were not of sufficient magnitude to breach the anticlines. It is likely that in the Amadeus Basin the main effects of the Petermann Ranges Orogeny (Forman & Shaw 1984; Shaw et al. 1991a) were confined to the western part of the basin, where the Mount Currie Conglomerate and Cleland Sandstone now crop out. In the western part of the basin, Stewart et al. (1991) attributes folds to both the Petermann Ranges Orogeny and the Alice Springs Orogeny; good overprinting relationships can be observed in the outcrop patterns.

Minor movement on the Gardiner Thrust west of the BMR seismic traverse (Fig. 1) during the initial stages of the Alice Springs Orogeny (mid-Devonian Pertnjara Movement) is indicated by thinning of the Parke Siltstone across the structure (Jones 1991), and by the reduced depths of burial deduced there from low conodont alteration values (Gorter 1991). Most of the movement on the Gardiner Thrust postdates deposition of the Hermannsburg Sandstone which is faulted against the Mereenie Sandstone along most of the structure. Thrust movement there probably occurred in the Carboniferous, following deposition of the presently exposed Brewer Conglomerate, because the conglomerate shows evidence of derivation from a source in the Arunta basement to the north (Jones 1991) and not an upthrust source to the south. Thin-skinned thrust systems, such as that recognised here over such a large region in the central and southern parts of the Amadeus Basin, presumably developed in stages over an extended period (possibly of the order of 10-30 Ma).

Structural Synthesis of the Amadeus Basin Reflection Profile

When compared with the interpreted depths to stratigraphic units derived from the reflection-seismic profile (Figs. 7, 8 & 9), the magnetic 'basement' estimated by Wellman (1991) appears to rack the Bitter Springs Formation and may relate to spilite within the Loves Creek Member (cf. Shaw et al. 1991a). Secondary magnetic ridges outlined by Wellman (1991) correspond approximately to the Gardiner Thrust and to the region of tight folding south of the Walker

Anticline. Both of these belts contain the Bitter Springs Formation at shallow depth, which is locally exposed in tight anticline cores.

The tectonic situation shows how the different structural and geophysical features mentioned above relate to each other (Fig. 10). It is evident that the depth to magnetic basement tracks the depth to the sole thrust throughout the central and southern parts of the basin. Because the sole thrust is localised in the salt horizon within the Bitter Springs Formation, which is known to contain altered basalt (spilite) elsewhere in the basin (Wells et al. 1970; Wyatt 1983), the level above the sole thrust may be a major magnetic unit which is detected as 'magnetic basement'. This relationship explains why the depth to magnetic basement is shallowest within the thrust-cored anticlines, because the Bitter Springs Formation has been transported to higher levels. North of Tempe Downs (TD in Fig. 10), the depth to magnetic basement drops to the level of the lower

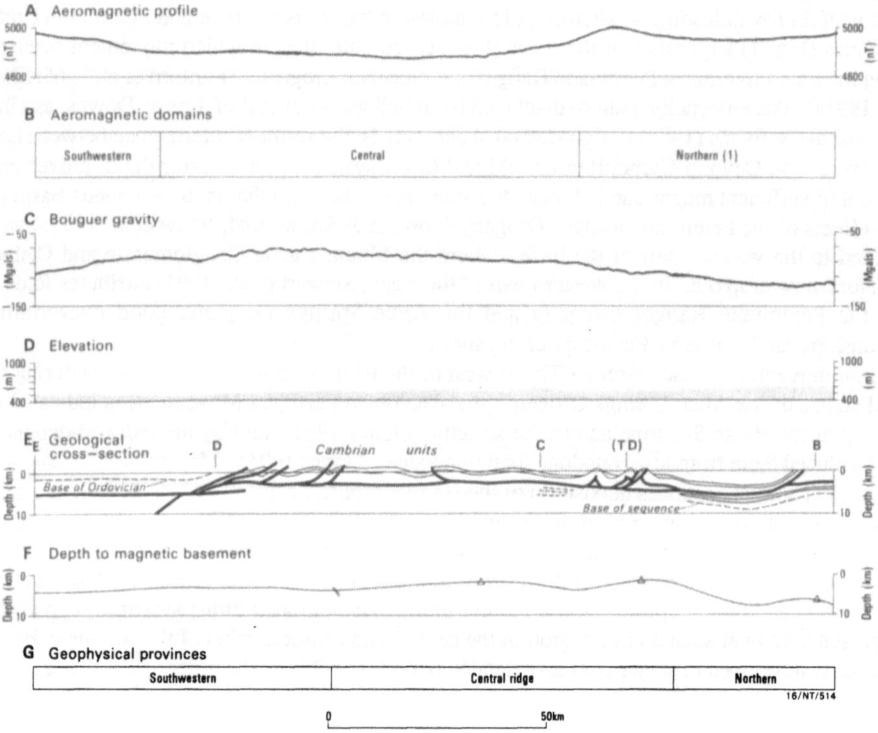

Figure 10. Composite section along the BMR seismic traverse showing its relationship to various geophysical and topographic features. A) Aeromagnetic profile (1985 survey along BMR seismic traverse), B) Aeromagnetic domains, C) Bouguer gravity (1km station spacing, 1985 BMR seismic survey), D) Elevation (a.s.l.), E) Generalised geological cross section (B = Missionary Plain; C = South of Petermann Creek Anticline; D = Detachment, Kernot Range region; E = end of traverse; TD = Tempe Downs, F) Depth to magnetic basement, triangles show magnetic ridges (after Wellman 1991), and G) Geophysical (combined gravity and magnetic) provinces of Wellman (1991). (after Shaw et al. 1991a).

Bitter Springs Formation under the flat section of the Gardiner Thrust, presumably because the Bitter Springs Formation in the overthrust sheet lacks spilite.

The simplified geological cross-section along the BMR reflection-seismic profile from the region of the Ngalia Basin southwards to the Kernot Thrust Zone in the southern Amadeus Basin is based on the interpretation of the reflection-seismic profile and interpolation of geological outcrop data in regions where the seismic image is poor (Fig. 11).

Discussion

Figure 12 is an attempt to synthesize our revised model for the crustal-scale structure of the central Australian region. Because the sub-horizontal sole thrust (Figs. 9, 10 & 12) cannot continue indefinitely to the south, a major structure is required somewhere towards the south of the basin to accommodate the shortening on a crustal scale. We postulate the existence of a major fault to the south of the Mount Currie Conglomerate (Fig. 12; near Uluru (Ayers Rock) north of the Woodroffe Thrust, Fig. 1). This fault would overthrust the basement onto sedimentary rocks, implying that sedimentary rocks may continue beneath the present-day basement exposures. This interpretation is supported by a series of sub-parallel magnetic and gravity gradients, evident in new detailed data, which have been interpreted as a complex set of major, anastomosing, north-directed thrusts (Edgoose et al. 1990).

Semi-continuous sedimentary sections can be traced underneath the Gardiner Thrust southwards from the Missionary Plain, implying duplication of the sedimentary section for at least 30 km. A thick (greater than 1.0 s TWT) layered succession imaged beneath the Bitter Springs Formation in the region of the Levi Syncline (Figure 1; CMP 10800-11050) may represent either a doubling of the Amadeus Basin succession or a relatively flat succession of pre-basin Proterozoic rocks. Although a thin-skinned structural model is favoured for the central-western and southwestern parts of the basin (Figs. 10 &12), we cannot completely discount the possibility that the basin structures are controlled by several deep crustal structures such as that imaged south of the Kernot Range (E in Fig. 9). Each of the closely-spaced fold and thrust packages may conceivably be controlled by local deep crustal structures. The distinct boundaries between magnetic domains are consistent with some degree of basement inheritance in the major structures of the basin (Shaw et al. 1991a). In order to explain why surface geological structures in the central and southwest parts of the basin are unrelated to the basement structural patterns inferred from detailed gradients evident in gravity and magnetic residual contours, Wellman (1991) suggested that a decollement underlay the bulk of the sedimentary succession at a relative shallow depth. However, Wellman (1991) found a closer relationship between surface geological structure and these geophysical patterns in the west and north of the basin, suggesting a greater degree of basement inheritance in the structural development of this area. Detailed magnetic and gravity modelling across the Kernot Range structures might help resolve the degree of basement inheritance there. An additional deep reflection-seismic survey or a teleseismic residual survey across the magnetic domain boundary farther south would help clarify the deep structure in the southern area.

Apatite-fission track results (Tingate 1991) indicate higher than average inheritance of pre-Alice Springs track ages in an easterly belt near Tempe Downs (Illamurta Structure and Petermann Creek Anticlines in Figure 1; near the northern margin of 'Central ridge'and 'TD' in Fig. 10). This implies that the rocks in this region were never buried by more than 1-2 km and were uplifted a similar amount during the Alice Springs Orogeny. If there was major duplication of the sedimentary section during the Alice Springs Orogeny in the region of Tempe Downs, as well as

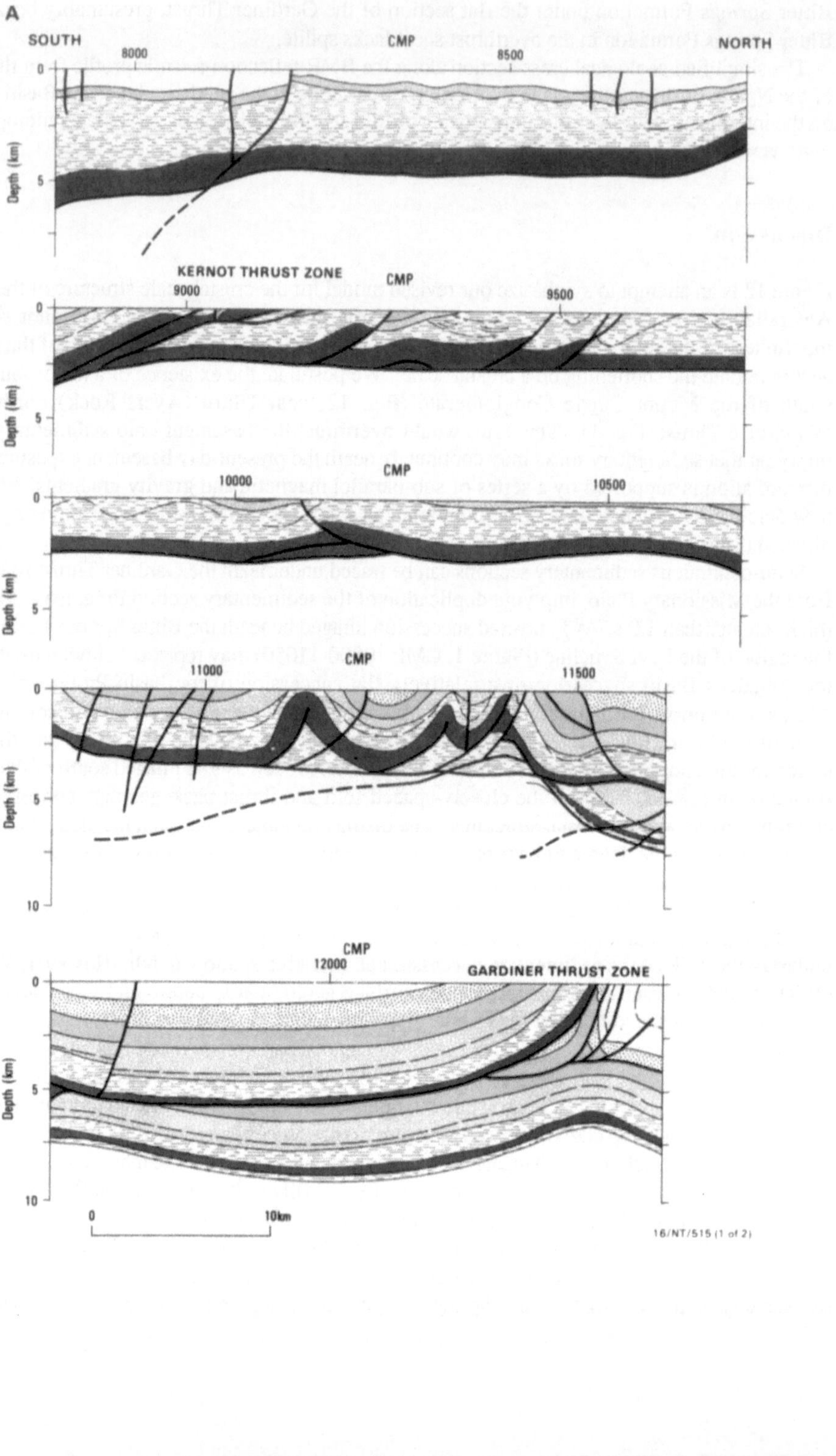
A
SOUTH
CMP
NORTH
8000
8500
0
5
Depth (km)
KERNOT THRUST ZONE
CMP
9000
9500
0
5
Depth (km)
CMP
10000
10500
0
5
Depth (km)
11000
CMP
11500
0
5
10
Depth (km)
CMP
12000
GARDINER THRUST ZONE
0
5
10
Depth (km)
0
10km
16/NT/515 (1 of 2)

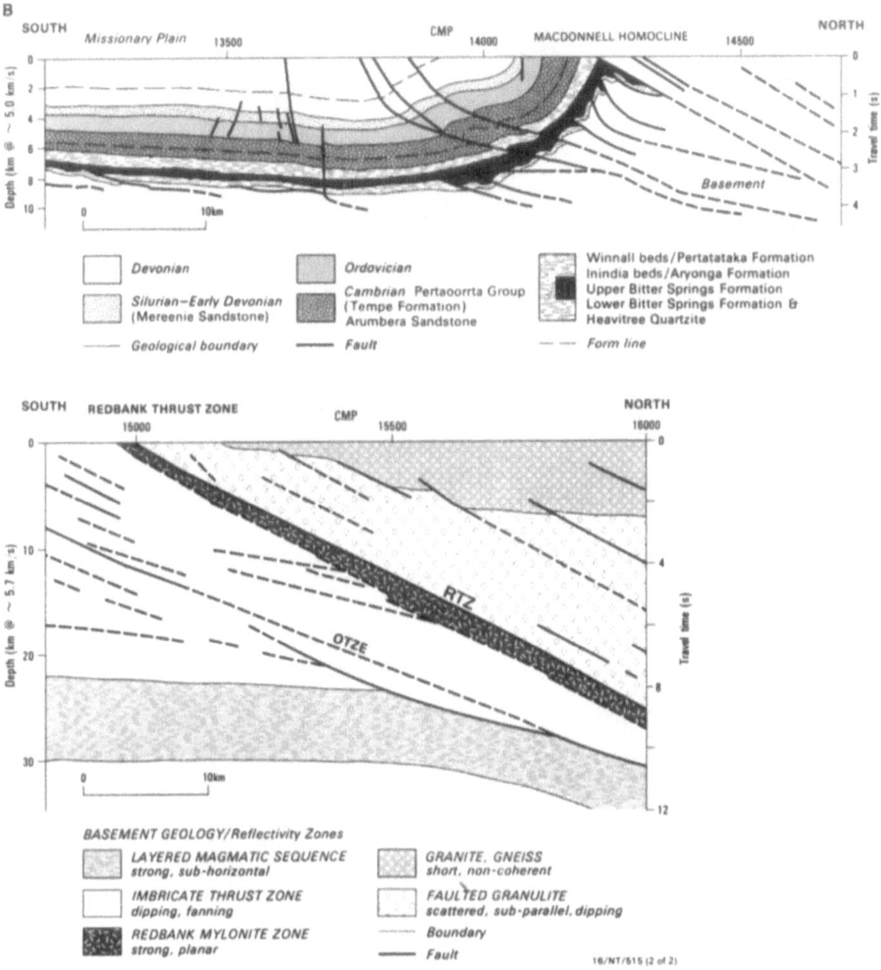

Figure 11. Simplified geological cross-section along the BMR reflection-seismic line (L1B, L1D, L1E in Figure 1) from the region of the Ngalia Basin southwards to the Kernot Thrust Zone in the southern Amadeus Basin. (Note that the seismic section recorded parallel to the Gardiner Thrust is omitted).
A) Geological interpretation of the southern end of the seismic line northward to the Gardiner Thrust, and
B) Geological interpretation of the Missionary Plain, MacDonnell Homocline, Ormiston Thrust Zone (ONTE), and Redbank Thrust Zone (RDZ; Panel for the Redbank Thrust Zone region shows a geological interpretation of seismic reflectivity zones). (Figure after Shaw et al. 1991a).

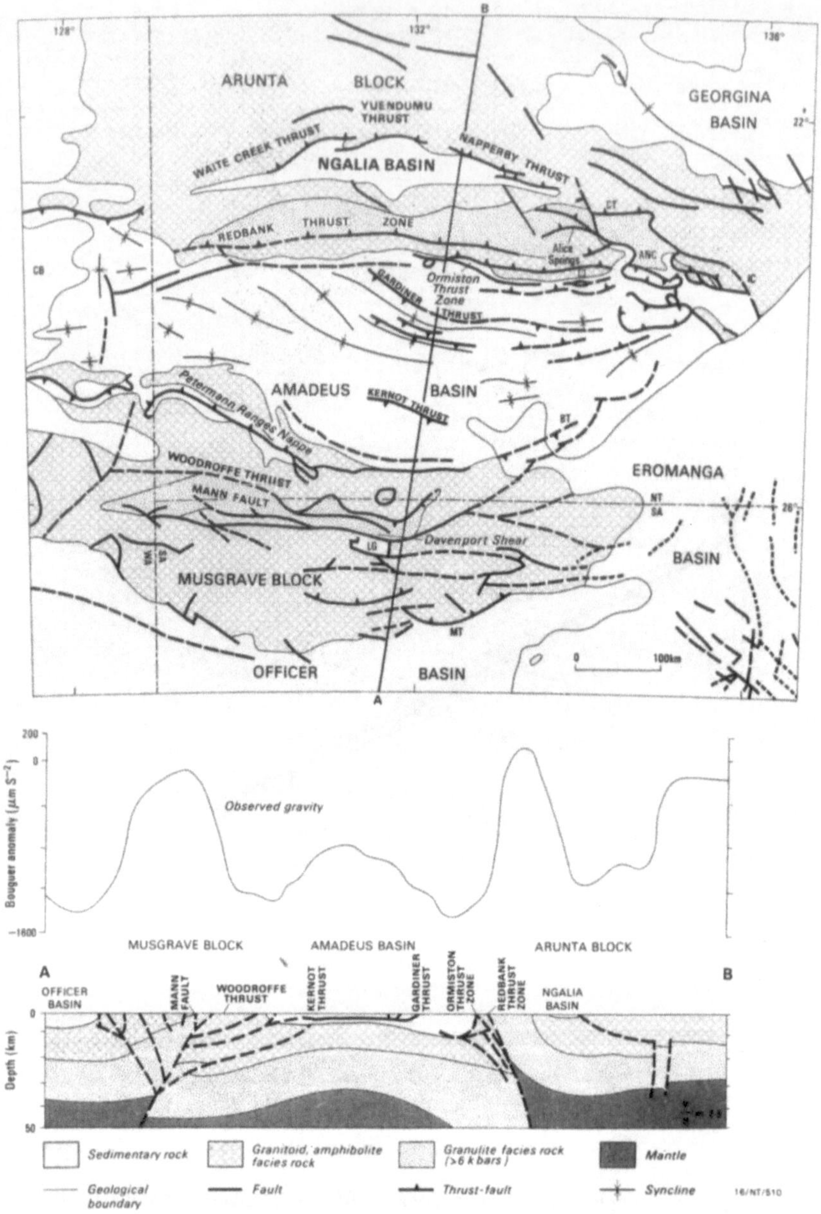

Figure 12. Major structural features of the Amadeus Basin and schematic north to south cross-section showing crustal-scale features. Additional structural features shown are: ANC = Arltunga Nappe Complex; CB = Canning Basin; CT = Cadney Thrust; MT = Munyadai Thrust . (Figure after Shaw et al. 1991a, b).

farther to the south, then uplift of more than 2 km might be expected to have occurred over this southern region, and this would have given rise to apatite fission tracks of Late Devonian or younger age (i.e. syn-or post-Alice Springs Orogeny; cf. Tingate 1991). For this reason, we do not favour a structural model involving substantial duplication of the sedimentary succession in the central and southwest of the basin, south of Tempe Downs.

The reason why upthrusting, rather than more sub-horizontal south-directed thrusting, took place at the northern margin of the Amadeus Basin becomes more understandable now that major north-directed cover thrusts such as the Gardiner Thrust and the Kernot Range Thrust System have been identified in the central and southwestern sectors of the Amadeus Basin (Figs. 10 and 11). The basement thrusting north of the basin margin involved the upward translation of a thrust wedge, rather than a more horizontal displacement, because easy thrust translation onto the foreland to the south was opposed by the north-directed, thin-skinned thrust complex within the basin. Although the basement-thrust systems resemble 'pop-up' structures, strike-slip movement is unlikely to have accommodated much of the crustal shortening because stretching lineations in the shear zones are orthogonal to the strike of the main thrusts (Teyssier 1985; Shaw & Black 1991). The relatively narrow trough (Brewer Trough) in the Missionary Plain, which contains synorogenic sediment (Hermannsburg Sandstone and Brewer Conglomerate), has been mechanically modelled as a crustal-scale synclinal footwall depression in front of the thick-skinned thrust system comprising the Ormiston and Redbank Thrust Zones (Shaw 1987; Shaw et al. 1991a&b). The dominantly thin-skinned thrust system of the central and southern Amadeus Basin, here termed the Gardiner-Kernot Thrust Complex, has over-ridden the synorogenic Brewer Trough resulting in overturning of the Mereenie and Hermannsburg Sandstones on the southern flank of the trough.

The convergence of thrust systems in the western sector of the basin explains why the sedimentary succession under the Missionary Plain escaped substantial tectonic activity, with the exception of the marked upturning of the succession at its northern and southern ends. Forward northward progression of the sole thrust of the thin-skinned southern thrust system was hindered by the upthrusting of basement wedges at the northern margin of the basin. The sedimentary succession under the Missionary Plain was probably thrust northwards to a limited extent over the basement at the MacDonnell Homocline. We suggest that the basement block underlying the Missionary Plain became tilted downwards and northwards as it attempted to underthrust the rising basement-thrust wedges which were linked to the crust-cutting Redbank Thrust Zone. This region of block tilting correlates with the region where the crust may be thicker than normal (~ 50 km). The balance of mechanical forces results in the present structure being out of isostatic equilibrium in the region between the overthrust crust centred on the Redbank Thrust Zone to the north and the underthrust crust to the south (Lambeck et al. 1988). Although a component of crustal thickening is due to the block-tilting, other additional causes for crustal thickening in this region are tectonic underplating by lower crustal underthrust slivers and/or crustal thickening inherited from earlier Proterozoic tectonism.

The upturned northern margin of the Amadeus Basin (MacDonnell Homocline) is considered to have developed as a triangular zone of strain accommodation (Lowell 1985; Soper & Higgins 1990) rather than as a simple antiformal thrust stack in the basement (cf. Vann et al. 1986). Scaled analogue modelling of a similar structure using rock models deformed under confining pressure (Chester et al. 1988) has shown that ramp anticlines similar to the MacDonnell Homocline may develop in thrust systems at the change from crystalline basement to the weaker sedimentary succession. In these models, a veneer having a high degree of mechanical anisotropy at the base of the sedimentary succession channels the interlayer-slip and favours the development of kink-folds.

The decollement developed in the Bitter Springs Formation near the base of the Amadeus Basin succession apparently accentuated the channelling of these forces. The geometry of mountain front monoclines, such as that developed at the northern part of the Amadeus Basin, depends on the interplay between crustal blocks with differing mechanical properties as well as the dominant dip of the controlling basement-thrust structures (Fig. 1; cf. Chester et al. 1988).

The Redbank Thrust Zone has now been confirmed in two across-strike profiles as a major crust-cutting feature which separates tectonic blocks with distinctly different character in reflection-seismic images (Wright et al. 1991a). The apparent lack of a listric geometry in the deeper sections of the Redbank Thrust Zones and its apparent penetration of the Moho are consistent with the 'thick-skinned' models for the tectonic evolution of the Arunta Block proposed by Shaw (1987) & Goleby et al. (1989).

The debate concerning the general applicability of 'thick-skinned' and 'thin-skinned' thrusting models, what has been taken up in several regions, notably in the Pyrenees (McCaig 1988; Roume et al. 1989) should be further considered in the light of our results in central Australia. Thick-skinned and thin-skinned thrust models are two end-members of a range of thrusting styles. We have demonstrated that these two thrusting styles can operate simultaneously and are not mutually exclusive.

Estimates of crustal shortening across the linked northern thrust belt, made up principally of the Redbank and Ormiston Thrust Zones, vary from as low as 25-35 km (15-25%) based on a thick-skinned model (Shaw et al. 1991b) to as high as 66 km (50%) for the models emphasising thin-skinned features (e.g. Teyssier 1985). Again, in the northeast of the Amadeus Basin thrusting shows both thin-skinned and thick-skinned elements, with the amount of crustal shortening estimated (30 km versus 60-70 km) depending on details of the structural model adopted (Stewart et al. 1991). Thus, one important outcome (Teyssier 1985; cf Fig. 2A) of the thick-skinned – thin-skinned debate is that it has drawn attention to the fact that the balancing of cross-sections is model dependant. A principal assumption in the balancing of sections on the scale of the crust will, necessarily, be the initial configuration of the crustal blocks. Other important considerations for the structural models are the extent of subcrustal underthrusting (which may involve delamination and/or A-subduction) and the amount of erosion at the thrust front. Shaw et al. (1984) and more recently Shaw & Black (1991) have demonstrated that the Redbank Thrust Zone represents a reactivated Proterozoic province boundary with a long and complex history. Consequently, we cannot assume that the crust was initially the same thickness on either side of the thrust zone in any model adopted for section balancing. Again, in the central and southern parts of the Amadeus Basin, we would need to make assumptions on the amount of duplication of the sedimentary succession and the geometry of thrusts in the root zone in relation to the thrust system before section balancing could be attempted.

We should now move our attention away from debates about whether the thrust system in any area is thick-skinned or thin-skinned, and try to look in more detail at the wide range of thrusting styles. In re-examining the deformation style in the classic region of so-called thick-skinned tectonics of the Laramide uplifts in the Wyoming foreland (USA), Brown (1989) and Spang & Evans (1989), have emphasised the wide variety of structural styles that have developed depending on the orientation of pre-existing structures with respect to the regional stress field and the mechanical properties of the basement and sedimentary rocks. In the Pyrenees, a variety of thrusting styles are also recognised. In central Australia, we also have evidence for a variable dip to the master-thrust systems and control of them by pre-existing deep crustal structure as in the Pyrenees (Roume et al. 1989).

Conclusions

The following features have been identified from the deep reflection-seismic profile.

1) The Redbank Thrust Zone is a relatively steeply dipping (around 45°) planar feature which cuts through the crust to mantle depths.
2) A thrust complex, which is equivalent to the regional Ormiston Thrust Zone, appears to be linked at depth to the north-dipping, crust-cutting Redbank Thrust Zone.
3) The MacDonnell Homocline is the preserved part of a mountain front monocline which is considered to result from rotation of the Late Proterozoic-Devonian Amadeus Basin succession, together with localised northward back-thrusting on a decollement at the base of the sedimentary succession.
4) A gently north-dipping para-autochthonous sedimentary section is present in the Missionary Plain region between the Gardiner Thrust and the MacDonnell Homocline.
5) Thrust repetition of part of the sedimentary succession (Bitter Springs Formation) by the Gardiner Thrust is indicated by identification of the seismic-reflection sections at a depth of 6.9-8.0 km beneath Cambrian-Devonian sediments. The seismic sections suggest that surface exposures south of the Gardiner Thrust are allochthonous, having been deposited at least 30 km to the south before transportation to their present position, and to have been uplifted at least 6 km.
6) Farther south, the seismic-reflection section shows a roll-over structure, possibly related to backthrusting, and a north-dipping repetition of the Late Proterozoic to Early Cambrian succession, possibly developed on the northern flank of a duplex-thrust zone.
7) In the southern part of the basin, there are tight macroscopic anticlines, some with overturned northern limbs. These anticlines are interpreted to be cored by blind to emergent thrusts that branch off the main sub-horizontal thrust at a depth of about 4.5 km. The best example of such an imbricate complex imaged in the seismic-reflection profiling is the Kernot Thrust.
8) South of the Gardiner Range, the Alice Springs Orogeny appears to have affected the sedimentary succession to the southernmost margin of the Amadeus Basin.
9) The two contrasting converging thrust systems, illustrated by the results of the reflection-seismic traverse across central Australia, both have thick-skinned and thin-skinned elements; with the northern thrust system being more obviously thick-skinned. Thus, these two styles of thrusting are not mutually exclusive. Whether one style or the other tends to develop depends on the interplay between a variety of factors including the orientation of pre-existing structures and on the differing mechanical properties of crystalline basement and sedimentary cover rocks.

Acknowledgments

The authors acknowledge the contributions from all members of the BMR Seismic Party. We thank John Lindsay for critically reading early versions of this manuscript. The figures were drafted by Lindell Emerton. This paper is published with the permission of the Executive Director, Bureau of Mineral Resources.

References

Allegre, C.J., Courtillot, V., Tapponnier, P., Hirn, A., Mattauer, M., Coulon, C., Jaeger, J.J., Achache, J., Scharer, U., Marcoux, J., Burg, J.P., Girardeau, J., Armijo, R., Gariepy, C., Gopel, C., Li. Tindong, Xiao Xuchang, Chang Chenfa, Li Guangqin, Lin Baoyu, Teng Jiwen, Wang Naiwen, Chen Guoming, Han Tonglin, Wang Xibin, Den Wanming, Sheng Huaibin, Cao Yougong, Zhou Ji, Qiu Hongrong, Bao Peisheng, Wang Songchan, Wang Bixiang, Zhou Yaoxiu, & Ronghua Xu 1984. Structure and evolution of the Himalaya-Tibet orogenic belt. Nature **5**, 17-22.

Allmendinger, R.W., Sharp, J.W., Von Tish, D., Serpa, L., Brown, L., Kaufman, S., & Oliver, J. 1983. Cenozoic and Mesozoic structure of the eastern Basin and Range province, Utah, from COCORP seismic reflection data. Geology **11**, 532-536.

Bradshaw, J.D., & Evans, P.R. 1988. Palaeozoic tectonics, Amadeus Basin, central Australia. APEA Journal **28**, 267-282.

Brown, W.G. 1989. Deformational style of Laramide uplifts in the Wyoming forelands. In, Schmidt, C.J. & Perry, Jr, W.J. (eds) Interaction of the Rocky Mountain Foreland and the Cordillera Thrust Belt. Geological Society of America, Memoir **171**, pp. 65-74.

Butler, R.W.H. 1983. Balanced cross-sections and their implications for the deep structure of the northwestern Alps. Journal of Structural Geology **5**, 125-137.

Cao, S., Goleby, B.R., & Kennett, B.L.N. 1991. Modelling seismic reflections in the Amadeus Basin by 3D isochronal technique. Exploration Geophysics, (submitted).

Chester, J.S., Spang J.H., & Logan, J.M. 1988. Comparison of thrust fault rock models to basement-cored folds in the Rocky Mountain foreland. In, Schmidt, C.J. & Perry, Jr, W.J. (eds) Interaction of the Rocky Mountain Foreland and the Cordillera Thrust Belt Geological Society of America, Memoir **171**, 65-74.

Coward, M.P. 1983. Thrust tectonics, thin-skinned or thick-skinned, and the continuation of thrusts to deep in the crust. Journal of Structural Geology **5**, 113-123.

Coward, M.P., Butler, R.W.H., Asif Khan, M., & Knipe, R.J. 1987. The tectonic history of Kohistan and its implications for Himalayan structure. Journal of the Geological Society of London **144**, 377-391.

Edgoose, C.J., Camacho, A., Wakelin-King, G.,W., & Simons, B. 1990. Kulgera, Northern Territory. 1:250000 Geological Series. Northern Territory Geological Survey, Explanatory Notes, SG53-5 (in press).

Forman, D.J., & Shaw, R.D. 1973. Deformation of the crust and mantle in central Australia. Bureau of Mineral Resources, Australia, Bulletin **144**, 20 pp.

Goleby, B.R., Kennett, B.L.N., Wright, C., Shaw, R.D., & Lambeck, K. 1990. Results from deep seismic reflection profiling in the Proterozoic Arunta Block of central Australia: processing for testing models of tectonic evolution. Tectonophysics **173**, 257-268.

Goleby, B.R., Shaw, R.D., Wright, C., Kennett, B.L.N., & Lambeck, K. 1989. Geophysical evidence for thick-skinned crustal deformation in central Australia. Nature **337**, 325-330.

Goleby, B.R., Wright, C., Collins, C.D.N., & Kennett, B.L.N. 1988. Seismic reflection and refraction profiling across the Arunta Block and the Ngalia and Amadeus Basins. Australian Journal of Earth Sciences **35**, 275-294.

Gorter, J.D. 1991. Palaeogeography of Late Cambrian - Early Ordovician sediments in the Amadeus Basin, central Australia. In, Korsch, R.J. & Kennard, J.M. (eds) Geological and Geophysical Investigations in the Amadeus Basin, central Australia. Bureau of Mineral Resources, Australia, Bulletin 236.

Jones, B.G. 1991. Fluvial and lacustrine facies of the Middle and Late Devonian Pertnjara Group, Amadeus Basin, Northern Territory and their relationship to tectonic events and climate. In, Korsch, R.J., & Kennard, J.M. (eds) Geological and Geophysical Investigations in the Amadeus Basin, central Australia. Bureau of Mineral Resources, Australia, Bulletin 236.

Korsch R.J., Shaw, R.D., Wright, C., Goleby, B.R., & Collins, C.D.N. 1990. Constraints on the tectonic evolution of the Amadeus Basin, central Australia: Implications for hydrocarbon exploration. In, Pinet, B., & Bois, C. (eds) The Potential of Deep Seismic Profiling for Hydrocarbon Exploration, Editions Technip, Paris, 249-264.

Lambeck, K. 1984. Structure and evolution of the Amadeus, Officer and Ngalia Basins of central Australia. Australian Journal of Earth Sciences **31**, 25-48.

Lambeck, K. 1991. Teleseismic travel-time anomalies and deep crustal structure of the northern and southern margins of the Amadeus Basin. In, Korsch, R.J., & Kennard, J.M. (eds) Geological and Geophysical Investigations in the Amadeus Basin, central Australia. Bureau of Mineral Resources, Australia, Bulletin 236.

Lambeck, K., Burgess, G., & Shaw, R.D. 1988. Teleseismic travel-time anomalies and deep crustal structure in central Australia. Geophysical Journal of the Royal Astronomical Society **94**, 105-124.

Lindsay, J.F., & Korsch, R.J. 1991. The evolution of the Amadeus Basin, central Australia. In, Korsch, R.J., & Kennard, J.M. (eds) Geological and Geophysical Investigations in the Amadeus Basin, central Australia. Bureau of Mineral Resources, Australia, Bulletin 236.

Lowell, J.D. 1985. Structural Styles in Petroleum Geology, Oil and Gas Consultants international Inc. Tulsa.(course notes)

Mathur, S.P. 1976. Relationship of Bouguer anomalies to crustal structure in southwestern and central Australia. BMR Journal of Australian Geology and Geophysics **1**, 277-286.

McCaig, A.M. 1988. Deep geology of the Pyrenees. Nature **331**, 480-481.

Ranford, L.C., Cook, P.J., & Wells, A.T. 1965. The geology of the central part of the Amadeus Basin, Northern Territory. Bureau of Mineral Resources, Australia, Report 86.

Rourne, F., Choukroune, P., Berastegui, X., Munoz, J.A., Villien, A., Matheron, P., Bareyt, M., Segunet, M., Camara, P., & Deramond, J. 1989. ECORS deep seismic data and balanced cross-sections: geometric constraints on the evolution of the Pyrenees. Tectonics **8**, 41-50.

Schroder, R.J., & Gorter, J. 1984. A review of recent exploration and hydrocarbon potential of the Amadeus Basin, Northern Territory. APEA Journal **24**, 19-41.

Shaw, R.D. 1987. Basin uplift and basin subsidence in central Australia. Ph.D. thesis, Australian National University, Canberra, Australia (unpublished).

Shaw, R.D., & Black, L.P. 1991. The history of the Redbank Thrust Zone in central Australia: based on structural, metamorphic and Rb-Sr isotopic evidence: Tectonic Implications. Australian Journal of Earth Sciences (in press).

Shaw, R.D., Korsch, R.J., Wright, C., & Goleby, B.R. 1991a. Seismic interpretations and thrust tectonics of the Amadeus Basin, central Australia, along the BMR regional seismic line. In, Korsch, R.J., & Kennard, J.M. (eds) Geological and Geophysical Investigations in the Amadeus Basin, central Australia. Bureau of Mineral Resources, Australia, Bulletin 236.

Shaw, R.D., Etheridge, M.A., & Lambeck, K. 1991b. Development of the Late Proterozoic to Mid-Palaeozoic, intracratonic Amadeus Basin in central Australia: a key to understanding tectonic forces in plate interiors. Tectonics (in press).

Shaw, R.D., Stewart, A.J., & Black, L.P. 1984. The Arunta Inlier: a complex ensialic mobile belt in central Australia, part 2: Tectonic history. Australian Journal of Earth Sciences **31**, 457-484.

Soper, N.J., & Higgins, A.K. 1990. Models for the Ellesmerian mountain front in northern Greenland: a basin margin inverted by basement uplift. Journal of Structural Geology **12**, 83-97.

Spang, J.H., & Evans, J.P. 1989. Geometrical and mechanical constraints on basement-involved thrusts in the Rocky Mountain foreland province. In, Schmidt, C.J., & Perry, Jr, W.J. (eds) Interaction of the Rocky Mountain Foreland and the Cordillera Thrust Belt, Geological Society of America, Memoir **171**, 65-74.

Stewart, A.J., Oaks, R.Q., Deckelman, J.A., & Shaw, R.D. 1991. 'Mesothrust' versus 'megathrust' interpretations of the structure of the northeastern Amadeus Basin, central Australia. In, Korsch, R.J., & Kennard, J.M. (eds) Geological and Geophysical Investigations in the Amadeus Basin, central Australia. Bureau of Mineral Resources, Australia, Bulletin 236.

Teyssier, C. 1985. A crustal thrust system in an intracratonic tectonic environment. Journal of Structural Geology **7**, 689-700.

Tingate, P.R. 1991. Apatite fission track analysis of the Pacoota and Stairway Sandstones, Amadeus Basin, central Australia. In, Korsch, R.J., & Kennard, J.M. (eds) Geological and Geophysical Investigations in the Amadeus Basin, central Australia. Bureau of Mineral Resources, Australia, Bulletin 236.

Vann, I.R., Graham, R.H., & Hayward, A.B. 1986. The structure of mountain fronts. Journal of Structural Geology **8**, 215-227.

Wellman, P. 1991. Amadeus Basin, Northern Territory: structure from gravity and magnetic anomalies. In, Korsch, R.J., & Kennard, J.M. (eds) Geological and Geophysical Investigations in the Amadeus Basin, central Australia. Bureau of Mineral Resources, Australia, Bulletin 236.

Wells, A.T., Forman, D.J., Ranford, L.C., & Cook, P.J. 1970. Geology of the Amadeus Basin, central Australia. Bureau of Mineral Resources, Australia, Bulletin 100, 222 pp.

Wright, C., Goleby, B.R., Shaw, R.D., Collins, C.D.N., Kennett, B.L.N., & Lambeck, K. 1991a. Two seismic profiles across a major thrust zone in central Australia. Earth and planetary Science Letters (submitted).

Wright, C., Goleby, B.R., Shaw, R.D., Collins, C.D.N., Korsch, R.J., Barton, T., Greenhalgh, S.A., & Sugiharto, S. 1991b. Seismic reflection and refraction profiling in central Australia: implications for understanding the evolution of the Amadeus Basin. In, Korsch, R.J., & Kennard, J.M. (eds) Geological and Geophysical Investigations in the Amadeus Basin, central Australia. Bureau of Mineral Resources, Australia, Bulletin 236.

GEOMETRY OF PERMIAN TO MESOZOIC SEDIMENTARY BASINS IN EASTERN AUSTRALIA AND THEIR RELATIONSHIP TO THE NEW ENGLAND OROGEN

R.J. KORSCH, K.D. WAKE-DYSTER, P.E. O'BRIEN, D.M. FINLAYSON AND D.W. JOHNSTONE
Onshore Sedimentary and Petroleum Geology Program
Bureau of Mineral Resources
GPO Box 378
Canberra ACT 2601
Australia

ABSTRACT. Sedimentary basins adjacent to the New England Orogen in eastern Australia contain a record of events from the Early Permian to the Late Jurassic that provide information on the later history of the orogen. The basins have complex geometries, as seen in BMR and company seismic data. The half-grabens of the Permian Taroom Trough (southern, sub-surface extension of the Bowen Basin) and Triassic Esk Trough formed by oblique extension, with the steep bounding faults having a significant strike-slip component. Localised thrusts, folds and positive flower structures in Jurassic sedimentary rocks resulted from transpression associated with reactivation of the strike-slip faults. Some previous workers postulated that the Bowen Basin, and also the Sydney Basin to the south, were initiated during a period of ENE-WSW oriented extension in the latest Carboniferous or earliest Permian. Nevertheless, these basins were separated by a zone in the southern Bowen and Gunnedah basins that was dominated by oblique extension and strike-slip. During subsidence of the basins, strike-slip fault movements also played a significant role in the adjacent basement, controlling the transport and accretion of displaced terranes and oroclinal bending of an accretionary wedge sequence. Hence the New England Orogen, as well as the adjacent basins, was greatly influenced by strike-slip faulting for a considerable part of its history.

Introduction

Several sedimentary basins in eastern Australia cover or flank the New England Orogen (Fig. 1) and contain sedimentary records that extend from the Early Permian to the Cretaceous. Of these, the Surat Basin and the Denison Trough of the Bowen Basin contain significant oil and gas reserves (Elliott & Brown 1988) and the other basins are considered to be potentially prospective for hydrocarbons.

In this paper we examine aspects of the geometry of the Bowen, Surat, Esk, Ipswich and Clarence-Moreton basins, mainly using Bureau of Mineral Resources (BMR) deep seismic reflection data, to constrain mechanisms for the development of the basins, and to examine the implications for the history of the surrounding basement from latest Carboniferous to the Cretaceous.

M. J. Rickard et al. (eds.), Basement Tectonics 9, 85–108.

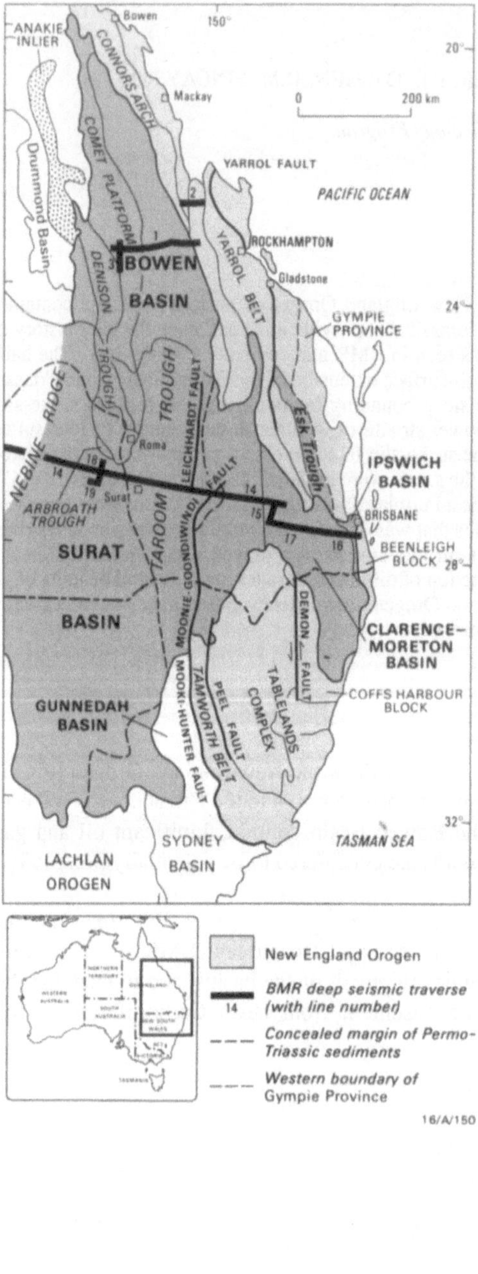

Figure 1. Locality map of eastern Australia showing locations of the Permian to Cretaceous sedimentary basins and positions of the BMR deep seismic lines. Lines 14 to 19 refer to data recorded in 1984 (BMR84.14-16) and 1896 (BMR84.17-19) and lines 1 to 3 refer to data recorded in 1989 (BMR89.B01-B03).

Models for basin evolution

Several models for the formation of the Bowen-Gunnedah-Sydney basin system (hereafter referred to as the Bowen basin system) have been proposed recently (see also summary in Murray 1990):

1. Foreland (foredeep) models or foreland loading mechanisms have been proposed by Jones et al. (1984), Murray (1985) and Hobday (1987). A basin can only subside due to foreland loading for a relatively short period, that is, during the duration of thrust events and for a very limited period of time after cessation of thrusting. The Bowen basin system subsided over a period of at least 150 Ma, both before and after shortening events, and thus the subsidence must have been driven by other mechanisms. Nevertheless, foreland loading operated as a subsidence mechanism for brief periods in the subsidence history of at least parts of the basin system.

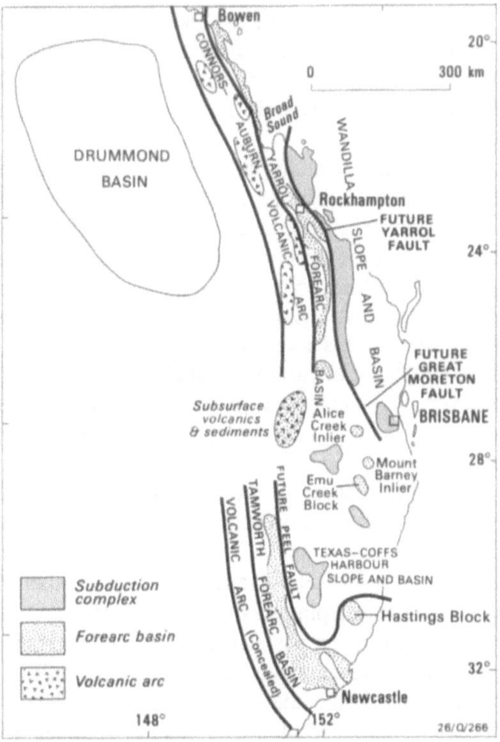

Figure 2. Map of the Late Devonian - Early Carboniferous palaeogeography (not restored palinspastically) showing the present-day distribution of the volcanic arc, fore-arc basin and accretionary wedge (after Murray et al. 1987).

2. Extensional origin: e.g. Denison Trough of the Bowen Basin (Paten et al. 1979; Bauer & Dixon 1981; Ziolkowski & Taylor 1985); whole Bowen Basin (Hammond 1987, 1988a); Surat Basin (Wake-Dyster et al. 1987a); early Permian extension in the Bowen-Sydney

system (Hobday 1987); Sydney Basin (Mallett et al. 1988); Gunnedah Basin (Tadros 1988). These models suggest that the extensional direction was approximately ENE.

3. Transtensional origin: e.g. Bowen basin system (Harrington 1982); Gunnedah Basin as a strike-slip basin (Harrington & Korsch 1985a); Taroom Trough of the Bowen Basin (Korsch et al. 1988, 1990a); Esk Trough (Korsch et al. 1989). The transtensional origin for the Taroom Trough is related to the north-south orientation of the Mooki Fault and is not incompatible with the extensional model for the Bowen Basin farther to the north.
4. Mixed-mode origin: This involves an early history of the basin system involving extension and/or transtension followed by a later history dominated by compression and/or transpression (e.g. Ziolkowski & Taylor 1985; Hobday 1987; Hammond 1987, 1988a; Korsch et al. 1988, 1990a; Elliott 1989; Fielding et al. 1990). Hence, all the models proposed previously are not mutually exclusive, and each could have operated at different times during the history of the basin.

Basement Geology

The western and southwestern parts of the Bowen basin system overlie the Lachlan and Thomson orogens, whereas to the east of a complex fault system the basins overlie the younger New England Orogen (Fig. 1). This fault system, the Mooki fault system *sensu lato* of Harrington & Korsch (1985a), consists of the Hunter, Mooki, Goondiwindi, Moonie, Leichhardt and Burunga faults, and had a major influence on basin development.

As yet, a consensus has not been reached on the tectonic evolution of the Lachlan Orogen; for descriptions of the geology and possible tectonic models refer to Crook & Powell (1976), Scheibner (1978, 1986), Cas (1983), Powell (in Veevers 1984), Leitch & Scheibner (1987), Chappell et al. (1988) and Coney et al. (1990). However, deformation in the Lachlan Orogen effectively ceased in the mid-Carboniferous (Powell, in Veevers 1984), some time before the basins were initiated, and therefore had little effect on basin history. The Thomson Orogen is mostly concealed beneath the younger sedimentary basins and is not well understood.

To the east of the Lachlan Orogen, tectonic development of eastern Australia was dominated during the Devonian and Carboniferous by a convergent plate margin setting, with west-dipping subduction taking place (see summary by Korsch et al. 1990b). An accretionary wedge formed the main part of the New England Orogen (the Tablelands Complex) and a fore-arc basin corresponded to the Tamworth and Yarrol belts (Fig. 2). In the southern part of the orogen in the New England Province, the volcanic arc is now missing, either largely concealed beneath the Gunnedah Basin or Tamworth Belt, or removed by erosion, by strike-slip faulting or both. To the north in the Yarrol Province, the magmatic arc is represented by the Connors and Auburn volcanic arches. Although the main part of the New England Orogen is related to west-dipping subduction, at least two units, the Gympie Province and Beenleigh Block (Fig. 1), have been postulated to be exotic terranes (e.g. Korsch & Harrington 1987; Murray 1988). There was a marked change in the tectonics of eastern Australia in the Late Carboniferous. Subduction ceased in the west, and jumped to the east of the present New England Orogen (excluding the Gympie terrane). A system of sedimentary basins was initiated in a back-arc setting. Following this, intense deformation on a regional scale occurred in zones within the orogen.

During the early Permian, the extensive Bowen basin system containing volcanics and shallow marine to terrestrial sediments developed along the western margin of the New England Orogen,

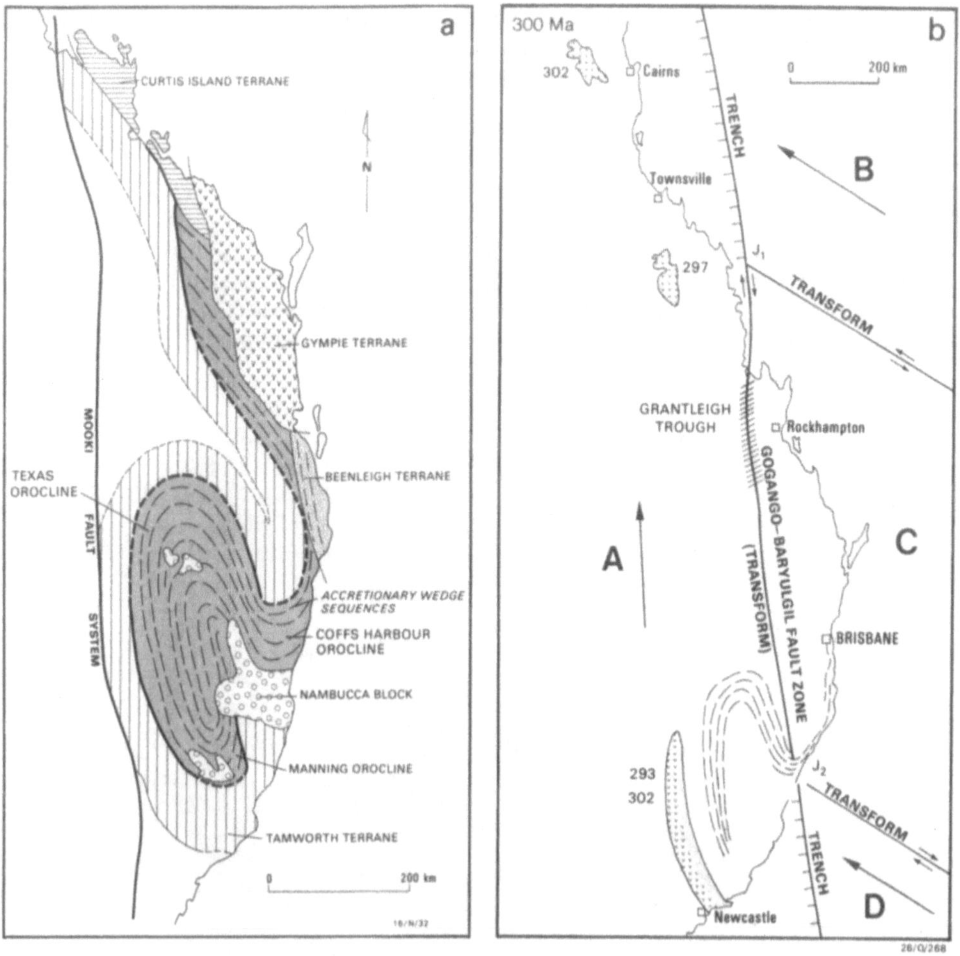

Figure 3. Orocline models of (a) Korsch and Harrington (1987) and (b) Murray et al. (1987). The major difference is in the location of the controlling strike-slip fault zone, which Korsch and Harrington infer to be their Mooki Fault System, whereas Murray et al. interpret it to be along their postulated Gogango-Baryulgil transform fault zone.

whereas to the east, some deep-water sedimentation occurred in localised basins, including the deposition of thick successions of marine diamictites (see Scheibner 1976; Korsch 1982).

After subduction ceased, the New England Orogen (or at least the accretionary wedge and fore-arc basin parts) was deformed by major oroclinal bending (Korsch & Harrington 1987; Murray et al. 1987) due to southwards movement of about 450-500 km by the fore-arc basin and accretionary wedge. There is some debate as to the mechanism of formation of the orocline; Korsch and Harrington (1987) and Harrington and Korsch (1987) favoured control on a western bounding fault, whereas Murray et al. (1987) inferred a major fault in the centre of the orogen (Fig. 3). To allow the southward movement of the orogen and formation of the orocline, Harrington and Korsch (1987) inferred the presence of a sub-horizontal, mid-crustal detachment surface, on which the oroclinal material was allowed to slide (Fig. 4). After the oroclinal bending, the Beenleigh and Gympie terranes accreted by predominantly strike-slip movements.

Seismic Data

In 1984 the BMR, in conjunction with the Geological Survey of Queensland, shot a 6-fold, common-depth-point explosive-source, 20 s record-length, seismic reflection profile across southeast Queensland (Wake-Dyster et al. 1987b), and in 1989 the BMR recorded 254 km of deep seismic data (8-fold CMP, explosive-source, 20 s record-length) across the exposed part of the Bowen Basin (Fig. 1). Here we use parts of Traverses BMR84.14 and BMR84.16 across the

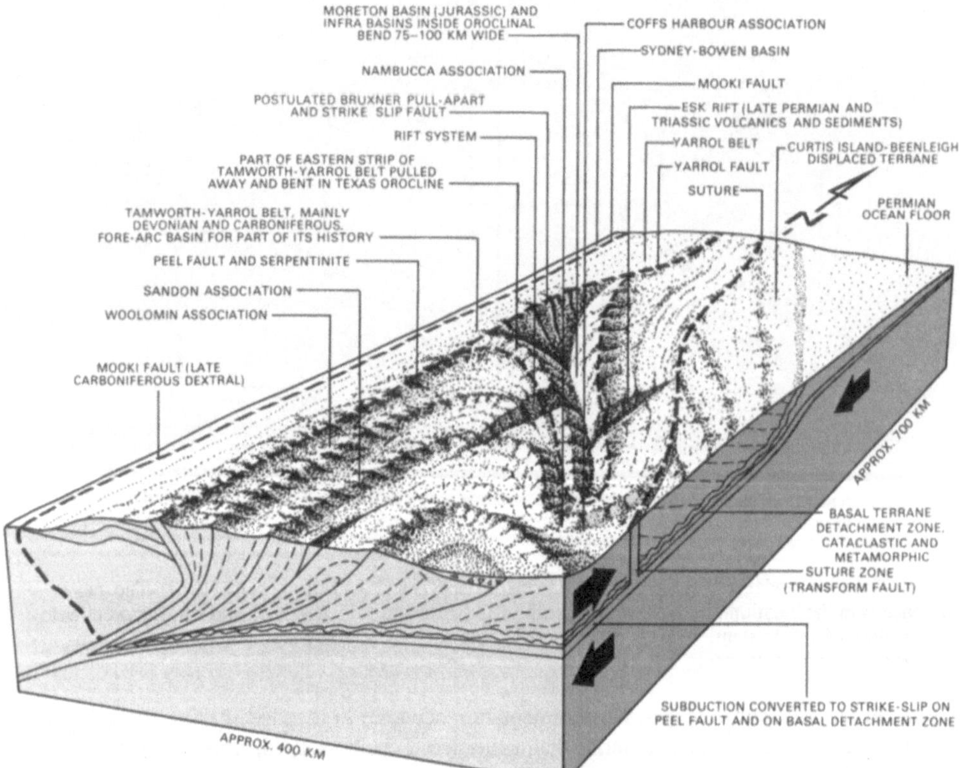

Figure 4. Block diagram interpretation of the oroclinal bending (after Harrington & Korsch, 1987) showing sliding and bending of the accretionary wedge above a detachment zone which is postulated to be the top of the subducted oceanic plate. Note that the Peel Fault and adjoining serpentinite belt are bent in the orocline. The Mooki Fault is postulated to be the main western boundary of the orocline region.

Taroom Trough (southern, subsurface part of the Bowen Basin) and line BMR89.B01 across the Bowen Basin to examine the geometries of the basins, remembering that single lines will not provide complete information on basin geometry.

Bowen Basin

The Bowen Basin is a complex sedimentary basin of latest Carboniferous or earliest Permian to Early Triassic age and consists of several structural units (Fig. 1). It is overlain, sometimes with marked unconformity, by the Early Jurassic to Cretaceous Surat Basin. Here we focus on two areas in the vicinity of the BMR deep seismic profiles.

TAROOM TROUGH AND SURAT BASIN

Part of seismic traverse BMR84.14 crosses the Surat Basin (Fig. 1), and the seismic data (Figs 5 & 6) show that the Surat Basin is predominantly a flat-lying blanket of sediments covering the Bowen Basin at depth. It covers the Nebine Ridge in the west and two asymmetric sub-basins in the east. These sub-basins have the geometry of half-grabens, are bounded by normal or subvertical faults (Leichhardt and Moonie faults) along their eastern margins, and are the subsurface part of the Bowen basin system. The geometry of the basin system is complicated in the vicinity of the seismic line because of the presence of the Undulla embayment (equivalent to the eastern half-graben on the seismic line, Fig. 6), which is an eastward offshoot of the western half-graben, the longitudinal Taroom Trough.

Below the Taroom Trough, the crust is characterised on the seismic line by numerous discontinuous reflections (Fig. 6). The base of the crust is interpreted as the bottom of the reflective zone; farther west in the Eromanga Basin, the base of the reflective zone correlates with the Moho as defined by seismic-refraction profiling (Finlayson & Collins 1987). A band of W dipping reflections that penetrates almost the entire crust (Fig. 6) separates sections of crust with quite different reflection characteristics, and is very similar to major terrane boundaries and sutures that have been imaged by deep seismic methods in other parts of the world (e.g. Klemperer et al. 1990). Farther east, numerous reflections project upwards to the east (Fig. 6) where they define a W dipping zone, again separating crust of differing reflection characteristics. This zone could be a terrane boundary but its relationship to the Leichhardt and Moonie faults is not obvious on the seismic section. The faults could detach on this boundary or continue to the Moho.

Under the Surat Basin, the Moho has considerable topography (Fig. 6) and a band of strong reflections represents the transition from the crust to upper mantle. It is deepest in the west but shallows on the eastern side of the Nebine Ridge. The crust beneath the Permian and Mesozoic sediments is thickest at the Nebine Ridge (over 40 km thick) and thinnest beneath the Taroom Trough (about 30 km thick) suggesting that there has been thinning of the crust beneath the trough by a ß factor of about 1.3.

Different interpretations have been presented for seismic lines in the vicinity of line BMR84.14. A previous BMR interpretation (Wake-Dyster et al. 1987a) favoured a bounding fault system that dipped steeply W. On the other hand, an industry interpretation (Elliott & Brown 1988; Elliott 1989) shows the bounding fault dipping steeply E, invoking high-angle thrusting. Here, we favour a very high angle to subvertical basement fault (Fig. 6). All interpretations place geometrical constraints on a foreland basin model because it is extremely difficult to generate much shortening on a steeply dipping fault.

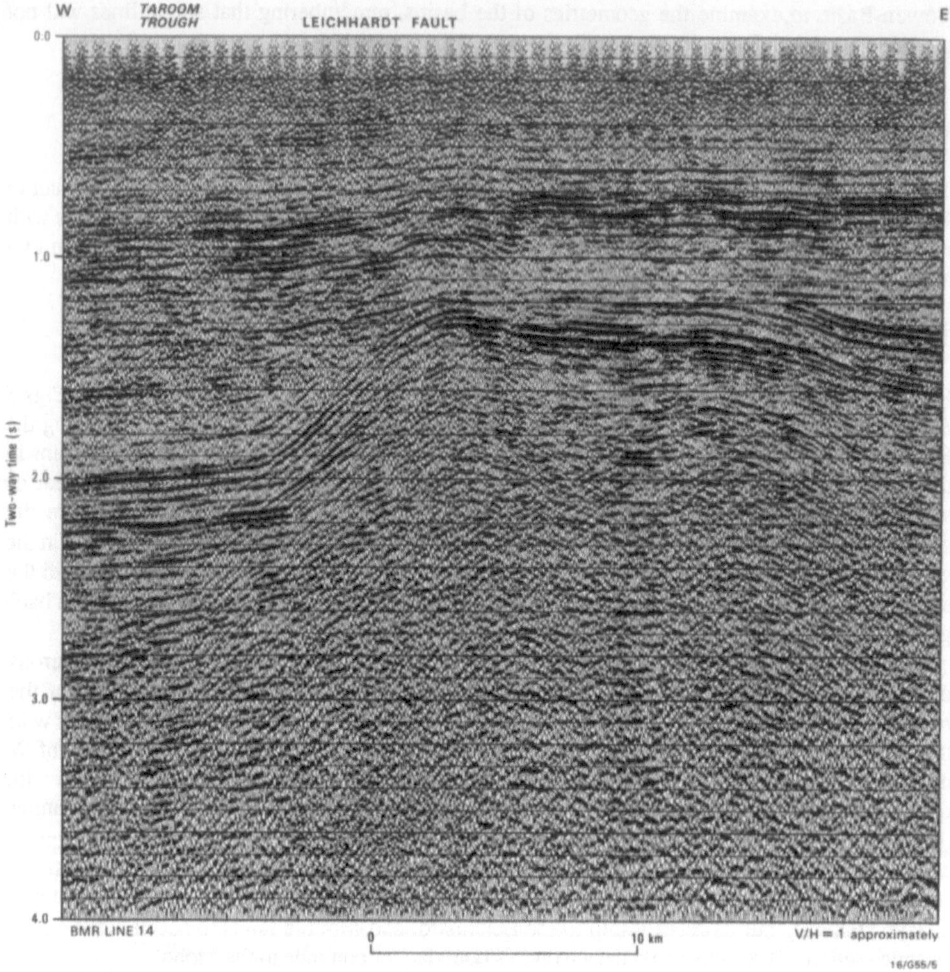

Figure 5. Unmigrated portion of seismic line BMR84.14 across the eastern Taroom Trough. Note the steep bounding Leichhardt Fault and transpressional structures in the Jurassic sedimentary rocks above the bounding fault.

The geometry of the Taroom Trough seen in the seismic section can be used to calculate the depth to the detachment that should link with a bounding listric normal fault (method of Gibbs 1983). This results in a predicted depth to the detachment in excess of 50 km, placing the detachment well within the mantle. As this is geologically unlikely, it implies that the trough did not develop by normal extension within the plane of the section. Therefore, there must have been movement out of the plane of section, and the extension had a strong oblique component. Thus, movement on the Leichhardt Fault had a significant strike-slip component to accommodate the observed geometry, and the basin-forming mechanism was transtension.

Seismic data over the eastern Taroom Trough and Leichhardt Fault show contractional structures in the Jurassic part of the succession (down to about 1.3 s TWT, Fig. 5). These structures are localised in extent and are best interpreted as positive flower structures (Harding 1985) due to transpressional reactivation above a strike-slip fault. In this case, the Jurassic strata have acted as a carpet being rucked up above a basement that was being displaced by faulting. Flower structures also occur above the Moonie Fault to the east (Fig. 6). Because Jurassic sediments are involved in the flower structures, the faults must have been reactivated in the Cretaceous or Tertiary.

The Mooki fault system *sensu lato* (Harrington & Korsch 1985a) has traditionally been mapped as a thrust system (e.g. Carey 1934; Voisey 1959) leading to suggestions by some workers (e.g. Murray 1985) that the Bowen basin system is a foreland loading basin. Harrington and Korsch (1979, 1985a) recognised that the Mooki fault system extends along the eastern sides of the Bowen and Gunnedah basins and southwards into the Sydney Basin. They emphasised a postulated (but not proved) major strike-slip component of movement on the system and later they emphasised also that there were also compressional and tensional components of movement (e.g. Harrington 1982; Korsch 1982; Korsch et al. 1988, 1990a). Two possibilities arise with regard to the thrust interpretation: firstly, the thrusting is related to a shortening event along the eastern margin of the Permian basins, or secondly, much of the observed thrusting is related to transpression accompanying more deeply rooted strike-slip faulting.

The thrusts place older rocks over Permian as well as Permian over Early Triassic sedimentary rocks, and hence are too young to be involved in initiation of the basin (Murray 1985). This raises the question of whether the basin originated as a foreland basin or as another kind of structure. Scheibner (1973), Korsch (1982), Harrington (1982) and Harrington and Korsch (1985a) all suggested the possibility of an extensional or transtensional origin for the basin. Paten et al. (1979) and Bauer and Dixon (1981) showed the existence of half-grabens beneath the Denison Trough farther north in the western Bowen Basin. This interpretation was followed by Ziolkowski and Taylor (1985) who suggested that the Denison Trough was initiated during a period of extension in the latest Carboniferous or earliest Permian, which resulted in the formation of a series of half-grabens along the western margin. This interpretation was developed further by Hammond (1987) into a model for the development of the whole of the Bowen Basin by extension in the upper crust. Hammond's extensional model predicts the opposite geometry to that observed in BMR84.14 and company seismic lines (e.g. Thomas et al. 1982) across the Taroom Trough. For a simple foreland-loading basin, thrusting should thicken the crust, but the crust under the Taroom Trough is much thinner than that under the Nebine Ridge, which also suggests the possibility of crustal extension. The preferred extension direction of Hammond (1987) is ENE-WSW which is oblique (65°) to the Leichhardt Fault. If this extension direction is correct, then movement on the Leichhardt Fault would be oblique (that is, there would be a component of strike slip). It is interesting to note that Katz (1986) demonstrated an ENE-WSW extension direction and inferred a transtensional character in the Early Permian for the Peel Fault in the southern New England Orogen (Fig. 1).

The blanket of Jurassic-Cretaceous sedimentary rocks forming the Surat and Clarence-Moreton basins was then deposited during subsidence interpreted to represent the thermal-recovery phase after cessation of major faulting (Wake-Dyster et al. 1987a).

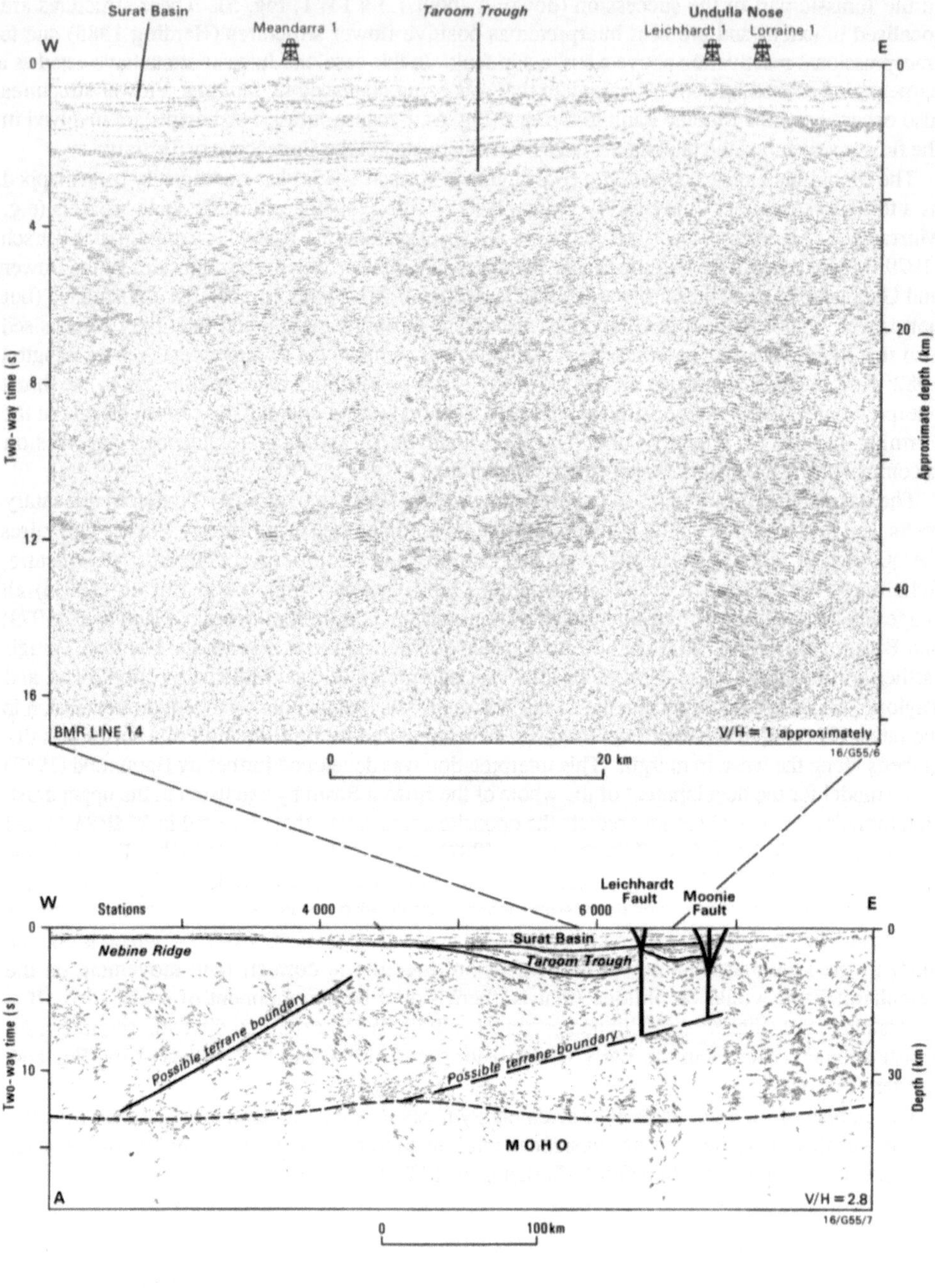
Surat Basin
Meandarra 1
Taroom Trough
Undulla Nose
Leichhardt 1
Lorraine 1
W
E
Two-way time (s)
Approximate depth (km)
BMR LINE 14
V/H = 1 approximately
16/G55/6
0
20 km
Stations
4 000
6 000
Leichhardt Fault
Moonie Fault
Surat Basin
Nebine Ridge
Taroom Trough
Possible terrane boundary
Possible terrane boundary
MOHO
Depth (km)
A
V/H = 2.8
16/G55/7
0
100km

CENTRAL BOWEN BASIN

The stratigraphy of the Early Permian to Middle Triassic Bowen Basin is reasonably well known but, as discussed above, there is some conflict in the models for its development. To test the various tectonic and structural models, the BMR conducted a deep seismic reflection survey across the basin in the vicinity of Blackwater during the latter half of 1989.

Seismic Line BMR89.B01 (Fig. 7) was positioned to follow a corridor between two transfer faults postulated by Hammond (1988a) and to test the extensional tectonics model. In the west, Early Triassic sedimentary rocks exposed at the surface are essentially non-reflective, but excellent reflections were recorded from the Late Permian coal measures immediately below the Triassic rocks (Korsch et al. 1990c, 1990d). The data suggest that the Permo-Triassic sedimentary wedge thickens to the east, a result which is similar to that seen in line BMR84.14 over the Taroom Trough farther to the south. This contrasts with westward thickening predicted by the model of Hammond (1988b, p.108).

The seismic line is dominated by (? mid-Triassic) deformation in the sedimentary succession controlled by thin-skinned thrusting on a series of listric faults which dip to the E and root in a major detachment that also dips to the E (Fig. 7). In the east, the detachment appears to flatten in the ductile zone in the middle crust (about 7 s TWT). Nevertheless, the Blackwater, Yarrabee and Dawson structural zones recognised by Hobbs (1985) exhibit distinct structural styles.

The Blackwater Zone, to the west of the Jellinbah Thrust, is dominated by subhorizontal reflections, interpreted as mainly sedimentary layering (Fig. 7). Displacements of up to 5 km on flat lying thrusts were reported by Hobbs (1985) from this zone. The Yarrabee Zone, between the Jellinbah and Yarrabee thrusts, is characterised by dome and basin structures (Hobbs 1985). The seismic data presented here (Fig. 7) indicate that the structures are controlled by listric thrust faults that have an imbricate fan geometry. Minor displacements occur at the boundary between the non-reflective and reflective sedimentary layers in the upper 1 s of the data. The Dawson Fold Zone, to the east of the Yarrabee Thrust, consists of tightly folded Late Permian sedimentary rocks. Because of the steep dips, reflectivity is low, but faults with moderate dips to the E occur, and are best seen at the eastern end of line BMR89.B01 (Fig. 7). The major E dipping detachment defines the base of this zone.

Thus, the BMR deep seismic data show that the central and eastern Bowen Basin contains an eastward-thickening wedge (partly post-dating most of the fill in the Denison Trough). If early extension occurred in this region, sedimentary relationships suggest the opposite polarity to that proposed by previous workers. The seismic data show the dominant role of thrust faults in controlling the deformation of the sedimentary succession. A series of listric faults root in a major E dipping detachment which appears to flatten in the middle crust.

Figure 6. A. Deep seismic line BMR84.14 across the Taroom Trough shown to a depth of nearly 18 s TWT. Note the lack of evidence for listric normal faults bounding the basin. The Moho is interpreted as the base of lower crustal reflections slightly deeper than 12 s. B. Interpreted line drawing of seismic line BMR84.14 showing the Nebine Ridge, Surat Basin, Taroom Trough, Leichhardt Fault, Moonie Fault and possible terrane boundaries. Note that the crust beneath the sedimentary succession is thinnest beneath the deepest part of the Taroom Trough.

Esk, Ipswich and Clatrence-Moreton Basins

The Early Triassic Esk Trough, Middle Triassic Ipswich Basin and Late Triassic to Cretaceous Clarence-Moreton Basin (Fig. 1) represent a linked system of basins that are genetically related. Seismic traverse BMR84.16 crossed the Clarence-Moreton Basin, and only the easternmost shot points were recorded in basement of the Beenleigh Block. The line also crosses the Esk Trough in the subsurface south of its southern outcrop limit and provides a good image of the geometry of the trough (Fig. 8). This is discussed in detail by Korsch et al. (1989, 1990a) and salient points only will be covered here.

The structures and sediments of the Esk Trough (Fig. 8) have the pronounced asymmetry typical of half-grabens and most rift basins. This implies that the basin developed by an extensional mechanism with sedimentation being controlled by a W dipping normal fault (West Ipswich Fault of the Great Moreton fault system).

Poor time control on a previously undetected packet of sedimentary rocks beneath the Esk Trough (Fig. 8, unit 2) and on the sedimentary rocks of the Esk Trough and Clarence-Moreton Basin, hampers construction of subsidence curves for the basins, and Figure 9 (curve A) uses data from the seismic line BMR84.16 to estimate depths to the unit boundaries. Nevertheless, the shape of the curve is typical for a sedimentary basin that was initiated by extension and followed by a protracted period of subsidence driven by thermal cooling of the lithosphere (e.g. McKenzie 1978). The curve indicates that the most rapid subsidence took place during initial sedimentation and that the subsidence rate decreased with time; the subsidence rate of the Clarence-Moreton Basin being less than that of the Esk Trough. The deep seismic data (Fig. 10) show that the crust beneath the sedimentary successions is thinnest below the deepest part of the Esk Trough, suggesting that there has been thinning of the crust by a ß factor of about 1.3.

A raw subsidence curve for the area east of the South Moreton Anticline, constructed using depths obtained from industry seismic and well data, has a pattern similar to that west of the South Moreton Anticline, again suggesting that deposition was controlled mainly by thermal subsidence after the extensional event which formed the Ipswich Basin (Fig. 8, curve B).

The structures and sediments of the extension and thermal relaxation phases show a pronounced asymmetry in cross-section, as expected during normal extension, but on deep seismic records (Figs 8 & 10), a low-angle detachment fault extending to depth from the bounding fault of the Esk Trough (the West Ipswich Fault) has not been detected. It is possible to calculate the depth to an inferred detachment following the method outlined by Gibbs (1983). Using the interpreted geometry in Figure 8, and assuming a velocity of 6 km sec¯1, the detachment should be located at about 11 s TWT (about 33 km depth) which is about 2 s shallower than the depth of the Moho defined by seismic reflections beneath the Esk Trough. For the fault to extend this deep it must cut through the brittle-ductile transition. Geologically, it is more reasonable to expect the detachment to flatten within the ductile zone. Thus, the steep dip of the fault and lack of evidence for flattening into a detachment suggests that the rift did not form by normal extension in the plane of the cross section. Extension was therefore oblique to the trough margin (West Ipswich Fault). To accommodate the geometry in the seismic section (Fig. 8), a horizontal (strike-slip) component of

---

Figure 7. Unmigrated deep seismic-reflection profile and preliminary geological interpretation of part of line BMR89.B01 showing the major E dipping detachment. The Duaringa Basin is an Early Tertiary oil-shale bearing basin that probably formed by reactivation of older faults in the eastern part of the Bowen Basin. Vertical scale equals horizontal scale for a velocity of 6 km s-1.

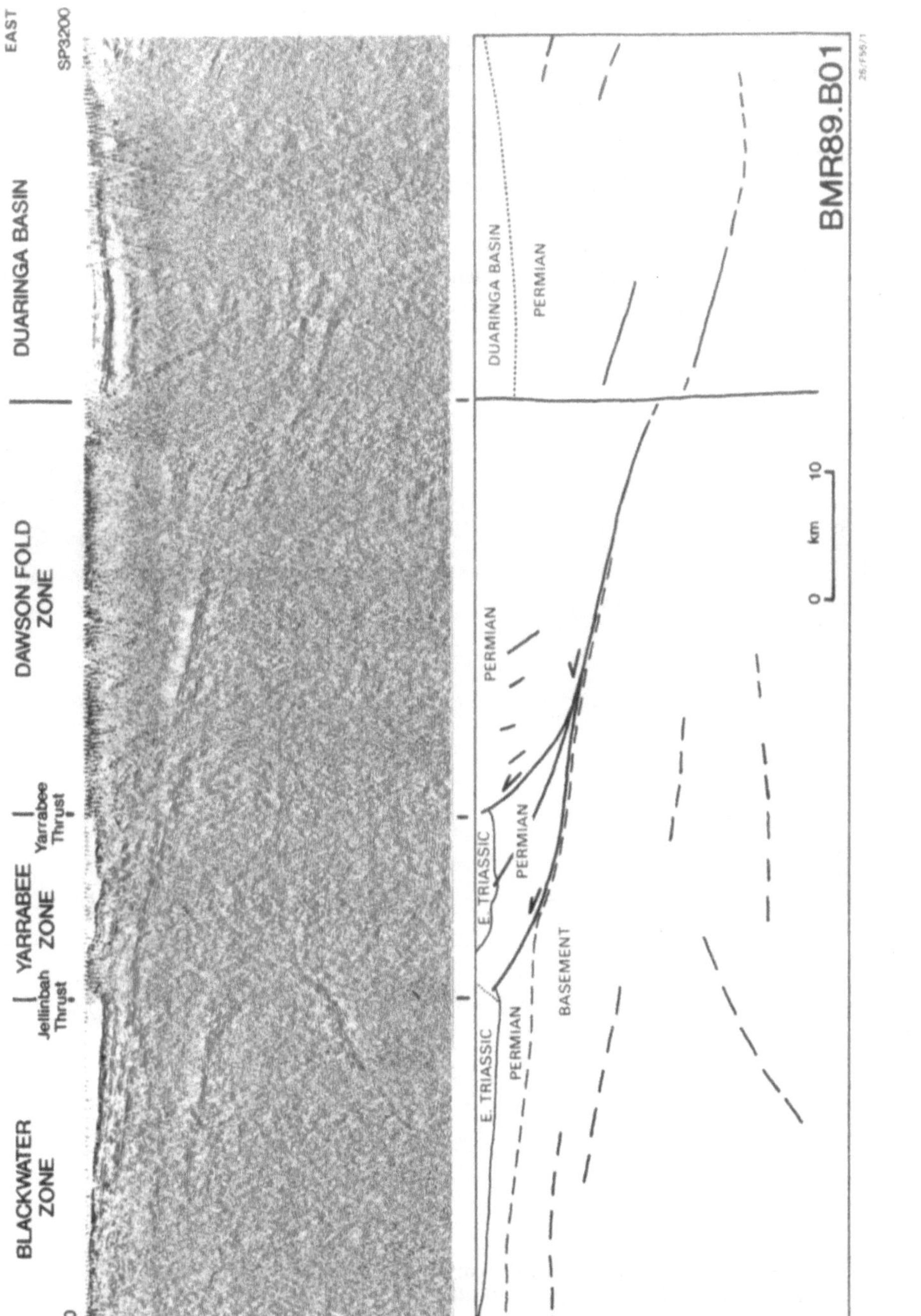
WEST
EAST
SP1500
SP3200
BLACKWATER ZONE
Jellinbah Thrust
YARRABEE ZONE
Yarrabee Thrust
DAWSON FOLD ZONE
DUARINGA BASIN
Two-way Time (s)
Approx. Depth (km)
E. TRIASSIC
PERMIAN
BASEMENT
DUARINGA BASIN
PERMIAN
0 km 10
BMR89.B01

movement is required parallel to the trough margins; the basin is thus transtensional (Harland 1971) in character. Therefore, the Esk Trough is not a pure pull-apart basin associated with either two parallel strike-slip faults or a major bend in a single strike-slip fault (Crowell 1974) because this geometry is not supported by the field or seismic data (Fig. 8).

The necessity for oblique extension suggests that the faults of the Great Moreton system (including the West Ipswich Fault) are surface traces of a major strike-slip fault system at depth. Low-angle reverse faults and folds on the crest of the South Moreton Anticline (see Fig. 5 in Korsch et al. 1989) are interpreted as positive flower structures on the crest of the strike-slip system. It is not known to what depth the major strike-slip fault system extends, but on the deep seismic image there appears to be a detachment, with a very shallow apparent dip to the west, at mid crustal levels (about 8 s TWT) beneath the Esk Trough (Fig. 10), and it is unlikely that the fault system extends beyond this depth.

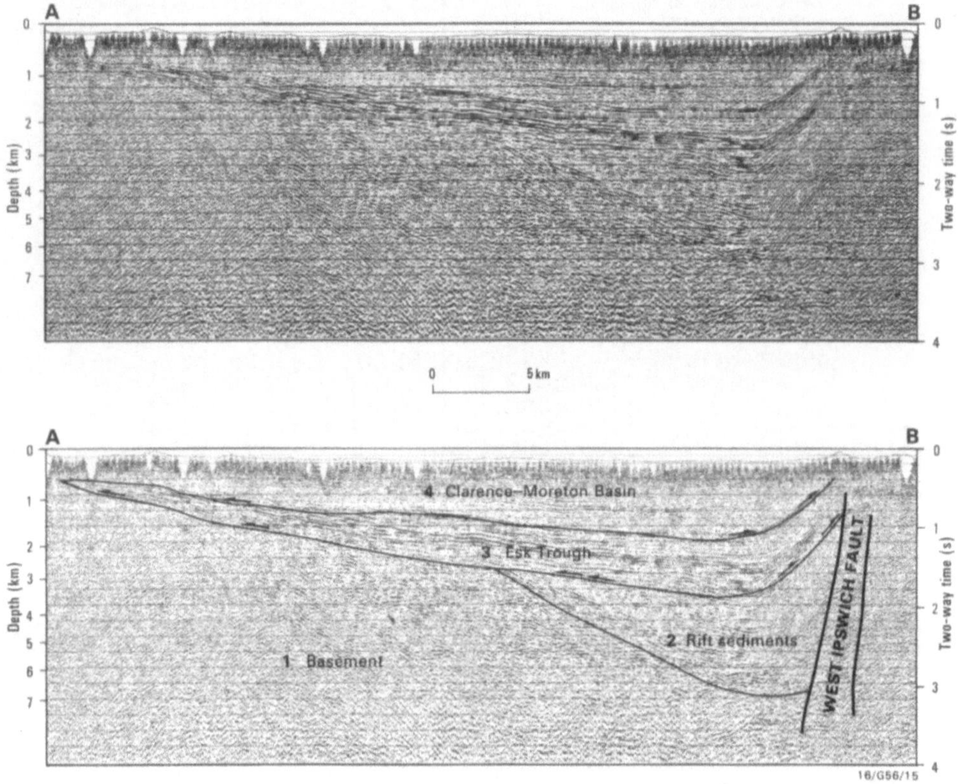

Figure 8. Unmigrated seismic-reflection profile across the Clarence-Moreton Basin and the Esk Trough. Profile shown is a portion of seismic traverse BMR84.16 (Fig. 1). Arbitrary seismic datum is approximately 250 m above sea level and irregular profile at top of section is ground elevation. The succession labelled 'rift sediments' is a previously unknown succession below the Esk Trough.

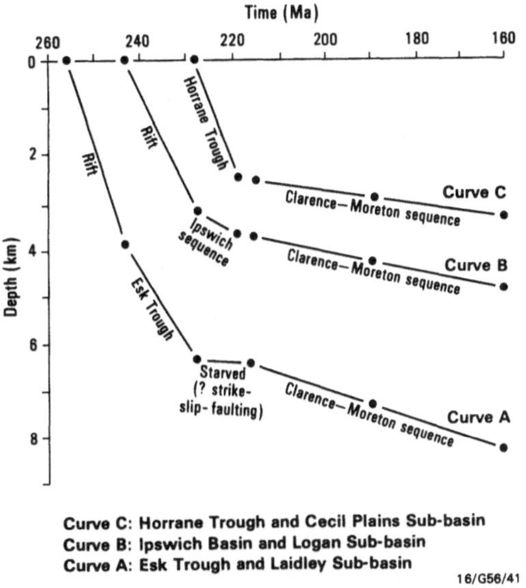

Figure 9. Raw subsidence curves showing cumulative thickness of sediment versus time for the fill of the Esk Trough (curve A), Ipswich Basin (curve B) and Horrane Trough (curve C) which occurs to the west of the Esk Trough.

The unconformity between the Esk Trough sediments and the Clarence-Moreton succession represents a break of about 12 Ma, during which time sediments and volcanics of the Ipswich Basin were deposited farther to the east. This unconformity represents a period of time when movement on the West Ipswich Fault was transpressional and the Esk Trough was elevated.

The geometry of the Ipswich Basin is not displayed in any single seismic section, and that shown in Figure 11 is composite and speculative. Data from the Rathdowney seismic survey (Milner & Milner 1981) suggest that the fault on the eastern margin of the South Moreton Anticline is the bounding fault to the rift sequence (Korsch et al. 1989). If this is the case, the rift sequence probably thins gradually towards the east away from the fault, and laps out rapidly to the west on to the footwall block, which is the South Moreton Anticline. Sediments on the eastern margin of the basin lap onto the Beenleigh Block, as shown by the eastern end of seismic line BMR84.16 (Fig. 11; see also Fig. 2 in Korsch et al. 1986).

The Ipswich Coal Measures wedge out on to the South Moreton Anticline but the Clarence-Moreton succession continues across it (Fig. 11). By the time deposition of the Clarence-Moreton succession commenced, thermal relaxation is considered to have been the dominant driving force for subsidence, allowing sediments to spread out over both rift sequences and basement highs. Note however, that the amount of subsidence generated over the South Moreton Anticline is significantly less than that generated above the two rift sequences (Fig. 11).

Strike-slip movement continued, at least intermittently, into the Jurassic, Cretaceous and possibly Tertiary. Contractional structures such as anticlines near Ipswich (Cranfield et al. 1981; Korsch et al. 1986), and thrust faults such as those at Redbank (Fig. 9 in Korsch et al. 1989) are best explained as transpressional features developed by an oblique component of the movement during strike-slip faulting on fault segments that have near linear surface traces; these folds and faults imply a dextral sense of movement. Thus, late stage transpressional, or possibly reverse, movement on the West Ipswich Fault indicates a long and complicated movement history, with

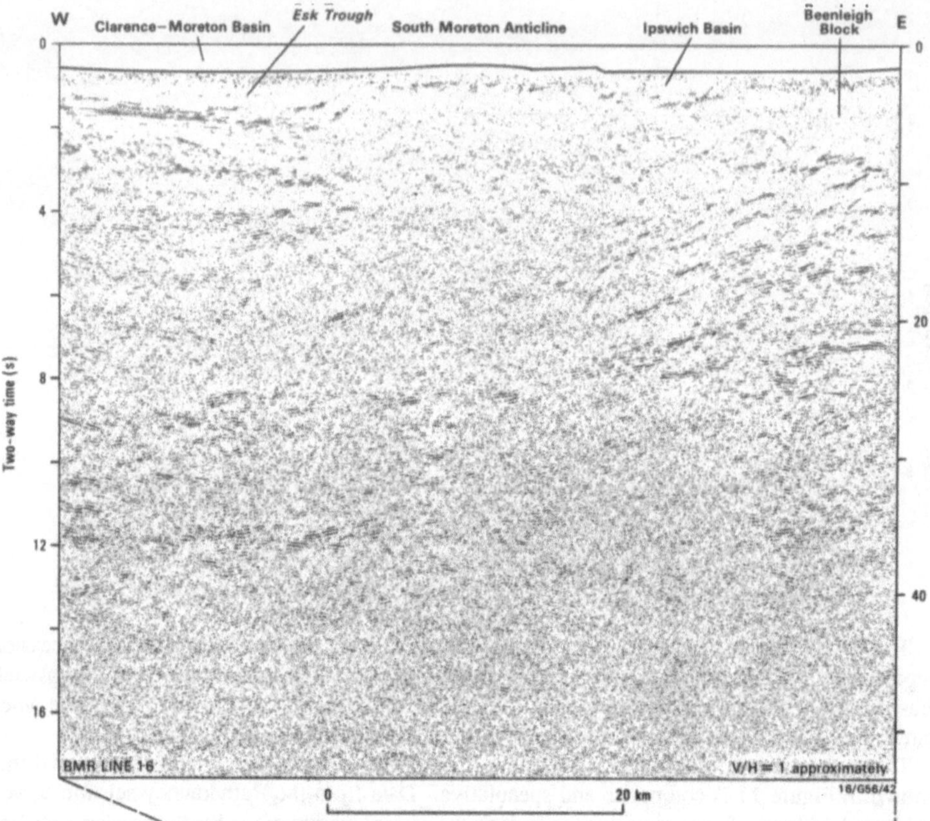
W
Clarence–Moreton Basin
Esk Trough
South Moreton Anticline
Ipswich Basin
Beenleigh Block
E
Two-way time (s)
Approximate depth (km)
BMR LINE 16
V/H = 1 approximately
0
20 km
16/G56/42

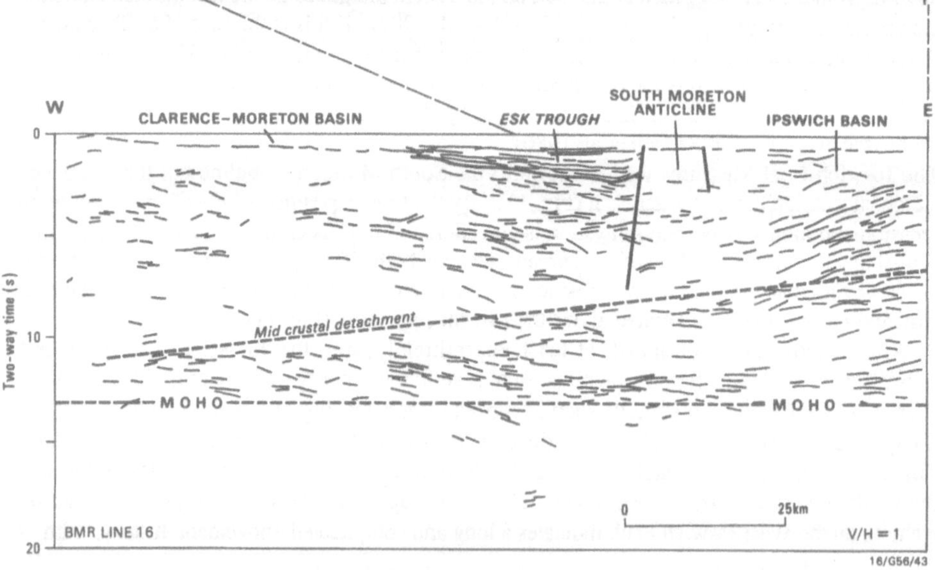
W
CLARENCE–MORETON BASIN
ESK TROUGH
SOUTH MORETON ANTICLINE
IPSWICH BASIN
E
Two-way time (s)
Mid crustal detachment
MOHO
MOHO
BMR LINE 16
0
25km
V/H = 1
16/G56/43

reactivation and accompanying tectonic inversion after deposition of the Clarence-Moreton succession.

Farther southeast in the Clarence-Moreton Basin, positive flower structures along a NNE trending zone also indicate strike-slip faulting (O'Brien et al. 1990). The orientation of these flower structures suggests that they formed on restraining bends or side steps in a dextral fault system (Fig. 12). Thrusting on the eastern side of the South Moreton Anticline and near the eastern margin of the basin can also be related to dextral transpression (O'Brien et al. 1990). These structures affect the youngest (Late Jurassic) sediments preserved in the basin. Because the opening of the Tasman Sea along this part of the the Australian margin took place by sinistral strike-slip in the Late Cretaceous (Shaw 1990), it is likely that the last dextral movements occurred in the Early Cretaceous (Gleadow & O'Brien, in press).

Discussion

TAROOM AND ESK TROUGHS

In the vicinity of seismic lines BMR84.14 and BMR84.16, we have demonstrated that the geometries of the Taroom Trough and Esk Trough support a mechanism of formation by transtension, that is, oblique extension coupled with strike-slip motion on their eastern bounding faults. It has been inferred previously (O'Brien et al. 1990) that most of the areas of deformation within the Clarence-Moreton Basin are related to strike-slip faults in the basement. When the structures are mapped out, they are confined to relatively narrow N-S or NNE-SSW trending zones (Fig. 12). Thus we interpret these structures as positive flower structures above strike-slip faults in the basement. These were produced by a period or periods of transpression when the faults were reactivated in post-depositional times.

In the uppermost crust, strike-slip faults can have significant thrusting associated with them (e.g. Woodcock & Fischer 1986). This type of thrusting often produces positive flower structures and splay faults (e.g. Harding 1985) which have been reported from many strike-slip fault zones with a component of reverse dip-slip movement, including the San Andreas Fault zone and the New Zealand Alpine Fault.

Therefore, in this part of the New England Orogen, strike-slip faulting was active periodically throughout basin development, and the major faults stepped progressively to the east, as did the locus of volcanism (Harrington & Korsch 1985b). This faulting succeeded the strike-slip regime initiated in the Late Carboniferous (Murray et al. 1987) when there was a major plate reorganization (Korsch 1982) and subduction ceased progressively in the New England Orogen. The locus of subduction moved east and a thick accretionary wedge was built up in New Zealand in the Permian and Mesozoic. New Zealand was adjacent to eastern Australia at this time, prior to Late Cretaceous rifting to form the Tasman Sea. Thus, the England Orogen in the Permian and Mesozoic was in a back-arc tectonic setting, dominated by a strike-slip regime.

Figure 10. A. Unmigrated deep seismic-reflection profile beneath the Esk Trough and Ipswich Basin (part of line BMR84.16). A possible mid-crustal detachment is between 8 and 9 s TWT. B. Line diagram of the eastern end of BMR84.16 (unmigrated). Esk Trough is bounded on the east by a steep fault which possibly links into a mid-crustal detachment zone.

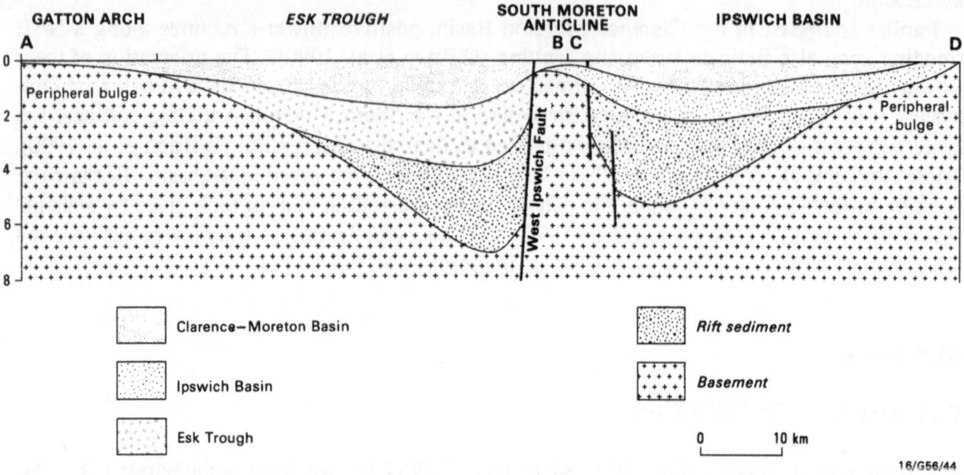

Figure 11. Composite cross section (4:1 vertical exaggeration) across the Esk, Ipswich and Clarence-Moreton basins. AB corresponds to the seismic line across the Esk Trough (Fig. 8) and CD is based on shallow industry data and is speculative (see Korsch et al. 1989).

Harrington and Korsch (1985a) proposed major strike-slip movement on the Mooki Fault (*sensu lato*) in the latest Carboniferous to Early Permian. There is controversy over movement directions on the Peel Fault to the east, particularly in the Permian. Evidence for sinistral strike-slip has been presented by Corbett (1976) and Offler andWilliams (1985), in contrast to dextral transtension determined by Katz (1986). Blake and Murchey (1988) have suggested that the Peel Fault formed in the Early Permian as an E dipping thrust.

In the New England Orogen, strike-slip faulting continued into the Tertiary, as shown by the displacements recorded in Surat and Clarence-Moreton basin sediments. Evidence for strike-slip faulting from surface geology includes movement on the Demon Fault, which had a dextral movement of about 20 km in post Early Triassic time (Korsch et al. 1978) and the structures of the Clarence-Moreton Basin which indicate that minor dextral movement continued until the Early to Mid Cretaceous (Gleadow & O'Brien, in press). Movements on strike-slip faults have played a significant role in the development of this part of the New England Orogen in that they controlled the transport and accretion of displaced terranes such as the Gympie and Beenleigh terranes and the oroclinal bending of the accretionary wedge sequence (Korsch & Harrington 1987), as well as the initiation of the transtensional basins during oblique extension.

CENTRAL BOWEN BASIN

The central Bowen Basin and the subsurface Taroom Trough farther to the south are part of a single sedimentary basin system, yet show totally different present-day geometries. Seismic line BMR89.B01, across the exposed part of the basin system, is dominated by listric thrust faults that root onto a detachment that dips gently E (Fig. 7). In this area, the deformation is so intense that indications of the original geometry of the basin before thrusting are all but obliterated. To the east of the Bowen Basin, in the vicinity of seismic line BMR89.B01, is the Gogango Overfolded Zone

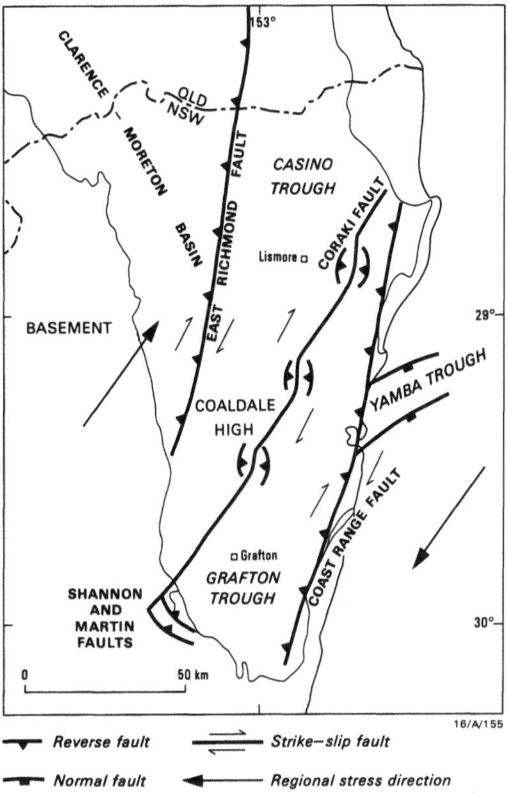

Figure 12. Map showing location of inferred strike-slip faults beneath the Clarence-Moreton Basin based on locations of contractional and extensional structures confined to narrow zones.

(Malone 1964; Fergusson 1990). This zone is dominated by thrust faults and highly cleaved, tight to isoclinal folds in rocks of Early Permian or older age. The Gogango Overfolded Zone is intruded by post-orogenic Late Permian to Mesozoic plutons and hence the age of the thrusting is constrained to being younger than the Early Permian but older than latest Permian. Farther to the west, in the Dawson and Yarrabee zones, the thrusts are younger, because Late Permian rocks are thrust over Early Triassic ones. This is consistent with observations in other thrust belts around the world where the thrusts propagate into the foreland sequentially.

MODEL

The thrust faults in the Bowen Basin place Permian rocks over Triassic rocks and hence are too young to be involved in initiation of the basin. This raises the question of whether the basin originated as a foreland basin, an extensional basin or a transtensional basin. Bauer and Dixon (1981) and Ziolkowski and Taylor (1985) suggested that the Denison Trough (western Bowen Basin) initiated during a period of extension in the latest Carboniferous or earliest Permian, resulting in the formation of a series of half-grabens along the western margin. This led Hammond (1987, 1988a) to propose a model for the development of the whole Bowen Basin by upper crustal

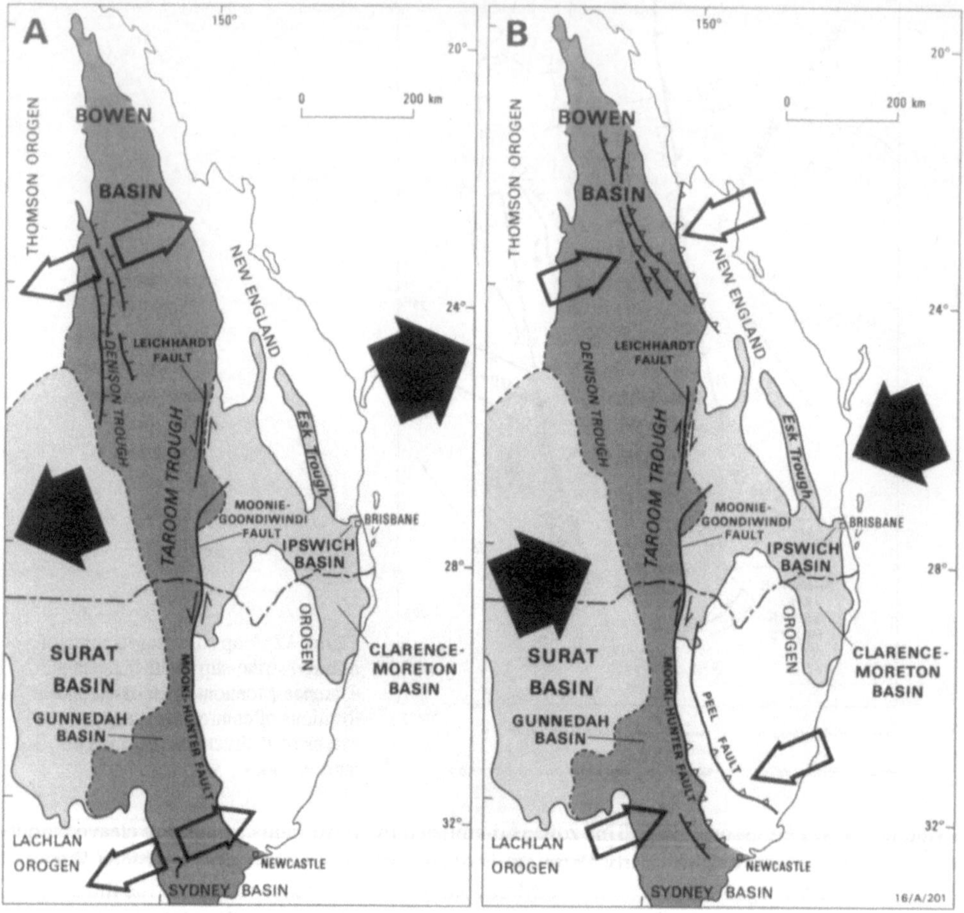

Figure 13. Model showing (A) extensional and transtensional areas during the early part (latest Carboniferous and earliest Permian) of the evolution of the Bowen basin system, and (B) contractional and transpressional areas during the later (mid Permian to Middle Triassic) history.

extension. Hammond's model has the opposite geometry to that observed on BMR and company seismic lines across the Taroom Trough and across the central Bowen Basin.

The contrast between the two BMR seismic profiles suggests that development of the Bowen Basin system was complex, with extension or transtension operating initially. The preferred extension direction of Hammond (1987) is ENE-WSW which is oblique (65°) rather than normal to the Leichhardt Fault (Fig. 13a). In the exposed part of the Bowen Basin, the basin system trends NNW-SSE and extension was close to normal, with possibly only limited strike-slip faulting. Movement on the Leichhardt Fault would be oblique (that is, there would be a component of sinistral strike slip during the extensional event). During the later shortening event, the basin again behaved differently along its length. In the Bowen Basin the effects of the shortening event are widespread, possibly because the basin was able to invert on earlier structures. This led to the development of a major thrust belt (Fig. 13b). In the southernmost Sydney Basin part of the basin system, major thrusting both in the basin and the Tamworth Belt to the east (Fig. 13b) has been documented (e.g. Roberts & Engel 1987; Glen & Beckett 1989). In the southern, subsurface part of the Bowen Basin (Taroom Trough), the eastern bounding fault was N-S, leading initially to oblique extension coupled with strike-slip (Fig. 13a). During later shortening, the deformation was again controlled by the fault, leading to a narrow zone of transpression near the fault (e.g. Fig. 5) and not the widespread deformation seen to the north (e.g. Fig. 7).

In conclusion, movements on strike-slip faults have played a significant role in the tectonic development of eastern Australia since the latest Carboniferous. The geometries observed in BMR and company seismic data place constraints on the interpretation of the development of late Palaeozoic to Mesozoic sedimentary basins in eastern Australia, and indicate that a significant strike-slip component was involved throughout much of its history, at least in the southern part of the Bowen basin system. As well as controlling the development of transtensional basins during oblique extension, the strike-slip faults also controlled the transport and accretion of displaced terranes such as the Gympie and Beenleigh Terranes and the oroclinal bending of the accretionary wedge sequence.

Acknowledgements

We wish to thank H.J. Harrington, M.J. Sexton, R.D. Shaw, J.M. Totterdell and A.T. Wells for helpful discussions and/or for their comments on the manuscript. Published with permission of the Executive Director, Bureau of Mineral Resources.

References

Bauer, J.A., & Dixon, O. 1981. Results of a seismic survey in the southern Denison Trough, Queensland, 1978-79. Bureau of Mineral Resources, Journal of Australian Geology and Geophysics **6**, 213-222.

Blake, M.C. Jr., & Murchey, B.L. 1988. A California model for the New England fold belt. In, Kleeman, J.D. (ed.) New England Orogen: tectonics and metallogenesis, University of New England, Armidale, p. 20-31.

Carey, S.W. 1934. The geological structure of the Werrie Basin. Proceedings of the Linnean Society of New South Wales **59**, 351-374.

Cas, R. 1983. Palaeogeographic and tectonic development of the Lachlan Fold Belt, southeastern Australia. Geological Society of Australia Special Publication **10**, 104 p.

Chappell, B.W., White, A.J.R., & Hine, R. 1988. Granite provinces and basement terranes in the Lachlan Fold Belt, southeastern Australia. Australian Journal of Earth Sciences **35**, 505-521.

Coney, P.J., Edwards, A., Hine, R., Morrison, F., & Windrim, D. 1990. The regional tectonics of the Tasman orogenic system, eastern Australia. Journal of Structural Geology **12**, 519-543.

Corbett, G.J. 1976. A new fold structure in the Woolomin Beds suggesting a sinistral movement on the Peel Fault. Journal of the Geological Society of Australia **23**, 401-406.

Cranfield, L.C., Hutton, L.J., & Green, P.M. 1981. Ipswich 1;100 000 geological map, first edition, Queensland Department of Mines, Brisbane.

Crook, K.A.W., & Powell, C.McA. 1976. The evolution of the southeastern part of the Tasman Geosyncline. 25th International geological Congress, Sydney, Excursion 17A Field Guide, 122 p.

Crowell, J.C. 1974. Origin of Late Cenozoic basins in southern California. In, Dickinson W.R. (ed.) Tectonics and Sedimentation, Society of Economic Paleontologists and Mineralogists, Special Publication **22**, 190-204.

Elliott, L.G. 1989. The Surat and Bowen Basins. APEA (Australian Petroleum exploration Association) Journal **29**, 398-416.

Elliott, L.G., & Brown, R.S. 1988. The Surat and Bowen Basins. In, Petroleum in Australia: The First Century, Australian Petroleum Exploration Association, Melbourne, p.120-138.

Fergusson, C.L. 1990. Thin-skinned thrusting in the Bowen Basin and northern New England Orogen, central Queensland. In, Beeston J.W. (compiler) Bowen Basin Symposium 1990, Proceedings, Geological Society of Australia (Queensland Division), Brisbane, p. 42-44.

Fielding, C.R., Gray, A.R.G., Harris, G.I., & Salomon, J.A. 1990. The Bowen Basin and overlying Surat Basin. Bureau of Mineral Resources, Australia, Bulletin **232**, 105-116.

Finlayson, D.M., & Collins, C.D.N. 1987. Crustal differences between the Nebine Ridge and the central Eromanga Basin from seismic data. Australian Journal of Earth Sciences **34**, 251-259.

Gibbs, A.D. 1983. Balanced cross-section construction from seismic sections in areas of extensional tectonics. Journal of Structural Geology **5**, 153-160.

Gleadow, A.J.W., & O'Brien, P.E. (in press). Apatite fission track thermochronology and tectonics in the Clarence-Moreton Basin of eastern Australia. Bureau of Mineral Resources, Australia, Bulletin.

Glen, R.A., & Beckett, J. 1989. Thin-skinned tectonics in the Hunter Coalfield of New South Wales. Australian Journal of Earth Sciences **36**, 589-593.

Hammond, R.L. 1987. The Bowen Basin, Queensland, Australia: an upper crustal extension model for its early history. Bureau of Mineral Resources, Australia, Record 1988/51, 131-139.

Hammond, R.L. 1988a. The geological structure of the Bowen Basin. In, Mallett, C.W., Hammond, R.L., Leach, J.H.J., Enever, J.R., & Mengel, C. (eds) Bowen Basin—stress, structure and mining conditions: assessment for mine planning. CSIRO Division of Geomechanics, NERDDC Project No. 901, Final Report, 10-43 (unpublished).

Hammond, R.L. 1988b. Geological structure of the Curragh lease area. In, Mallett, C.W., Hammond, R.L., Leach, J.H.J., Enever, J.R., & Mengel, C. (eds) Bowen Basin—stress, structure and mining conditions: assessment for mine planning. CSIRO Division of Geomechanics, NERDDC Project No. 901, Final Report, 105-129 (unpublished).

Harding, T.P. 1985. Seismic characteristics and identification of negative flower structures, positive flower structures, and positive structural inversion. American Association of Petroleum Geologists Bulletin **69**, 582-600.

Harland, W.B. 1971. Tectonic transpression in Caledonian Spitsbergen. Geological Magazine **108**, 27-42.

Harrington, H.J. 1982. Tectonics and the Sydney Basin. In, 16th Symposium on Advances in the Study of the Sydney Basin, University of Newcastle, Newcastle, p. 15-19.

Harrington, H.J., & Korsch, R.J. 1979. Structural history and tectonics of the New England-Yarrol Orogen. Bureau of Mineral Resources, Australia, Record, 1979/2, 49-51.

Harrington, H.J., & Korsch, R.J. 1985a. Tectonic model for the Devonian to middle Permian of the New England Orogen. Australian Journal of Earth Sciences **32**, 163-179.

Harrington, H.J., & Korsch, R.J. 1985b. Late Permian to Cainozoic tectonics of the New England Orogen, Australian Journal of Earth Sciences **32**, 181-203.

Harrington, H.J., & Korsch, R.J. 1987. Oroclinal bending in the evolution of the New England-Yarrol Orogen and the Moreton Basin. In, Pacific Rim Congress 87, Australasian Institute of Mining and Metallurgy, Melbourne, p.797-800.

Hobbs, B.E. 1985. Interpretation and analysis of structure in the Bowen Basin. Geological Society of Australia, Abstracts **17**, 151.

Hobday, D.K. 1987. Gondwana coal basins of Australia and South Africa: tectonic setting, depositional systems and resources. Geological Society of London, Special Publication **32**, 219-233.

Jones, J.G., Conaghan, P.J., McDonnell, K.L., Flood, R.H., & Shaw, S.E. 1984. Papuan Basin analogue and a foreland basin model for the Bowen-Sydney Basin. In, Veevers, J.J. (ed.) Phanerozoic earth history of Australia, Oxford University Press, Oxford, pp. 243-261.

Katz, M.B. 1986. Tectonic analysis of the faulting at Woodsreef Asbestos Mine and its possible relationship to the kinematics of the Peel Fault. Australian Journal of Earth Sciences **33**, 99-105.

Klemperer, S., Hobbs, R.W., & Freeman, B. 1990. Dating the source of lower crustal reflectivity using BIRPS deep seismic profiles across the Iapetus suture. Tectonophysics **173**, 445-454.

Korsch, R.J. 1982. Early Permian events in the New England Orogen. In, Flood, P.G., & Runnegar, B.N. (eds) New England Geology, Proceedings of the Symposium on the Geology of the New England Region, University of New England, Armidale, p.35-42.

Korsch, R.J., & Harrington, H.J. 1987. Oroclinal bending, fragmentation and deformation of terranes in the New England Orogen, eastern Australia. In, Leitch, E.C., & Scheibner, E. (eds) Terrane acretion and orogenic belts, American Geophysical Union Geodynamics Series **19**, 129-139.

Korsch, R.J., Archer, N.R., & McConachy, G.W. 1978. The Demon Fault. Journal and Proceedings of the Royal Society of New South Wales **111**, 101-106.

Korsch, R.J., Harrington, H.J., Murray, C.G., Fergusson, C.L., & Flood, P.G. 1990b. Tectonics of the New England Orogen. Bureau of Mineral Resources, Australia, Bulletin **232**, 35-52.

Korsch, R.J., Harrington, H.J., Wake-Dyster, K.D., O'Brien, P.E., & Finlayson, D.M. 1988. Sedimentary basins peripheral to the New England Orogen: their contribution to understanding New England tectonics. In, Kleeman, J.D. (ed.) New England Orogen—tectonics and metallogenesis, University of New England, Armidale, p. 134-140.

Korsch, R.J., Lindsay, J.F., O'Brien, P.B., Sexton, M.F., & Wake-Dyster, K. 1986. Deep crustal seismic reflection profiling, New England Orogen, eastern Australia: Telescoping of the crust and a hidden deep layered sedimentary sequence. Geology **14**, 982-985.

Korsch, R.J., O'Brien, P.E., Harrington, H.J., Wake-Dyster, K.D., Finlayson, D.M., & Johnstone, D.W. 1990a. Constraints from deep seismic profiling on models for the evolution of Permian-Mesozoic sedimentary basins in eastern Australia. In, Pinet, B., & Bois, C. (eds) The potential of deep seismic profiling for hydrocarbon exploration. Éditions Technip, Paris, p. 275-290.

Korsch, R.J., O'Brien, P.E., Sexton, M.J., Wake-Dyster, K.D., & Wells A.T. 1989. Development of Mesozoic transtensional basins in easternmost Australia. Australian Journal of Earth Sciences **36**, 13-28.

Korsch, R.J., Wake-Dyster, K.D., & Johnstone, D.W. 1990c. A Mapping Accord Project: Sedimentary Basins of eastern Australia, with comments on 1989 Bowen Basin Deep Seismic Profiles. In, Muir, W.F. (ed.) Queensland 1990 Exploration and Development. 12th Annual PESA(Qld) - ODCAA - SPE Petroleum Symposium, Brisbane **5**, 78-86.

Korsch, R.J., Wake-Dyster, K.D., & Johnstone, D.W. 1990d. Deep seismic profiling across the Bowen Basin. In, Beeston, J.W. (compiler) Bowen Basin Symposium 1990 Procedings, Geological Society of Australia (Queensland Division), Brisbane, p. 10-14.

Leitch, E.C., & Scheibner, E. 1987. Stratotectonic terranes of the eastern Australian Tasmanides. American Geophysical Union, Geodynamics Series **19**, 1-19.

Mallett, C.W., Hammond, R.L., & Sullivan, T.D. 1988. The implications for the Sydney Basin of upper crustal extension in the Bowen Basin. Proceedings of the 22nd Symposium on Advances in the Study of the Sydney Basin, University of Newcastle, Newcastle, p. 1-8.

Malone, E.J. 1964. Depositional evolution of the Bowen Basin. Journal of the Geological Society of Australia **11**, 263-282.

McKenzie, D. 1978. Some remarks on the development of sedimentary basins. Earth and Planetary Science Letters **40**, 25-32.

Milner, J.E., & Milner, C.J. 1981. Rathdowney seismic survey, A.T.P. 266P, Queensland: Final report for Bridge Oil Limited. Geological Survey of Queensland Open File Report CR 9725.

Murray, C.G. 1985. Tectonic setting of the Bowen Basin. Geological Society of Australia Abstracts **17**, 5-16.

Murray, C.G. 1988. Tectonostratigraphic terranes of southeast Queensland. In, Hamilton, L.H. (ed.) Field excursions handbook for the Ninth Australian Geological Convention, Geological Society of Australia (Queensland Division), Brisbane, p.19-51.

Murray, C.G. 1990. Tectonic evolution and metallogenesis of the Bowen Basin. In, Beeston, J.W. (compiler) Bowen Basin Symposium 1990 Procedings, Geological Society of Australia (Queensland Division), Brisbane, p. 201-212.

Murray, C.G., Fergusson, C.L., Flood, P.G., Whitaker, W.G., & Korsch, R.J. 1987. Plate tectonic model for the Carboniferous evolution of the New England Fold Belt, Australian Journal of Earth Sciences **34**, 213-236.

O'Brien, P.E., Korsch, R.J., Wells, A.T., Sexton, M.J., & Wake-Dyster, K.D. 1990. Mesozoic basins at the eastern end of the Eromanga-Brisbane Geoscience Transect: strike-slip faulting and basin development. Bureau of Mineral Resources, Australia, Bulletin **232**, 117-132.

Offler, R., & Williams, A.J. 1985. Evidence for sinistral movement on the Peel Fault System in serpentinites, Glenrock Station, N.S.W. Geological Society of Australia, Abstracts **14**, 171-173.

Paten, R.J., Brown, L.N., & Groves, R.D. 1979. Stratigraphic concepts and petroleum potential of the Denison trough, Queensland. APEA (Australian Petroleum Exploration Association) Journal **19**, 43-52.

Roberts, J., & Engel, B.A. 1987. Depositional and tectonic history of the southern New England Orogen. Australian Journal of Earth Sciences **34**, 1-20.

Scheibner, E. 1973. A plate tectonic model of the Paleozoic tectonic history of New South Wales. Journal of the Geological Society of Australia **20**, 405-426.

Scheibner, E. 1976. Explanatory notes on the tectonic map of New South Wales. Geological Survey of New South Wales, Sydney, 283p.

Scheibner, E. (ed.) 1978. The Phanerozoic structure of Australia and variations in tectonic style. Tectonophysics **48**, 153-427.

Scheibner, E. (ed.) 1986. Metallogeny and tectonic development of eastern Australia. Ore Geology Reviews **1**, 147-412.

Shaw, R.D. 1990. Development of the Tasman Sea and easternmost Australian continental margin—a review. Bureau of Mineral resources, Australia, Bulletin **232**, 53-66.

Tadros, N.Z. 1988. Structural subdivision of the Gunnedah Basin. Quarterly Notes of the Geological Survey of New South Wales **73**, 1-20.

Thomas, B.M., Osborne, D.G., & Wright, A.J. 1982. Hydrocarbon habitat of the Surat/Bowen Basin. APEA (Australian Petroleum Exploration Association) Journal **22**, 213-226.

Veevers, J.J. (ed.) 1984. Phanerozoic earth history of Australia. Clarendon Press, Oxford.

Voisey, A.H. 1959. Tectonic evolution of northeastern New South Wales, Australia. Journal and Proceedings of the Royal Society of New South Wales **92**, 191-203.

Wake-Dyster, K.D., Drummond, B.J., Korsch, R.J., Finlayson, D.M., Sexton, M.J., Johnstone, D.W., & Bracewell, R. 1987a. An extensional model for the formation of the Surat Basin, eastern Queensland, Australia, based on deep seismic profiling. Bureau of Mineral Resources, Australia, Record 1987/51, 140-142.

Wake-Dyster, K.D., Sexton, M.J., Johnstone, D.W., Wright, C., & Finlayson, D.M. 1987b. A deep seismic profile of 800 km length recorded in southern Queensland, Australia. Geophysical Journal of the Royal Astronomical Society **89**, 423-430.

Woodcock, N.H., & Fischer, M. 1986. Strike-slip duplexes. Journal of Structural Geology **8**, 725-735.

Ziolkowski, V., & Taylor, R. 1985. Regional structure of the North Denison Trough. Geological Society of Australia Abstracts **17**, 129-135.

FACTORS AFFECTING THE ACQUISITION OF STRUCTURAL DATA FROM REMOTELY-SENSED IMAGES OF EASTERN AUSTRALIA

C.R. NASH
Australian Photogeological Consultants
48 Jacka Crescent
Campbell ACT 2601
Australia

ABSTRACT Systematic structural analysis based upon the spatial information contained in remotely-sensed images depends upon the accurate interpretation and annotation of the attitudes and surface traces of fundamental structural elements (bedding and foliation planes, faults and joints). In eastern Australia, as in other regions characterised by humid weathering and mid-latitude forests, these elements are largely manifested as structural landforms produced through the agency of fluviatile erosion. The interpretation of structural landforms may however be compromised by several inter-related factors which include image scale, solar illumination direction and elevation, terrain roughness and erosional history. The effects of these factors and suggested techniques to overcome the subjectivity which they introduce into photogeological studies are illustrated from the results of research in eastern Queensland.

Introduction

Structural analysis based upon spatial information contained in remotely-sensed images differs from conventional field analysis not only in scale of observation but, more importantly, in the fact that the observer is generally limited to mapping structural landforms rather than actual structures. This constraint is particularly relevant in regions of humid weathering and mid-latitude forest such as eastern Australia, where a strong element of subjectivity may be introduced into photogeological studies. Successful outcomes in these areas depend ultimately upon the skill and experience of the interpreter in recognising landforms associated with (a) the surface traces of faults and joints, and (b) the traces and attitudes of bedding or foliation planes formed by primary or secondary lithological layering. All subsequent macroscopic and megascopic structural analysis depends upon the correct annotation of these fundamental structural elements (Nash 1990).

A number of previous contributions in the fields of Basement Tectonics and remote sensing have focussed upon image interpretation procedures and their inherent problems. In particular, the effects of differential illumination direction and azimuth have been discussed by Wise (1969, 1976) and by Siegal (1977) while the effects of seasonal variations between alternative images have been discussed by Viljoen et al. (1975) and by Larson (1982). These topics were further analysed by the writer during research in southeastern Queensland (Nash 1987a), which demonstrated that effective analysis of both landforms (Nash 1988) and regional structure (Nash 1987b) could be obtained from properly designed image-interpretation techniques, notwithstanding widespread regolith development and afforestation. The present paper seeks to bring together the results of studies in the portions of eastern Queensland depicted in Figure 1 and is intended to demonstrate how subjectivity in interpretation of landforms associated with fundamental structural elements

M. J. Rickard et al. (eds.), Basement Tectonics 9, 109–121.

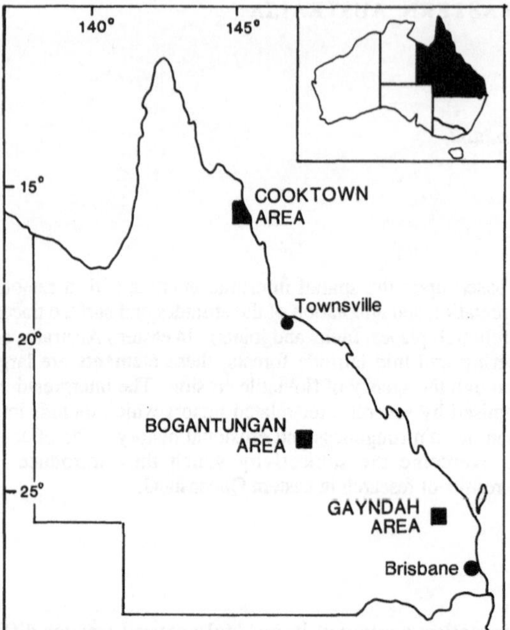

Figure 1. Locations of areas described in text

may be minimised through careful data selection, terrain analysis and consideration of geomorphic evolution.

Image Scale

The selection of appropriate image scale for a particular geological application requires careful consideration. This is demonstrated by comparative lineament interpretations of Landsat MSS image 22298-23031 acquired on 8/5/1981, using image enlargements at 1:1,000,000, 1:500,000, 1:250,000 and 1:100,000 scale respectively (Nash 1987a). The study area covers the GAYNDAH 1:100,000 topographic sheet area located in southeastern Queensland. The geology of the area has been described most recently by Cranfield (1986) and by Stephens (1986), who have shown the dominant rock types to be Triassic ignimbrites and associated subvolcanic granites.

Results of the comparative study are depicted in Figure 2. Lineaments from each of the different scale interpretations have been digitised and plotted as frequency histograms using standard lineament analysis software of the type described by McGuire and Gallagher (1976). The resulting histograms reveal a systematic variation between the number of lineaments and scale, ranging from 69 observed lineaments at 1:1,000,000 scale to 983 lineaments at 1:100,000 scale. The average lengths of observed lineaments are also related in a proportional way to image scale, confirming the generally held view that systems of sub-parallel lineaments seen at larger scales tend to merge into single large structures at smaller scales. Confirmation of the 1:100,000 scale Landsat lineament-interpretation exercise has been provided by a comparative fracture-trace analysis of the GAYNDAH 1:100,000 sheet area using 1:80,000 scale aerial photographs, from which 919 fracture traces with an average computed length of 2.99km were obtained (Nash 1987a).

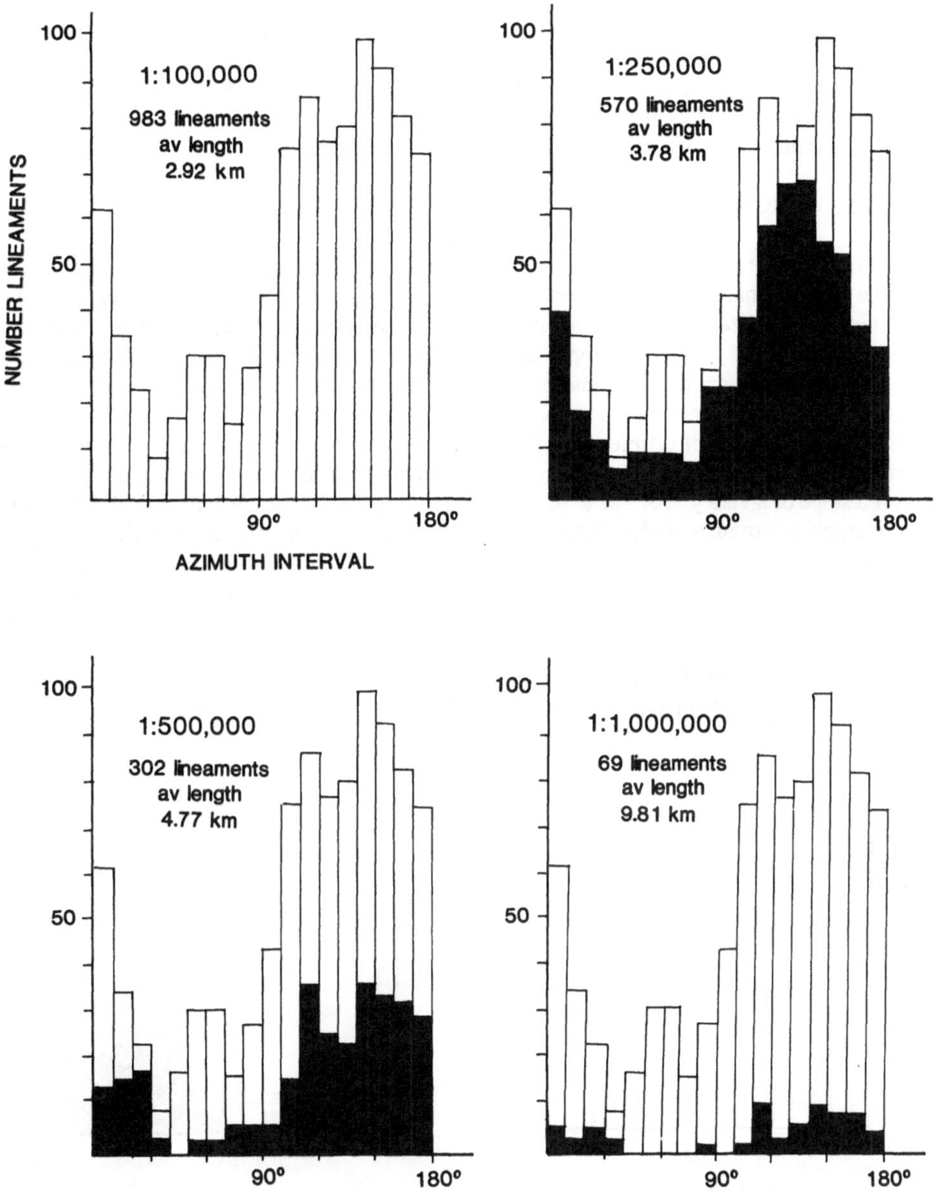

Figure 2. Frequency histograms (10° azimuth interval) of lineaments in the GAYNDAH sheet area interpreted from Landsat MSS imagery at scales of 1:1,000,000, 1:500,000. 1:250,000 and 1:100,000 respectively. Lineament interpretation based upon enlargements of Landsat MSS image 22298-23031, acquired 8/5/1981.

The implications of the foregoing data are significant in image selection prior to embarking upon applications of remotely-sensed data to structural geology. In particular, it is extremely important to select imagery which will permit detection of smaller structural landforms such as bedding and foliation features. In general it will be found that small-scale aerial photographs (1:50,000-1:100,000) and enlargements of satellite imagery to similar scales, where resolution permits, will provide an optimum trade-off between areal coverage and resolution.

SOLAR ELEVATION AND AZIMUTH

In order to fully appreciate the effects of solar elevation and azimuth during image acquisition, it should be realised that in dissected terrains virtually all structural information is relayed to the interpreter through differential illumination of structural landforms produced by fluviatile erosion. In particular, faults and joints tend to be etched out preferentially by stream erosion, while resistant beds tend to form remnant hogback landforms. These structural landforms are most easily seen in stereoscopic airphoto models; on monoscopic satellite imagery however their recognition is largely governed by the degree of differential solar illumination. What the interpreter observes as a 'lineament' is generally the linear boundary between an illuminated facing slope and the opposing shadowed slope of a valley. For this to occur, it is essential that (a) the angle of solar illumination is low enough to produce shadowing, and (b) that the azimuth of solar illumination diverges sufficiently from that of the eroded lineament. Geometric relationships between solar azimuth, slope angle and solar elevation have been discussed by Wise (1969); the effects of solar elevation on structural interpretation of monoscopic Landsat imagery have been described by Larson (1982) and Williams (1983).

In southeastern Queensland a comparative study of low and high sun-angle Landsat MSS imagery has been carried out over the GAYNDAH 1:100,000 sheet area by Nash (1987a). This

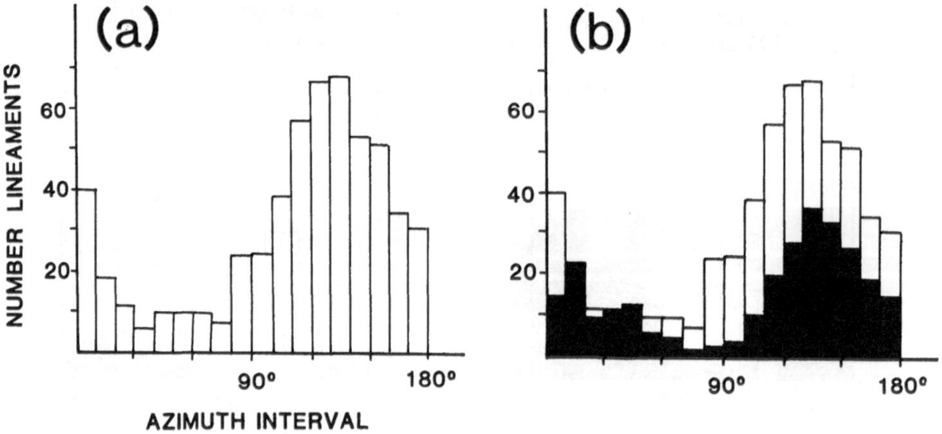

Figure 3. Frequency histograms (10° azimuth interval) of lineaments observed in the GAYNDAH sheet area, illustrating effects of solar elevation. Histogram (a) shows lineaments interpreted from a winter image (sun-angle 30°); lineaments shown as solid colour in histogram (b) are interpreted from a summer image of the same area (sun-angle 49°).

study made use of a summer, high sun-angle image acquired on 30/10/1975 (image 2264-23041, solar elevation 49°) and a winter, low sun-angle image acquired on 8/5/1981 (image 22298-23031, solar elevation 30°). Results of the comparative interpretations show that almost twice as many structural lineaments are recognisable on the low sun-angle images compared to the high sun-angle data (Fig. 3).

The effects of solar azimuth are equally significant, and were first noticed by the writer in a comparative study of Landsat and airphoto data over the Monto region of southeastern Queensland (Nash 1984). Subsequent detailed analysis (Nash 1987a) has shown that structures parallel to the direction of solar illumination at the time of image acquisition tend to be under-represented. This is well demonstrated in Figure 4, which depicts comparative circular frequency histograms of lineaments, interpreted from aerial photographs and Landsat MSS imagery respectively, over the GAYNDAH 1:100,000 scale sheet area. Landsat interpretation was carried out at the conventional scale of 1:250,000 while the fracture trace data were obtained from interpretation of 1:80,000 scale aerial photographs.

The data presented in Figure 4 reveal a coincidence between Landsat lineament and airphoto fracture trace trends oriented in the azimuth range 120°-140°. The 140°-160° lineament trends visible in the Landsat data are due to large Tertiary landform features which were not included in the photogeological fracture-trace data (Nash 1988). Of particular significance however is the presence of a prominent 050°-070° frequency peak in the airphoto data which is conspicuously absent from the Landsat lineament histogram. This anomaly is best explained by the local acquisition time for Landsat MSS imagery over southeastern Queensland (0930hrs), when solar

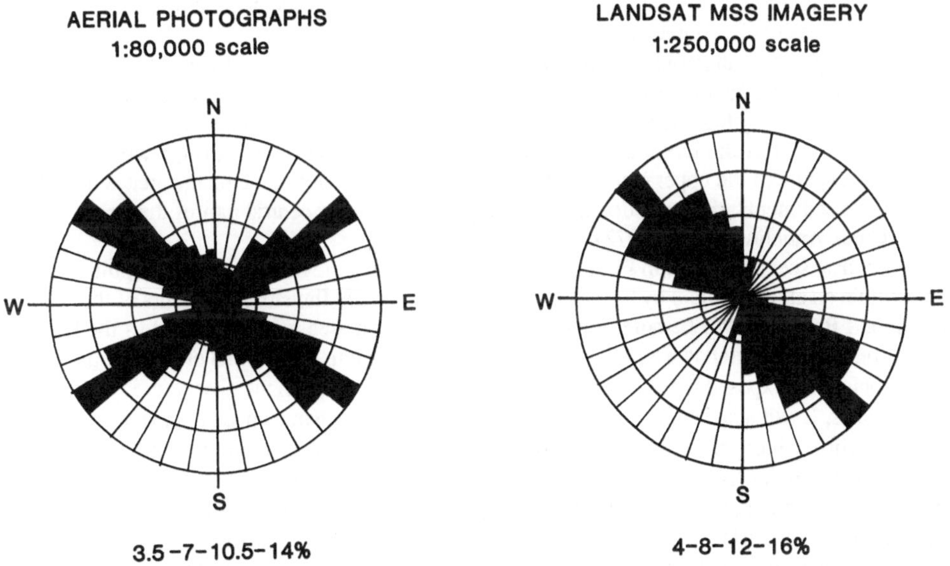

Figure 4. Comparative circular frequency histograms (10° azimuth interval) of fracture-trace and lineament data from the GAYNDAH sheet area, illustrating effects of solar azimuth. Fracture-trace data depicted in left-hand diagram reveals a strong 050°-070° preferred orientation which is absent from Landsat lineament data shown in right-hand diagram.

illumination is from the east or northeast depending upon season. Since the majority of observed lineaments are due to differential illumination of incised drainage features, it follows that lineaments aligned with the NE solar-illumination direction will not have one face shadowed and will consequently be invisible on satellite imagery. By contrast, the small-scale aerial photographs acquired when the sun is directly overhead show no such bias.

TERRAIN ROUGHNESS

The foregoing sections have emphasised the importance of fluviatile dissection in the enhancement of fundamental structural features. It is therefore to be anticipated that the degree and frequency of stream incision will bear a relationship to the overall topography of a given region. In particular, it is to be expected that the greatest density of stream-enhanced lineaments will be found in areas of roughest terrain. This concept will be familiar to all photogeologists who have experienced the difficulty of locating structural lineaments in the flat, deeply-weathered regions of central and western Australia compared to the ease with which these features may be discerned in the deeply dissected mountainlands of Papua New Guinea and comparable areas of recent uplift.

An excellent example of the effects of terrain roughness is provided by the Bogantungan area, located in the Drummond Basin region of Central Queensland (Fig. 5). The basin is a large intracratonic feature which developed during Late Devonian to Early Carboniferous times (Hutton 1989). The main part of the depository contains two cycles of sedimentation (Olgers 1972) which are dominated by feldspathic and quartzose arenite and rudite with interbedded mudstone and shale, and by volcanolithic sediments and tuffs respectively. The contents of the basin are folded into gentle to tight folds, often doubly plunging, whose axes follow the general trends of the basin and which are easily discernible on Landsat imagery.

The physiography and geology of the Bogantungan area are illustrated by means of a Landsat MSS image in Figure 5a and by a geological sketch map in Figure 5b. The northeastern portion of the area is underlain by basement rocks which have a characteristic photogeological expression; the basal volcanic unit of the Drummond Basin is the prominent NW trending ridge which may be traced to the centre of the upper margin of the image. The remainder of the image area is underlain by fluviatile sediments; the large doubly-plunging structure in the central part of the image is the Pebbly Creek Anticline (Olgers 1972).

The Landsat scene depicted in Figure 5a suggests that two quite distinct physiographic terrains occur within an area underlain by the same sequence of deformed fluviatile sediments of the central Drummond Basin. It will be noted that the dissected upland area is not confined to the Pebbly Creek Anticline, but extends well to the south and to the east, thus differential erosion of the lower stratigraphic units exposed in the anticline cannot be invoked to explain the observed physiographic discrepancies. It will also be noted that the flat planation surface characterised by a veneer of pale-toned regolith which rests upon the Drummond Basin in the western part of the area is conspicuously absent from the dissected upland area. These observations strongly suggest that uplift of the Bogantungan area took place during late Cainozoic time, after the formation of the regional planation surface. It is probable that the roughly circular uplift was associated with widespread late Cainozoic basaltic volcanism with occurred throughout eastern Queensland (careful inspection of Figure 5a will reveal the presence of numerous small volcanic craters in the basement terrain to the east of the Drummond Basin margin).

Terrain variations such as that described above will obviously provide anomalous results unless they are taken into account during photogeological interpretation. In the cited example, fracture-controlled lineaments are clearly visible in the uplifted and dissected part of the image, but are

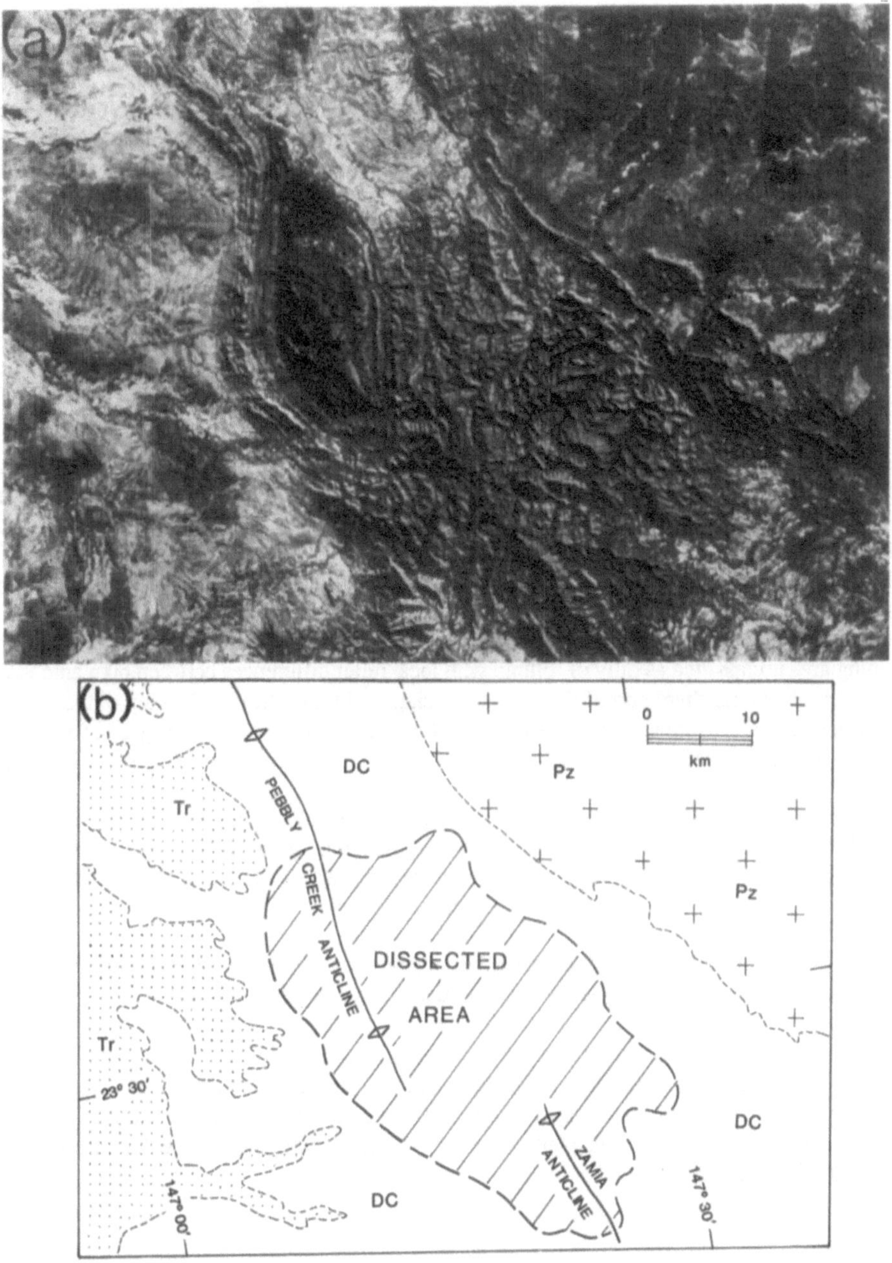

Figure 5. (a) Portion of Landsat MSS image covering Bogantungan area, central Drummond Basin. (b) Geological map of the Bogantungan area after Olgers (1972) showing areas underlain by Palaeozoic basement (Pz), Devonian-Carboniferous Drummond Basin rocks (DC) and Tertiary regolith (Tr).

difficult to see in the remainder because of the regolith cover. A spurious pattern of lineament density may therefore be expected to result from image interpretation, which is unrelated to the actual structure of the Drummond Basin. Similarly, while the patterns described by bedding traces in the dissected country are quite sufficient to define the major structural features, it would be impossible to decipher fold culminations in the area to the west, leading to misleading inferences regarding the style of deformation. It is useful in situations such as this to draw rough boundaries around the major landform types in a given region so that varying levels of confidence may be assigned to structural information depicted on photogeological maps.

From these considerations it may be surmised that interpreted lineament- and bedding-trace densities over a given area (although not absolute lineament densities) are likely to be related to some measure of terrain roughness. This relationship has been examined semi-quantitatively by Nash (1987a) in Gayndah region of southeastern Queensland, where lineament densities have been contoured into map form and compared to a terrain roughness map at a common scale.

In this study, lineaments were initially interpreted from 1:100,000 scale Landsat MSS imagery. Lineament densities were obtained by counting the number of lineament intersections per $2km^2$ cell. The terrain-roughness factors for the area were obtained by computing the elevation differences between stream courses and their adjacent interfluve ridges, as suggested by Galloway (1969). The resulting elevation-difference values, which may conveniently be measured from published 1:100,000 scale topographic maps, were contoured to provide a terrain roughness map for direct comparison with the lineament density map.

The results of the study are depicted in Figure 6. Four terrain categories were identified. In the eastern portion of the area, rugged country forming the Ban Ban and Bin Bin Ranges attains local relief differences in excess of 200m and may be classified as mountainous terrain. Most of the remaining area is classified as hilly or rolling, with local relief differences between 60m amd 80m. The flat plain in the northern part of the area is associated with the alluviated valley of the Burnett River where relief differences seldom exceed 40m. The area of undulating country in the southwest is formed upon relatively undeformed Triassic volcanic rocks where relief differences are generally less than 60m.

Superimposition of the two maps depicted in Figure 6 reveals a good degree of coincidence between lineament density and terrain roughness. Areas of greatest lineament density (>7 lineaments/$2km^2$ cell) occur in the mountainous eastern part of the study area, while lowest lineament densities (0-4 lineaments/$2km^2$ cell) correspond closely with the defined limits of the plain of the Burnett River.

An important application of terrain-roughness estimation lies in the field of lineament-and fracture-trace analysis. Many workers have endeavoured to relate the density of lineaments observed on aerial photographs and satellite images to specific geological situations such as local patterns of intense fracturing around intrusive stocks or oil prone anticlines. While there is nothing inherently wrong in this approach, it is essential that regional lineament or fracture-trace data be filtered for the effects of terrain roughness prior to the selection of anomalies. If this is not done, the resulting lineament density or isopleth maps may end up showing little more than the positions of elevated and dissected areas.

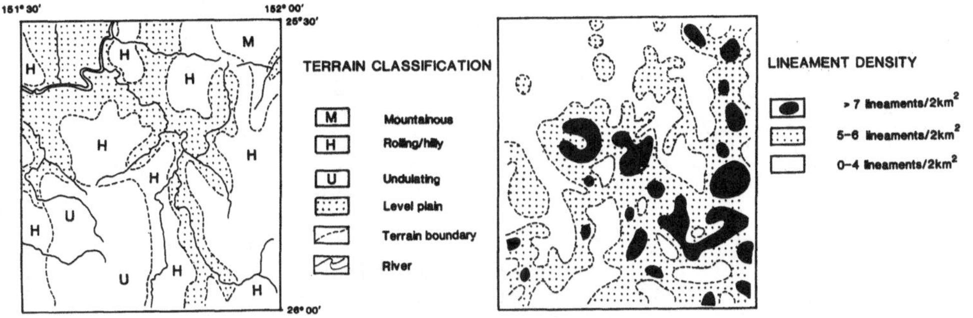

Figure 6. Comparative terrain-classification and lineament-density maps of the GAYNDAH sheet area. Terrain classification based upon method of Galloway (1969). Lineament-density map derived by counting lineament intersections per $2km^2$ cell.

Geomorphic Evolution

The agencies of fluviatile erosion which are responsible for structural landform evolution in eastern Australia require further mention. In most instances landform genesis may be explained in terms of a simple process-response model in which uplift or sea-level changes cause stream incision and the ultimate etching out of structurally controlled landforms. However this is not always the case. Inherited or superimposed drainage patterns are not uncommon and can give rise to spurious 'structural landforms'. An example is to be found in the Cooktown region of far northern Queensland where the writer carried out a detailed photogeological and field investigation during 1988 (Nash 1990; in prep.).

The geomorphology of the Cooktown region is dominated by the meridional Great Escarpment (Ollier 1982) located some 30km west of the present coastline. To the west of this feature are a series of highlands consisting of incised valleys and structural planation surfaces formed upon Mesozoic sandstones of the Laura Basin, whereas to the east of the escarpment a dissected lowland dominated by meridional hogback ridges, relict mesas and granite uplands is present. The Cooktown region also provides striking examples of drainage piracy by short, eastward flowing streams which have captured the headwaters of northwestward flowing rivers draining into the Laura Basin (de Keyser & Lucas 1968; Nash in prep). Figure 7 is a landform interpretation of the Cooktown hinterland area, based upon 1:250,000 scale Landsat MSS imagery, which clearly illustrates drainage capture by the Annan River, Oaky Creek and Endeavour River systems. Evidence of stream piracy is provided by the windgap in the extreme northwest of the diagram and by 'elbows of capture' located at three points along the course of Oaky Creek, including its junction with the Annan River.

The figure also illustrates the positions of positive topographic landforms (hogback ridges) in the area; these generally coincide with bedding trends in the Palaeozoic Hodgkinson Basin which underlies the country to the east of the Great Escarpment (de Keyser & Lucas 1968). Field inspection shows that the majority of the meridional hogback ridges in the Cooktown hinterland are composed of resistant chert interbeds within a strongly deformed flysch sequence.

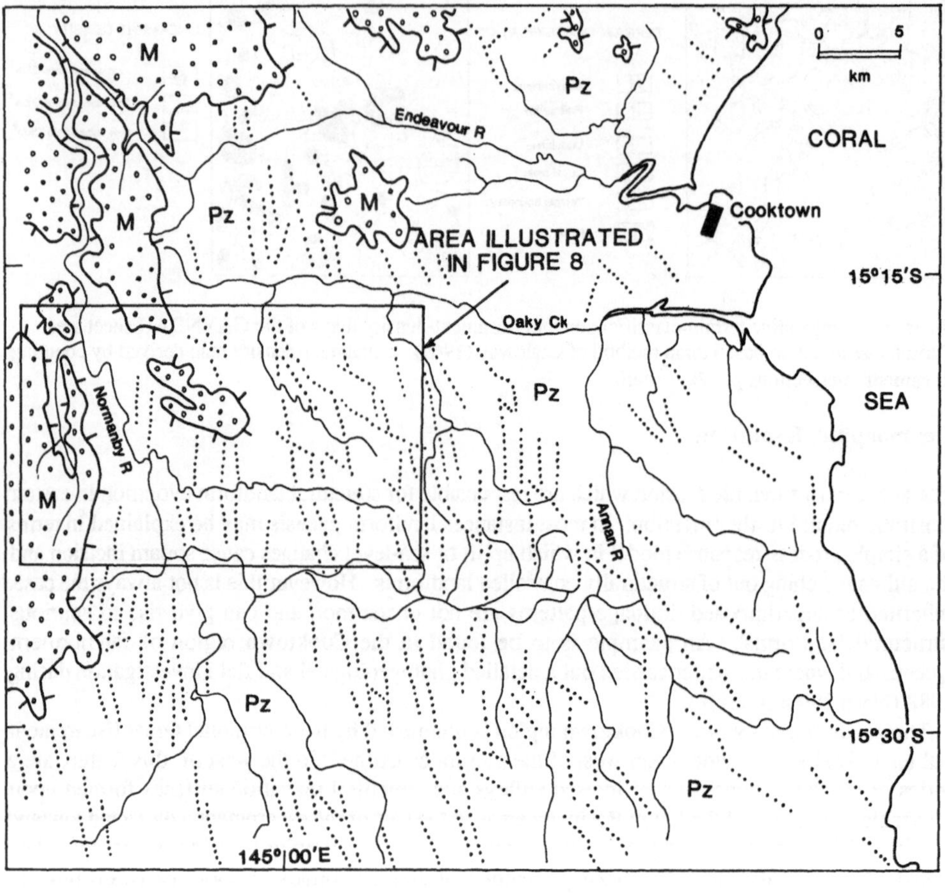

Figure 7. Structural landforms of the Cooktown hinterland area interpreted from 1:250,000 scale Landsat MSS images 22233-23411 and 22233-23045, acquired on 4/3/1981 (sun-angle 46°).

An inspection of Figure 7 however also reveals the presence of WNW oriented topographic lineaments which are at variance with the N-S structural trends of the Hodgkinson Basin. The nature of these anomalous trends has been investigated by means of a detailed photogeological and field study of the Mount McCormack area, located between Oaky Creek and the Normanby River, using 1:80,000 and 1:25,000 scale aerial photographs (Fig. 8). The detailed map of the Mount McCormack area shows WNW and ENE drainage interfluves; the former set correspond to the topographic lineaments identified from Landsat MSS imagery. It will be noted that the ENE trending interfluves identified from the airphoto study are not visible on satellite imagery, since this direction is parallel to the solar-illumination direction.

The airphoto interpretation also reveals numerous fracture or joint traces in the tableland region to the west of the Normanby River. These form deep linear valleys, presumably caused by the selective erosion of structurally weakened sandstone. The predominant directions of the joint traces (WNW and ENE) correspond quite closely with the directions of the drainage interfluves in the country to the east of the Normanby River where the Mesozoic cover appears to have been stripped away by erosion. The drainage interfluves may have evolved through a process of parallel scarp retreat adjacent to the courses of linear subsequent streams which developed selectively along joints in the Mesozoic sandstone. These streams possibly propagated downward into the underlying deformed Palaeozoic sequence following uplift of the Laura Basin, although it is also possible that their courses were controlled by fracture patterns in the Palaeozoic basement. Continuing scarp retreat caused progressive reduction of the sandstone capping upon the interfluve ridges, leaving behind small remnant mesas (e.g. Mount McCormack).

Further erosion removed all traces of the sandstone, leaving behind a series of linear interfluve ridges such as those which presently form Lookout Range and Barrons Range. These interfluve ridges are separated by superimposed streams which cross and are apparently unaffected by the regional system of steep chert hogback ridges of the Palaeozoic Hodgkinson Basin sequence.

Discussion

Landforms in dissected terrains such as those of eastern Australia are dominated by the effects of fluviatile erosion. Structural landforms which are of particular significance in photogeological studies are linear hogback ridges, which provide an indication of bedding and foliation directions and attitudes, and linear stream segments and drainage interfluves which commonly reveal fault and joint directions.

Selection of appropriate image scale should be matched carefully to the sizes of the target landforms to be identified. In most cases it will be found that small-scale (1:50,000-1:100,000) aerial photographs and high-resolution satellite imagery provide the ideal medium for regional landform studies. If monoscopic satellite imagery is used alone, the effects of solar elevation and azimuth must be considered. Low sun-angle winter images are always preferable for structural interpretation.

Terrain variation within a given area will cause considerable discrepancies in the density of recorded structural data from aerial photographs and particularly from satellite images. A recommended procedure to counter this problem is the depiction of major landform class boundaries upon photogeological maps.

The cited example of drainage superimposition from far northern Queensland will draw to the attention of the reader the importance of understanding the geomorphic evolution of an area before attempting to assign any structural significance to observed landforms. While the assumption that linear drainage systems and hogback ridges are respectively indicative of fractures and bedding traces in areas of fluviatile erosion is generally correct, there are many instances where this is not so, leading to interpretation of spurious structural elements. In the Cooktown area it is possible to observe primary hogback ridges in a direction parallel to bedding and a secondary set related to fracturing in either the basement or the cover. Without an adequate analysis of the geomorphic history of the area it would not be possible to evaluate the proper significance of these features.

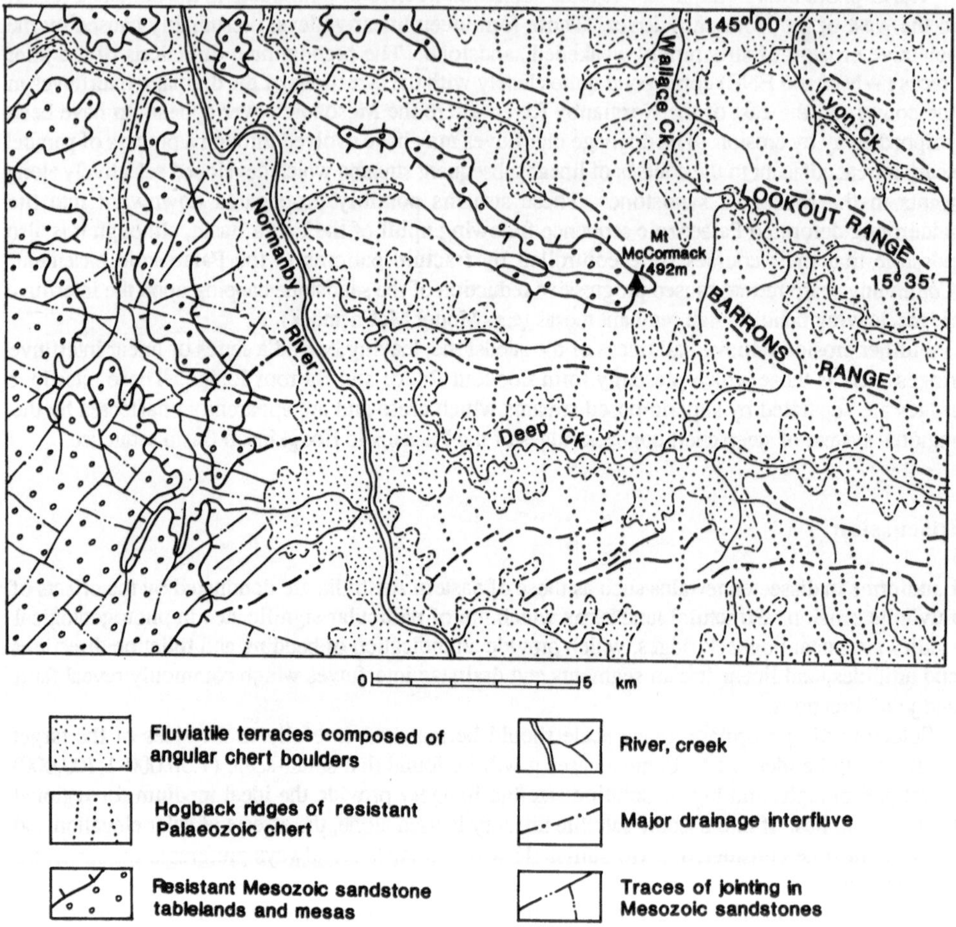

Figure 8. Detailed landform map of the Mount McCormack area, based upon interpretation of 1:80,000 scale aerial photographs and field traversing.

References

Cranfield, L.C. 1986. Geology of the South Burnett District. In, Willmott, W.F. (ed.) 1986 Field Conference, The South Burnett District. Brisbane. Geological Society of Australia, Queensland Division p.1-12.

De Keyser, F., & Lucas, K.G. 1968. Geology of the Hodgkinson and Laura Basins, north Queensland. Australian Bureau of Mineral Resources Geology & Geophysics Bulletin **84**.

Galloway, R.W. 1969. Geomorphology of the Queanbeyan-Shoalhaven area. Commonwealth Scientific & Industrial Research Organisation, Australian Land Research Series **24**, 76-91.

Hutton, L.J. 1989. A stratigraphy and tectonic model for the Drummond Basin and its relationship to gold mineralisation. Proceedings of North Queensland Gold '89 Conference, Townsville, p.31-40.

Larson, B.S. 1982. Examination of some factors used in selecting Landsat imagery for lineament interpretation. Proceedings International Symposium for Remote Sensing of the Environment, 2nd Thematic Conference on Remote Sensing for Exploration Geology. Fort Worth, p. 293-302.

McGuire, M.J., & Gallagher, J.J. 1976. Techniques for computer-aided analysis of lineaments. In, Podwysocki, M.H., & Earle, J.L. (eds) Proceedings 2nd International Conference on Basement Tectonics, Newark, p. 528-541.

Nash, C.R. 1984. Tectonic interpretation of the northern Tasman Orogenic Zone from Landsat structural data. Proceedings 3rd. Australasian Remote Sensing Conference, Gold Coast, Queensland, p. 433-441.

Nash, C.R. 1987a. Late Palaeozoic to Cainozoic evolution of the New England Orogen in southeastern Queensland—a photogeological investigation. Unpublished PhD thesis, Macquarie University, 255p.

Nash, C.R. 1987b. Late Palaeozoic and Early Mesozoic evolution of the New England Orogen in southeastern Queensland—a photogeological study. In, Leitch, E.C., & Scheibner E. (eds) Terrane Accretion and Orogenic Belts. American Geophysical Union, Geodynamics Series **19**, 153-160.

Nash, C.R. 1988. Delineation of structurally controlled landforms in southeastern Queensland using remotely sensed data. Earth Science Reviews **25**, 427-432.

Nash, C.R. 1990. Structural interpretation of remotely-sensed data. Adelaide, Australian Mineral Foundation, Course Notes, 668/90.

Nash, C.R. (in prep.) Evolution of superimposed drainage systems in the Cooktown area, north Queensland—a photogeological study. Earth Surface Processes and Landforms.

Ollier, C.D. 1982. The Great Escarpment of eastern Australia: tectonic and geomorphic significance. Journal Geological Society of Australia **29**, 13-23.

Olgers, F. 1972. Geology of the Drummond Basin, Queensland. Australian Bureau Mineral Resources Geology & Geophysics Bulletin **132**.

Siegal, B.S. 1977. Significance of operator variation and the angle of illumination in lineament analysis on synoptic images. Modern Geology **6**, 75-95.

Stephens, C. 1986. Late Triassic silicic volcanism near Gayndah. In, Willmott, W.F. (ed.) 1986 Field Conference, The South Burnett District, Brisbane. Geological Society of Australia, Queensland Division, p. 32-38.

Viljoen, R.P., Viljoen, M.P., Grootenboer, J., & Longshaw, T.G. 1975. ERTS-1 imagery: an appraisal of applications in geology and mineral exploration. Minerals Science &Engineering **7**, 132-168.

Williams, R.S. 1983. Geological applications. In, Colwell, R.N. (ed.) Manual of Remote Sensing (2nd ed.), Falls Church. American Society Photogrammetry, p.1790-1791.

Wise, D.U. 1969. Regional and sub-continental sized fracture systems detectable by topographic shadow techniques. In, Baer, A.J., & Norris, D.K. (eds) Research in Tectonics. Geological Survey Canada, p.175-199.

Wise, D.U. 1976. Sub-continental sized fracture systems etched into the topography of New England. In, Hodgson, R.A., Parker Gay, S., & Benjamins, J.Y. (eds) Proceedings 1st International Conference on The New Basement Tectonics, Salt Lake City, p. 416-422.

ELUSIVE TRAILS IN THE BASEMENT LABYRINTH

E.S.T. O'DRISCOLL
14 Renwick Street
West Beach S.A. 5024
Australia

ABSTRACT. The Australian continental crust is characterised by subtle pattern breaks corresponding to what E. Sherbon Hills called "zones of yielding" in the basement. These breaks can be identified in appropriate compilations of standard geological, geophysical and geographic data. Pattern breaks appear both as linear and arcuate (ring) discontinuities, and generally correspond to thresholds of change in the distribution of geological structures, attitudes and/or material compositions. Intersecting linear discontinuities form the systematic floor plan of the continental basement, and control the nature and distributions of surface geological patterns. They localise the propagation of crustal energies along crustal disjunctions and interruptions of one kind or another. Such disjunctions frequently correspond to fault and shear zones (including rifts), and to zones of anomalous conditions in terms of lithology, deformation, metamorphism and mineralisation.

From local applications at ore-deposit scale, the use of lineament tectonics extends to the comparison between lineament-related crustal patterns at continental scale and similar features visible in sea-floor topography and atmospheric patterns. A global lineament system is postulated which is spiral in nature, and which may be related to the remarkable criss-cross pattern of lumps on the earth's deep core-surface as revealed by seismic (*catscan*) tomography.

An outstanding example of the lineament-ore relationship at several different scales is provided by studies of Australia's latest and greatest copper-uranium deposit, Olympic Dam, in the discovery of which the application of lineament-tectonics played a leading role.

Introduction

Someone once remarked that elegant architecture is like frozen music. I think that the basement and crustal architecture of Australia may be contemplated in the same terms, and in this keynote address I will sound a separate note for each of a number of ideas that invite our deeper inquiry. As always, pictures provide the score for the lineament-tectonics theme, and I rely on them to support the theme with as little lyric content as practicable. In that respect my address is more of a tour d'oeil than a tour de force.

Continental Gravity Lineaments

The Bouguer gravity contour map of Australia (Bureau of Mineral Resources 1971, 1975) is a much used but still convenient vehicle for displaying a number of continental crustal megalineaments, some crossing from coast to coast. These are seen in Figure 1 (adapted from O'Driscoll 1986), which shows the internal trails of selected lineaments indicated by the directed tmarginal arrows pointing along them into the map area. The lineaments can be shown to be

M. J. Rickard et al. (eds.), Basement Tectonics 9, 123–148.

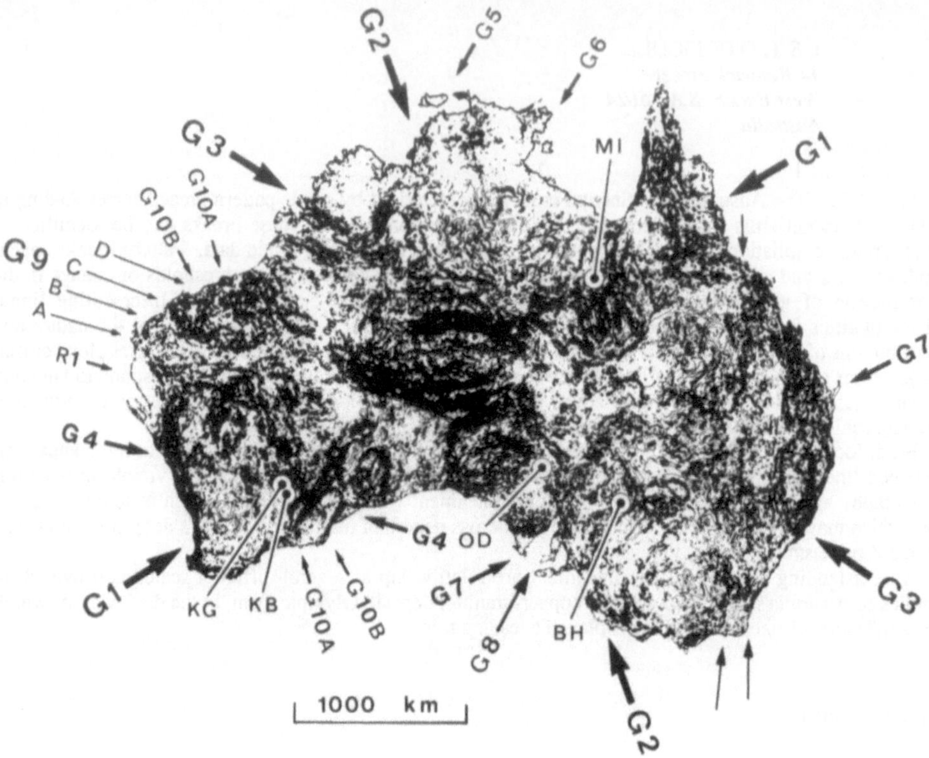

Figure 1. Marginal arrows point along continental lineaments visible in the Bouguer gravity map of Australia. (Adapted from O'Driscoll, 1986 after Bureau of Mineral Resources, 1971, 1975).

related to singular geological and physiographic features, and are characteristically associated with mineralised centres, especially at their intersections. Lineament GIOA in Western Australia is the track of the famous "gold line", and is also a trend of maximum nickel occurrence. Two major ore centres are shown circled at its intersection with the transverse WNW lineament G4, namely, Australia's greatest gold field, Kalgoorlie (KG), and its greatest nickel centre, Kambalda (KB). In a similar context 1400 km to the east, on lineament G2, is Australia's greatest copper-uranium deposit, Olympic Dam (OD), located where lineament G2 is crossed by the transverse lineament G9C. In analogous settings, the great mineral deposits, Mount Isa (copper-lead-zinc) and Broken Hill (silver-lead-zinc), are located respectively at "MI" on lineament G1, and at "BH" on lineament G7. Clearly, the greatest mineral deposits are associated with the most conspicuous continental lineaments, corresponding to basement fractures representing Hills' "zones of yielding" in the crust (Hills 1961). The character and associations of Australian regional and continental lineaments have been discussed in the texts and references given by Woodall (1984,1985,1990), Preiss (1987), O'Driscoll (1981,1989,1990), and Campbell & O'Driscoll (1989).

Antiquity of Major Crustal Lineaments

Evidence for the antiquity of present-day crustal lineaments may be seen in many contexts of ancient Australian stratigraphy. A useful example is available in the distribution pattern of Precambrian stratigraphy in the southern Adelaide Geosyncline (Preiss 1987), where three major lineament corridors transect its central area. The surface pattern of the Precambrian Burra Group sediments, shown as the black ingredient in Figure 2, illustrates a strikingly angular response to the tracks of all three linear corridors, particularly to lineaments G7 and G2. In the latter instance, the Burra Group geology effectively delineates the two edges of the NNW corridor G2.

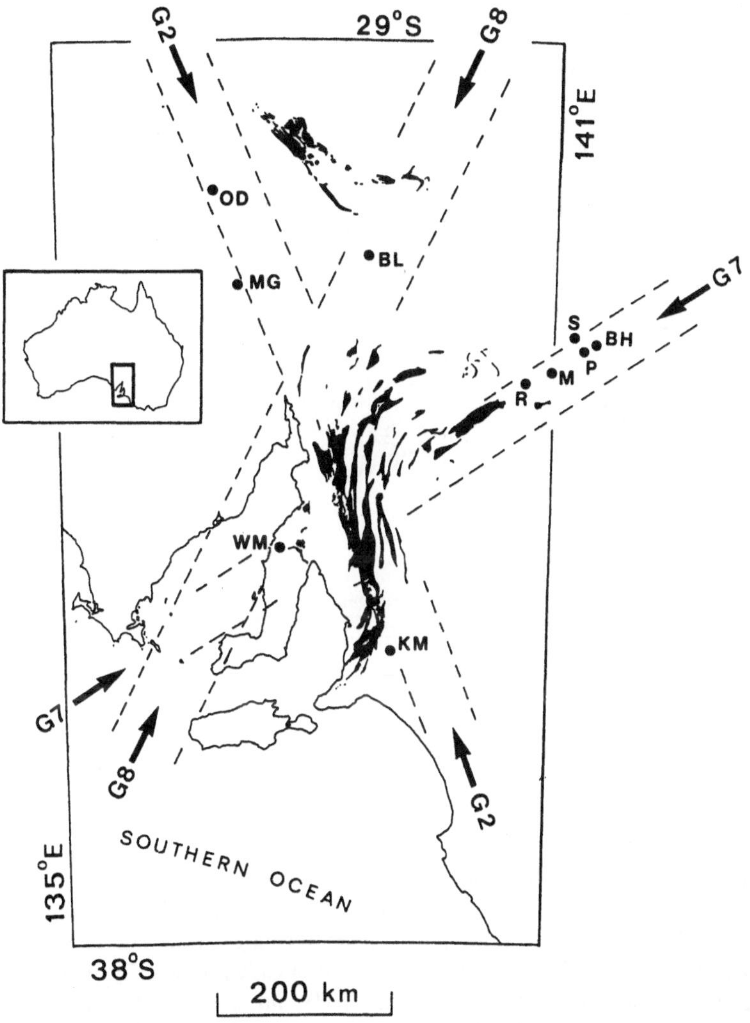

Figure 2. The angular response of Precambrian sediments (black) to the tracks of major lineaments in South Australia. (O'Driscoll 1983a; 1983b; 1986).

The ENE arm of the Burra Group sediments, running along corridor G7, is pointing to an ENE alignment of significant mineral deposits (black dots) identified as Radium Hill uranium (R), Mutooroo copper (M), Pinnacles Pb-Zn (P), Silverton Pb-Ag (S), and Broken Hill Pb-Ag-Zn (BH). Other deposits shown are Kanmantoo copper (KM), Wallaroo-Moonta copper (WM), Blinman copper (BL), Mount Gunson copper (MG), and Olympic Dam Cu-Au-U (OD). For reasons of cartographic clarity, the copper deposits at Burra and Kapunda are omitted, but they occur in the NNW corridor G2 (O'Driscoll 1986).

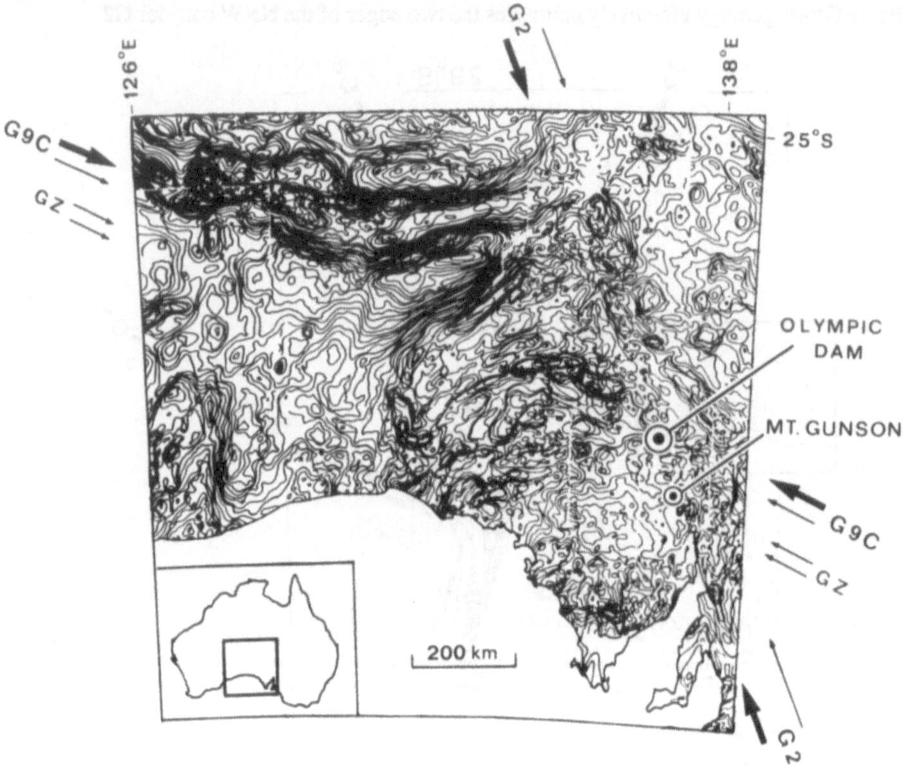

Figure 3. Bouguer gravity map of central southern Australia showing the relationship of Olympic Dam to the intersection of lineaments G2 and G9C. (After O'Driscoll, 1985).

The WNW Geological Corridor 4A-4B.

The Bouguer gravity contour map of the area in Figure 3 (enlarged from Fig. 1), shows the NNW gravity lineament, G2, to be a double-edged corridor, as already geologically attested in Figure 2. Figure 3 also shows the position of the transverse WNW lineament, G9C, (cf Fig. 1), which in greater detail, is seen as a narrow WNW corridor, accompanied by a parallel companion corridor, GZ, distant about 100 km to the south of it. These two WNW gravity trends, which sweep swath-like through the gravity pattern of South Australia, are also represented in the overall geological ingredient outline map of the Adelaide Geosyncline. There they are expressed as two WNW geological lineaments identified by O'Driscoll (1986, Fig. 15) as "4A" and "4B" respectively. When these two transverse lineaments are superimposed on the paleogeologic Precambrian maps of Preiss (1987, Figs. 38 & 75), it is apparent that their combined WNW system exercised a strong influence over Precambrian sedimentation and synsedimentary faulting in the northern Geosyncline.

The 4A-4B lineament influence is shown quite graphically in Figures 4 and 5. Figure 4 displays the gravity backdrop, as in Figure 3. Superimposed on it is Preiss's paleogeologic map (op. cit., Fig. 38) showing in hatched outlines the distribution of the Precambrian upper Burra Group sediments, together with the traces (heavy black lines) of faults active at the time of deposition. To this I have added the tracks of the transverse WNW lineaments 4A and 4B, which are then seen to have had a marked influence on the pattern of the Precambrian litho- structural distributions. The other lineaments shown are from Figures 1 and 2. As a matter of collateral interest, the positions of the most significant mineral deposits at that latitude have been added, numbered as (1), Olympic Dam Cu-Au-U; (2) Ediacara Pb-Ag; (3) Blinman Cu, (4) Honeymoon U; (5) Broken Hill Pb-Ag-Zn. The WNW alignment of these deposits along the 4A-4B corridor is notable.

A comparable example of lineament-controlled deposition is seen in Figure 5, indicating the continued influence of the 4A-4B corridor during the deposition of the Younger Precambrian Wilpena Group. This is clearly shown by the distribution of the Nuccaleena submerged dolomites (Preiss 1987, Fig. 75), identified by horizontal hatching, and the supra-tidal dolomites (cross hatched), coinciding with the 4A-4B/G8 lineament intersection. Thus, in combinations of Figures 2, 3, 4 and 5, are seen the interactions of four established lineament trends in controlling the pattern of ancient stratigraphy during its deposition.

The Role of Lineament-Tectonics in the Olympic Dam Discovery

In 1974, after a competitor had discovered a new and significant copper deposit in a northerly-trending Proterozoic pre-Adelaidean basement upwarp at Mt Gunson, on the Torrens 1/250,000 sheet in South Australia, Western Mining Corporation carried out a study of the lineament tectonics of that area and of the adjoining Andamooka 1/250,000 sheet to the north (O'Driscoll 1985,1986). The study revealed that the new Mt Gunson deposit related to a narrow WNW photo/magnetic basement corridor where it was intersected by a NNE orthogonal lineament. This was identified as the repetition of a specific lineament-ore signature that had already been recognised in other mineral deposits, and which had been prescribed as a criterion of primary importance in ore-search techniques (O'Driscoll 1986).

From its local outcropping culmination in the Mt Gunson area, the northerly trending basement dipped away under the younger sediments of the surrounding Stuart Shelf. The lineament-tectonic pattern of the Andamooka sheet to the north, was therefore subjected to a close study for signs of a further repetition of the prescribed lineament signature which would indicate a potential target in the buried basement. The Bouguer gravity lineaments of Figure 3 were also important factors in the assessment of potential target areas, because the western edge of the NNW G2 corridor was known to be ore-associated where it traversed the Adelaide Geosyncline to the south, and again where it tracked through the Peake & Denison Ranges to the north. In addition, the transverse G9C gravity trends (Fig. 3) represented the controlling system presiding over ore-associated WNW photo-magnetic lineaments.

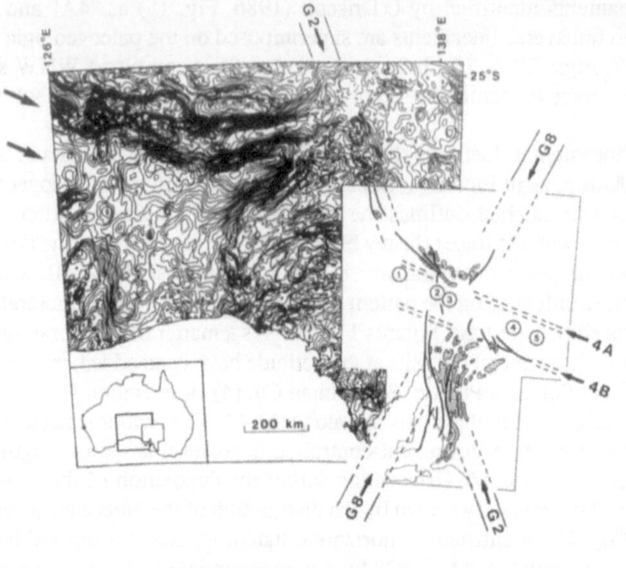

Figure 4. Stratigraphy and structure of Precambrian Burra Group sediments in the Adelaide Geosyncline in relation to major lineament corridors (after O'Driscoll 1986, and Preiss 1987). Circled numbers refer to important ore deposits identified in text.

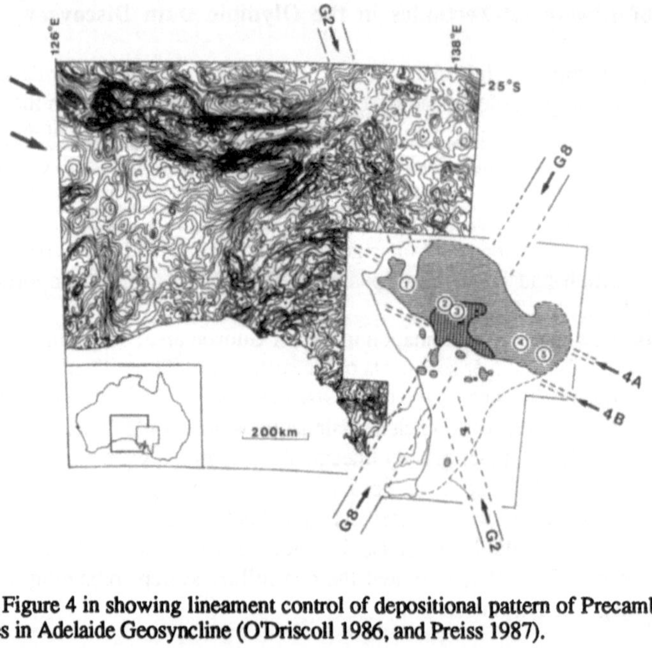

Figure 5. Similar to Figure 4 in showing lineament control of depositional pattern of Precambrian Nuccaleena dolomites in Adelaide Geosyncline (O'Driscoll 1986, and Preiss 1987).

The detailed study of the lineament-tectonics of the Andamooka sheet revealed four WNW photo/magnetic lineaments with the prescribed qualifications, one of them being a narrow corridor on which we recognised a repetition of the same lineament-tectonic signature as that seen earlier in the Mt. Gunson ore pattern. Subsequent drilling in this priority target intersected the Olympic Dam deposit in the Proterozoic basement at a depth of some 300 metres (Roberts & Hudson 1983). Figures 6, 7 and 8 show the transverse WNW lineaments identified on the Torrens and Andamooka sheets, and tell an interesting story of the lineament-ore relationship. Figure 6 shows the WNW photo/magnetic lineaments, including the two corridors, identified in 1974 before exploration drilling began. The associated NNW gravity corridor, G2, is also shown. At that time, almost nothing was known of the depth to pre-Adelaidean Proterozoic basement in the northern Andamooka sheet, and little of it also in the southerly Torrens sheet, except at the latitude of the culminating inlier containing the Mt Gunson (MG) area. Figure 7 shows the picture nine years later, after 61 exploration drill holes had penetrated the Proterozoic basement at depths ranging between 300 and 800 metres. Fifty four of these holes were on the northern Andamooka sheet, including the first to be drilled, which was the Andamooka discovery hole ("OD" in Fig. 7).

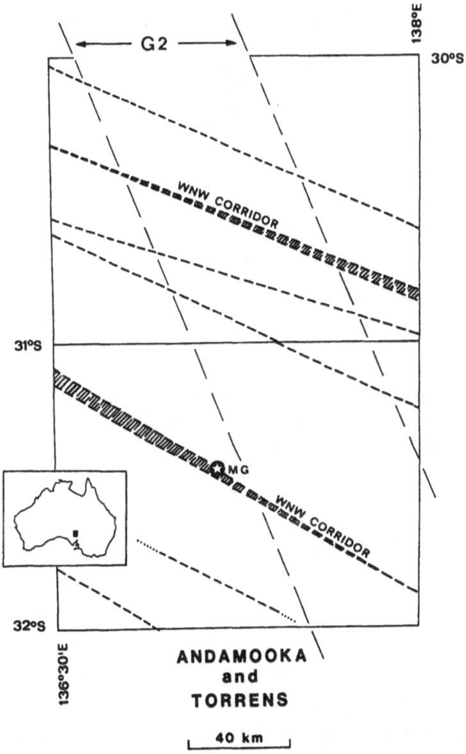

Figure 6. WNW lineaments on the Andamooka and Torrens sheets as defined in 1974 prior to drilling. The nature of the basement was largely unknown except around the area of Mt Gunson (MG).

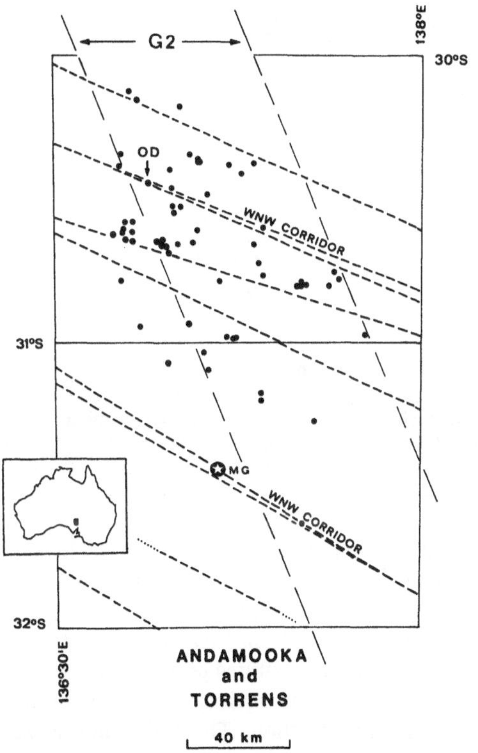

Figure 7. Similar to Figure 6, but nine years later, showing the positions (black dots) of sixty one drill holes that had penetrated to basement by 1983. They include Olympic Dam discovery hole, "OD".

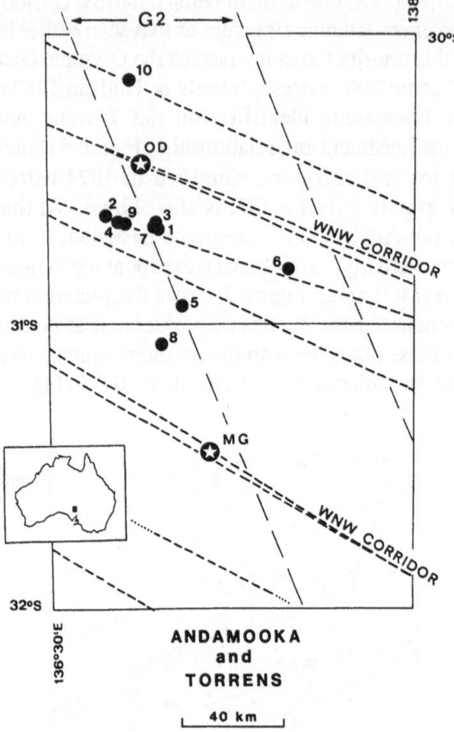

Figure 8. Similar to Figure 7, but showing only the drill holes (black dots) that intersected significant copper mineralisation (as at 1983). They include Olympic Dam ("OD") and another ten of the remaining 60 that were drilled. The association of ore with WNW lineaments is obvious.

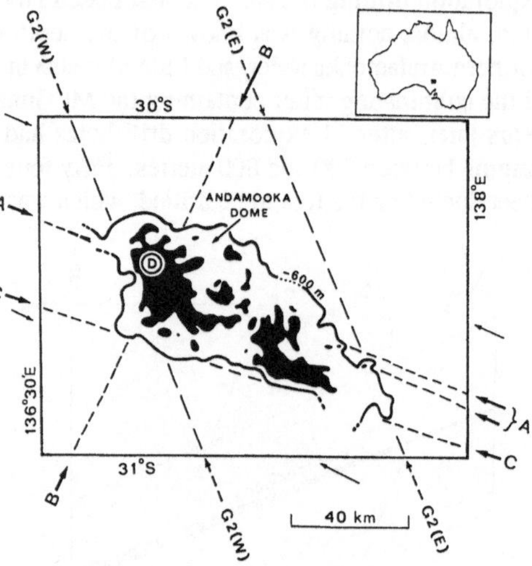

Figure 9. Structure of the Andamooka Dome within the lineament framework of Figures 6, 7 and 8, with Olympic Dam at "D" on the WNW corridor "A-A". The structural control of the dome by lineaments is clearly evident.

By 1983, Olympic Dam was fully established as a major Cu-Au-U ore deposit. Apart from Olympic Dam, the mineralised intersections in the remaining 60 drill holes (in Figure 7) were all analysed for economic significance according to their assay ratings in m.% copper. The ten best of these, plus Olympic Dam (OD), are shown (plotted as dots) in Figure 8 in relation to the transverse WNW lineaments identified prior to drilling. The association of mineralisation with the transverse WNW lineaments is obvious, and provides an illuminating substantiation of the basis for the tectonic target selection. The drill results also confirmed the lineament influence in shaping the Andamooka dome, a broad buried basement high, framed by the edges of the NNW G2 corridor and by the transverse WNW lineaments already discussed. Figure 9 shows the contour structure of the upper part of the dome derived from a contemporary magnetic interpretation by GEOTERREX. The outline of the dome is indicated by the basement structure contour at 600 m below surface, as marked. The interior ridge area, coloured in black wash, is bounded by the structure contour at 400 m below surface, whilst Olympic Dam (D) is at 300 m below surface at a peak on the ridge where it is intersected by the WNW corridor "A-A" (cf. Fig. 8). The NNE lineament "B-B" was also a diagnostic trend governing the selection of the Olympic Dam tectonic target.

The gross structure of the Andamooka Dome suggests an interference fold engineered by the combined tectonic contributions of the NNW and WNW lineament systems exemplified in Figure 3 by the two gravity lineament systems, G2 and G9C. The familial connection between the WNW photo/magnetic corridor through Olympic Dam (Figs. 8 & 9) and its great parent system, the continental G9C trend, becomes clearer in Figure 10, where the Andamooka 1/250,000 sheet is shown as a small rectangle set to scale against the gravity backdrop. The thin dashed line through "OD" (Olympic Dam) is the plot of the corresponding WNW corridor in Figure 8, and it can be seen in Figure 10 as a confluent representative of the G9C system.

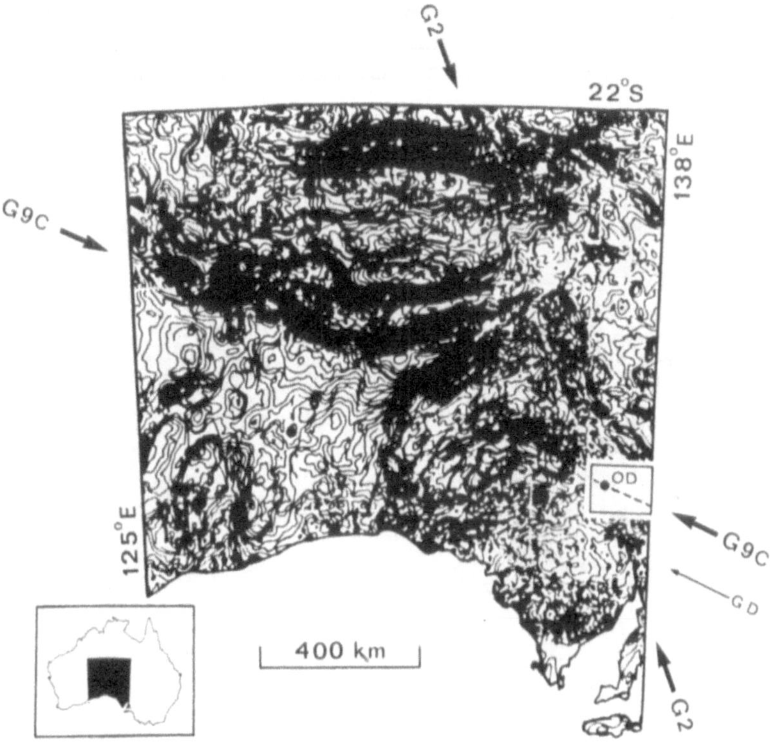

Figure 10. Gravity map (cf. Fig. 3) with inset of Andamooka sheet showing the Olympic Dam deposit at "OD", and the associated WNW photo/magnetic lineament (dashed line) as part of the continental G9C lineament system.

South Australian Ring Patterns

Crustal ring patterns of differing sizes can be seen in geological, gravimetric and aeromagnetic maps of various parts of the Australian continent. In central South Australia, the Woomera Ring is evident in the geological ingredient outline map shown in Figure 11. The ring is centred at

Woomera, at the tip of the arrow "WR", and its radius is about 160 km, which, for the convenience of the viewer, is the length of the accompanying scale bar. It will be seen that the ring has a distinctly polygonal outline, perhaps even pentagonal or hexagonal, depending on how one defines the edge of its northeasterly quadrant. The NNW corridor G2 is also visible in the geological outline data of Figure 11, and can be seen to cross through the Woomera Ring a little east of centre.

The significance of the Woomera Ring is unexplained, but it has some characteristics which are noteworthy. The presence of the ring can be detected not only in the geological ingredient data of Figure 11, but also in aeromagnetic, gravimetric, topographic and drainage data, as well as in satellite imagery. The ring, as a whole, rather neatly encompasses the Stuart Shelf in a way that is reminiscent of the twice-larger geological ring pattern in the U.S.A., which encompasses the Colorado Plateau. The fact that each of these rings contains its country's best uranium deposit is a feature to be noted. A suggested origin for the ring is by sub-crustal pressure, resulting in crustal upwelling, which would produce a circular outline at the crustal surface, subject to later erosional sculpture.

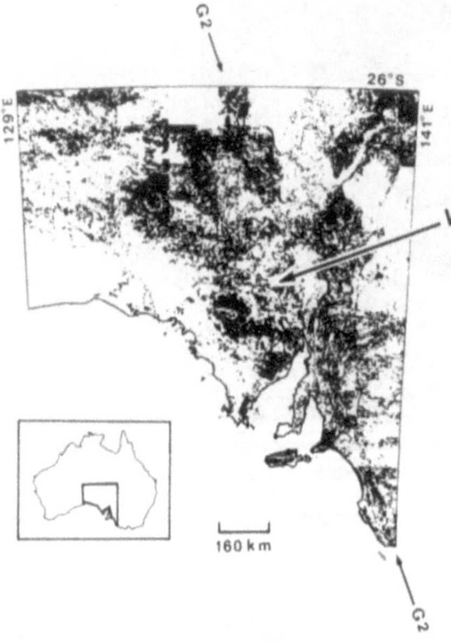

Figure 11. Geological ingredient outline map of South Australia, revealing the polygonal Woomera Ring (WR). Its radius is 160 km, and it is transected by the NNW lineament G2 seen in previous figures.

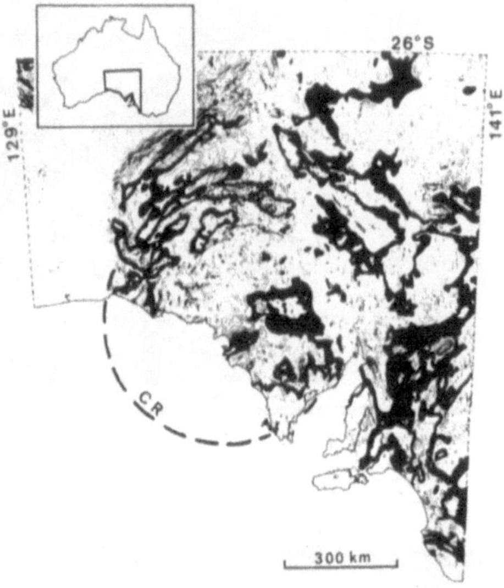

Figure 12. The Ceduna Ring (CR), with radius 300 km, visible in the Bouguer gravity map of South Australia. Black areas represent a gravity contour slice bounded by (0) mgl and (- 10) mgl contours (adapted from Fig. 3).

Adjacent to, and overlapping the Woomera Ring, is a much larger one, the Ceduna Ring, with a radius of some 300 km, and centred about 50 km easterly from Ceduna on the southwest coast of South Australia. It is visible in the Bouguer gravity pattern shown in Figure 12, where it is marked "CR". Its on-shore presence is corroborated by fault patterns and curved geological distribution patterns. Off-shore, its circumference can be traced in the aeromagnetic pattern. The relationship of the Woomera Ring (WR) to the Ceduna Ring (CR) is seen in Figure 13, against a background plot of the major lake systems in South Australia. These latter are shown in full black colour, and appear as a succession of arcuate forms, convex to the east, and evenly spaced along an east-northeasterly axis of alignment, as indicated in the figure. The coast-line where it straddles the alignment axis is also seen to be shaped in the same arcuate form, at a similar spacing interval. The meaning of this lake pattern has long been a subject of speculative interest, which is heightened by the observation that the Woomera Ring (WR) and Ceduna Ring (CR) are also positioned on this axis of alignment. A third small ring, marked "LAR" on the same ENE axis, is the Lake Acraman Ring (radius 50 km), encircling Lake Acraman which is visible as the small lake at the centre of the ring. Lake Acraman itself has already been described in literature as a Precambrian impact structure (Gostin et al. 1986; Williams 1986), from which wide-spread ejectamenta are found within the Woomera Ring and beyond.

Another noteworthy coincidence is shown in Figure 14 in the relation between the Ceduna Ring (CR) and the Gairdner dyke swarm (GDS). These NW-striking Proterozoic mafic dykes (Tucker et al. 1986) with slightly arcuate trends are concentrated within the northeasterly quadrant of the Ceduna Ring while their distributional boundary is closely controlled by the curved perimeter of the ring on the northeastern side in a way that suggests a structural relationship. It may be more than incidental that the dyke swarm also straddles the ENE axis of alignment shown in Figure 13, and its arcuate shape is similar to those of the major lakes which share the alignment. A similar coincidence outside the Ceduna Ring is given in Figure 15, illustrating how the arcuate pattern of earthquake epicentres for the period 1978-84 (Greenhalgh et al. 1986) shows a high degree of concentricity with the ring, apart from the deflective influence of the NNE corridor G8.

The overall alignment of rings and arcuate patterns of lakes, dykes and seismic epicentres raises an interesting question. For instance, is it possible that an impact (Lake Acraman?) directed east-northeasterly could have initiated resonant repercussions along a broad ENE corridor causing circular upwellings under Woomera and Ceduna, and other curved ancient basement structures

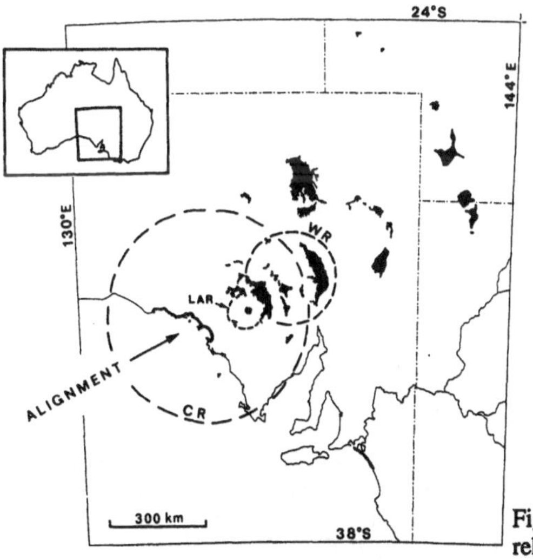

Figure 13. South Australian arcuate lake pattern in relation to the Ceduna (CR), Woomera (WR) and Lake Acraman (LAR) rings.

which have been inherited by present-day lakes? Whatever the explanation, this remarkable pattern calls for a serious inquiry which could have a tangible effect on regional tectonic interpretations.

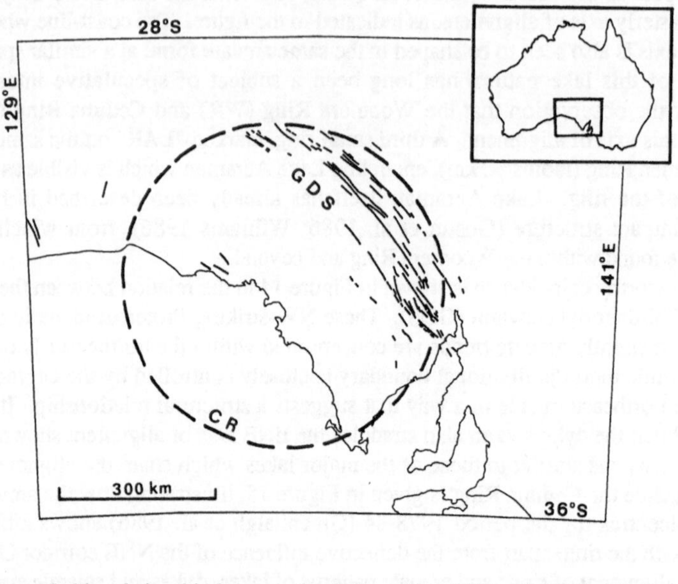

Figure 14. The remarkable relationship between the Ceduna Ring (CR) and the distribution of the mafic Gairdner Dyke Swarm (GDS).

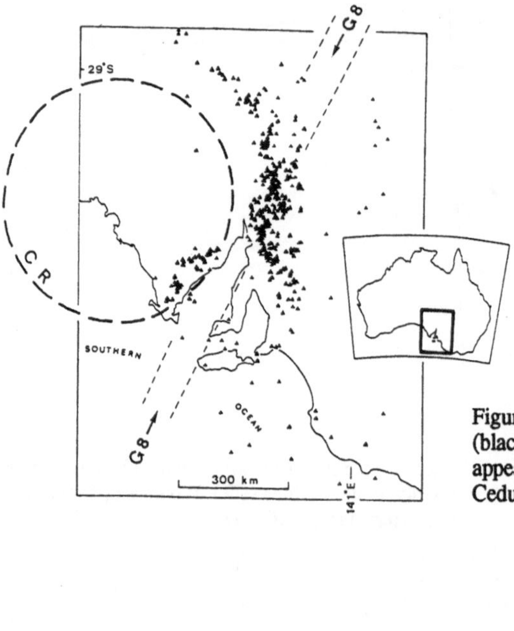

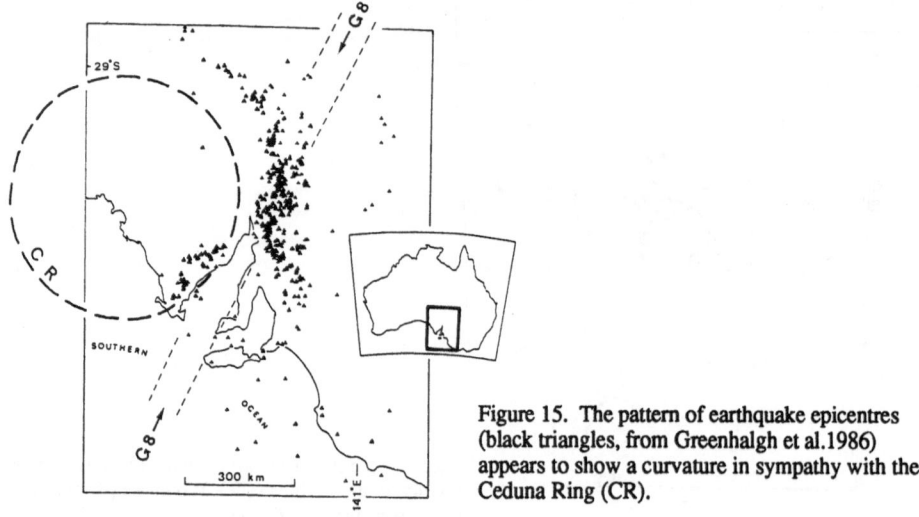

Figure 15. The pattern of earthquake epicentres (black triangles, from Greenhalgh et al.1986) appears to show a curvature in sympathy with the Ceduna Ring (CR).

Teamwork in Basement-Lineament Interplay

The constant integration of data from all and different sources readily testifies to the interdependence and interplay of lineament-tectonic criteria. In places where a structural signal from gravity may grow feeble, magnetics (or ingredient geology, or topography) may step in and take up the signal to provide the missing link, and vice versa. To receive the co- ordinated testimony of all these contributors, it is necessary to make constant comparisons of data, preferably by superimposing one set on another. One useful way of doing this is by using what I loosely call "patch maps" where a "patch" of one kind of data is correctly positioned on or within the display of another as has already been done in Figures 5 and 6. Figure 16 gives an example where a "patch" of aeromagnetics of the Yilgarn Block in southwestern Australia is positioned on the geological ingredient outline map of Australia. In this instance the two NNE magnetic corridors, MZ1 and MZ2, prominently displayed in the Yilgarn Block, have their northerly extensions taken up by the geological ingredient outline data, running through to the north coast to positions corresponding to gravity lineaments G5 and G6 marked thereon (cf Fig. 1). Other parallel NNE trends are visible in Figure 16, including the R12 system along the eastern coast. A "patch-map" version of this is given in Figure 17. Here a patch plot of significant mineral occurrences (black dots) in New South Wales (M. Aubrey, pers. comm., 1980), has been correctly positioned on the geological ingredient outline map. The patch shows not only a broad NNE belt (135 km wide) of maximum mineral

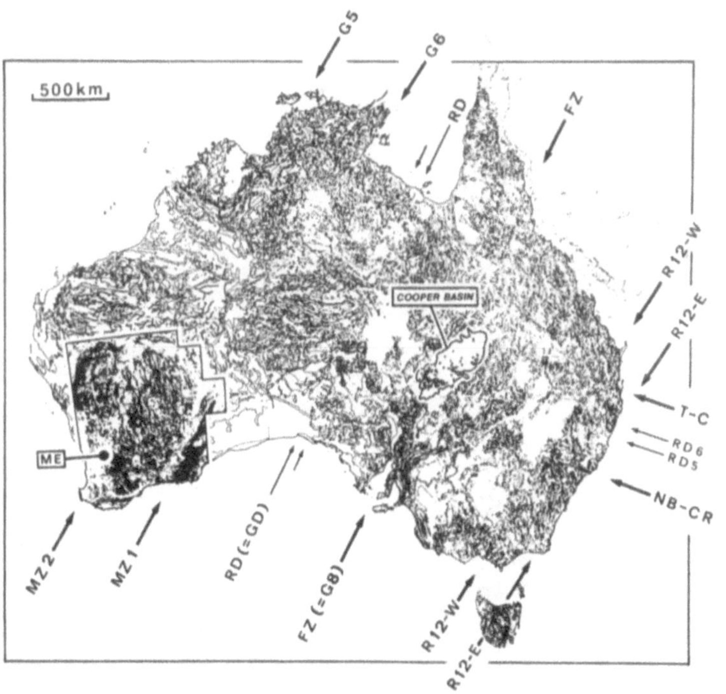

Figure 16. Lineaments visible in a geological ingredient outline map of Australia, with a"patch" of aeromagnetics superimposed on its southwest corner (after Campbell and O'Driscoll, 1989).

occurrence, reflecting the R12 lineament system, but also sharply defined internal trails belonging to the same system. These are not in competition with the obvious NNW trend of the gross pattern, nor with the obvious WNW cross-trends parallel to the Toowoomba-Charleville (T-C) lineament and its family members, nor with several meridional trails which are also detected in the gravity map in Figure 1. Rather do they exemplify the concerted interplay between different systems of basement lineaments maintaining their interactive teamwork through successive geological periods and domains.

A more diversified "patch-map" of Australia is shown in Figure 18. In this, the western and central areas of the continent, west of longitude 141°E, are represented by the gravity map of Figure 1, and the remaining eastern area by the geological ingredient map of Figure 16, together with the mineral occurrence map of Figure 17. The small geological map shown positioned in eastern South Australia (west of 141°E) has been transferred from Figure 5, and shows the paleogeologic distribution of the Precambrian Nuccaleena dolomites occupying the WNW 4A-4B belt. The black geological ingredient west of centre, and again on the west coast, shows the surface distribution of Permian sediments in the area covered by the gravity backdrop, west of 141°E.

The purpose of Figure 18 is to emphasise the prevailing nature and continuity of WNW continental lineaments when they are reported by a succession of different kinds of data. The lineaments as a whole are gently convex toward the equator and traceable right across the continent from west to east. The main body of the Permian ingredient aligned transversely in an approximate

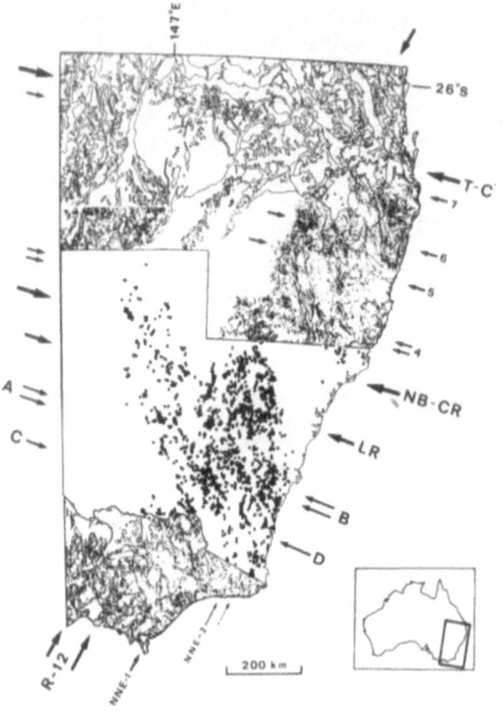

Figure 17. Southeast part of Figure 16, showing a "patch map" of mineral occurrences (black dots) registering NNE alignments parallel to lineament R12.

east-to-west direction, is certainly reflecting the WNW system of gravity lineaments in Western Australia, and the WNW Precambrian dolomite trend in South Australia, as well as similarly oriented trails in the mineral occurrence plot of southern New South Wales. In co-ordinated reporting, all of these data sources together corroborate the persistence of trends with a joint authority that none, alone, could offer.

Concentrations Within Corridors

We have seen various examples of material concentrations (e.g., ore deposits) on lineaments (cf. Figs. 1-5 and 8). These are also examples of concentrations of particular structures within corridors which are bounded by lineaments. The "patch-map" of Archaean Yilgarn Block aeromagnetics in Figure 16 is a suitable example to consider separately. For this purpose it is displayed accordingly in Figure 19.

The aeromagnetic pattern in the figure is quite intricate. The general NNW grain of the pattern is apparent, following the sinuous weave of NNW linear anomalies generally representing

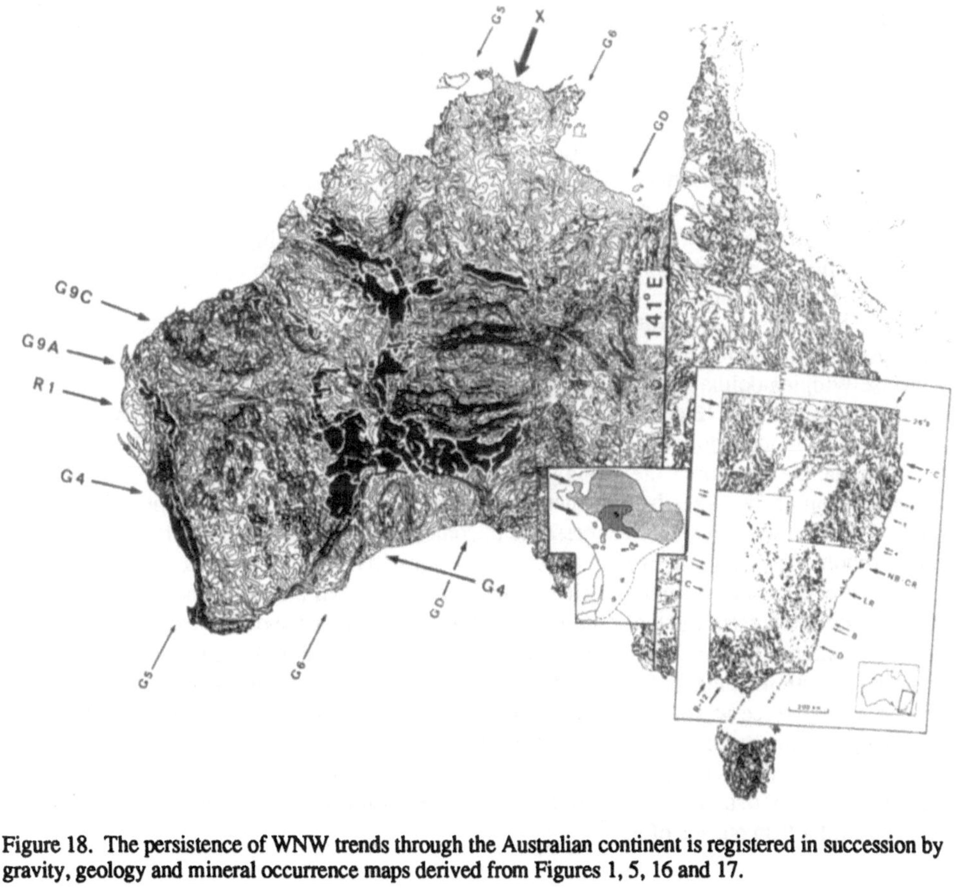

Figure 18. The persistence of WNW trends through the Australian continent is registered in succession by gravity, geology and mineral occurrence maps derived from Figures 1, 5, 16 and 17.

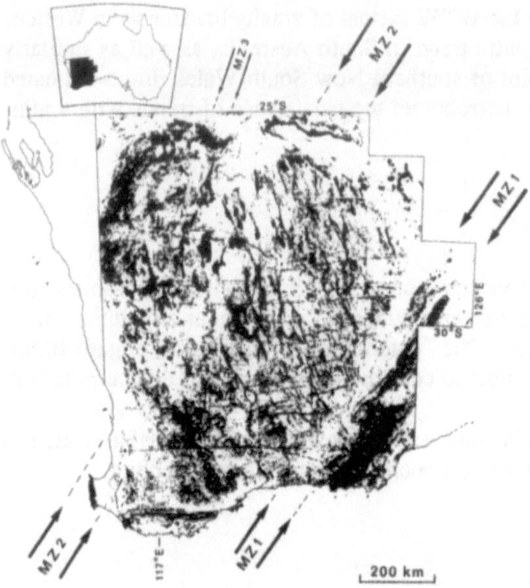

Figure 19. Aeromagnetics of Yilgarn Block in southwestern Australia, showing the great NNE magnetic corridors, MZ1 and MZ2, which control the terminal limits of the "gold line". (Adapted from Figure 16).

stratigraphy. Running transversely (E-W) across this weave are prominent members of the Proterozoic Widgiemooltha mafic dyke suite, occupying regional tensional fractures. Less visible at the scale of Figure 19, but quite evident in enlargements, is a pattern of multitudinous magnetic ring structures, ranging between 20 km and 120 km in diameter, possibly defining the margins of cylindrical plugs of diapiric granites which have risen through the main granitic body after the manner of rising salt domes.

The two NNE magnetic corridors, MZ1 and MZ2, appear to have exerted a powerful influence over these intra-block tectonic features. Figure 20 shows the most prominent members of the transverse Widgiemooltha mafic dyke suite traced off separately as a geological ingredient, and seen in relation to the NNE aeromagnetic corridors MZ1 and MZ2. It seems clear that either the block between the corridors has been moved vertically to expose a different level of the dyke pattern, or that the corridors were already present in Proterozoic times to control the introduction and distribution of the dykes by lineament-related movements. The latter interpretation seems preferred, since it would explain the presence of some non-conforming dykes running along the NNE corridors. It would also explain the transverse-tensional-dyke fractures as resulting from a right-lateral couple acting along the NNE corridor direction. In that case, the transverse-dyke fractures would represent the tensional-fracture direction along the short axis of the appropriate bulk-deformation ellipse, perpendicular to the long axis which would be extended NNW in the prevailing direction of magnetic anomalies. In addition, right- lateral movement along NNE corridors would agree with the movement observed on many NNE shear zones at local scales. Hence the NNE lineaments, MZ1 and MZ2 may be considered as structural domain boundaries against which the dykes either terminate or diminish, in a process which may be called cross-fracture segmentation, creating a pattern in which cross fractures are more intensively developed within a corridor than outside of it.

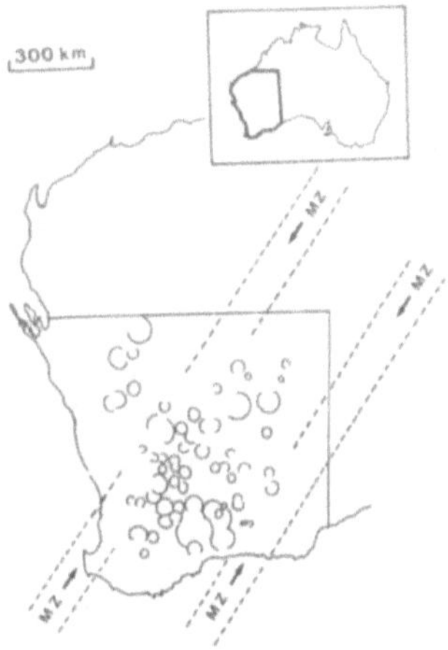

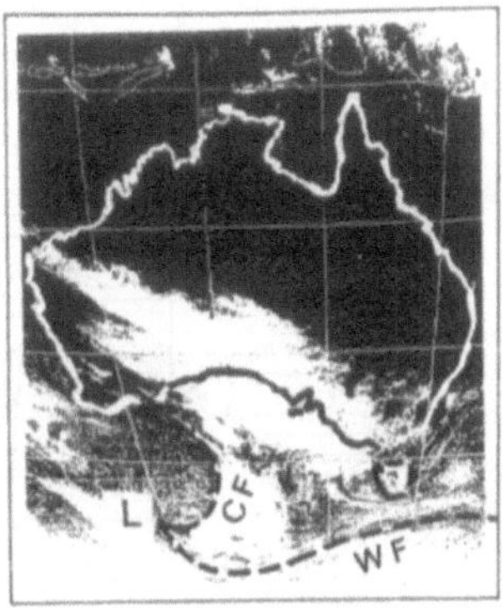

Figure 20. The distribution of the transverse Widgiemooltha mafic dykes in the Yilgarn Block (from Fig. 19) illustrates the influence of the NNE magnetic MZ corridors.

Figure 21. Ring structures detected in the Yilgarn aeromagnetic pattern (Fig. 19) show a distributional response to the NNE magnetic MZ corridors.

Figure 22. The cold front (CF) - warm front (WF) system in a satellite weather map of Australia.

Lithospheric Front Tectonics - An Atmospheric Analogy

Since the lithosphere and the atmosphere share the same system of terrestrial rotational dynamics, some comparisons may be drawn between their respective structural patterns. A geologist's cross section through a lithospheric subduction zone bears a remarkable resemblance to the cross section a meteorologist draws through an atmospheric cold front (O'Driscoll 1980). In plan, the typical atmospheric trough of the southern hemisphere, with its familiar cold-front/warm-front "y" structure, is a familiar sight in displays of satellite weather pictures of Australia. A suitable example is shown in Figure 22, where the cold front "CF", and the warm front "WF", converge southerly into the low-pressure centre, "L", at the foot of the "y", holding the cloud-filled sector between them. Despite the changing directions of cyclonic winds, these fronts, particularly the aggressive cold fronts, maintain their form and identity, in length and direction, with incredible persistence over hundreds of kilometres as they come in from the Indian Ocean and move easterly across the continent to the Pacific area.

The cloud-filled sector between the frontal arms of the atmospheric "y" has a counterpart in similar lithospheric troughs, of which the Adelaide Geosyncline provides an example, as in Figure 23(a). Here the "cold front" Torrens Hinge Zone, and the "warm front" Fleurieu-Nackara Arc (Rutland et al. 1981) form the fork of the "y" and converge southwesterly to its foot south of Spencer Gulf. The upwelling body of the geosyncline is held in the warm sector between them. In this context the Torrens Hinge Zone, which behaves like a subduction zone, is the lithospheric analogue of an atmospheric cold front. The diapiric intrusions rising through the core of the

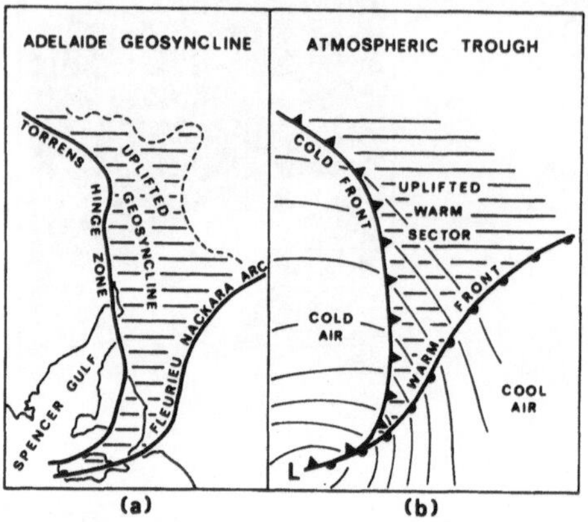

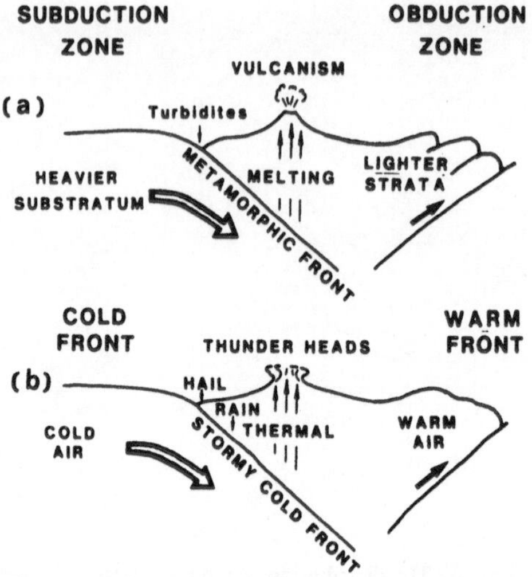

Figure 23. A comparison of the "y" structure of the Adelaide Geosyncline with that of a typical cold front - warm front weather trough in the southern hemisphere.

Figure 24. Comparison of the cross section through a lithospheric subduction/obduction zone, with that of an atmospheric cold front - warm front system.

Geosyncline (Preiss 1987) are analogous to the thermal upwellings hoist by a cold front in an atmospheric system.

For comparison, Figure 23(b) gives a plan view of a typical atmospheric trough, similar in form to Figure 22, with its two fronts converging south-westerly into the cyclonic low-pressure centre "L". Figure 24 shows the corresponding cross sections through (a) a lithospheric trough, and (b) an atmospheric trough, illustrating the correspondence of similar features in each.

It may be noted the wedge-shaped Yilgarn Block of the Western Australian Archaean shield has a similar "y" form to the Adelaide Geosyncline, but at a much broader scale. This is illustrated in Figure 25 where the north-trending Darling Fault (DF) and the northeasterly Bremer-Fraser Fault (BFF) form the arms of the "y", and converge southwesterly to the Naturaliste Plateau (NP). The complex network of linear features within the trough of the "y" are myriad magnetic anomalies attributed to basic dykes recognised in airborne magnetic data for onshore Australia plotted by computer (Tucker et al. 1986). The intensity of their development increases markedly toward the southwest corner of the wedge where the two arms converge. When translated to atmospheric equivalence, this position is one of inter-arm compression where an occluded front may develop, owing to the cold front overtaking and underthrusting the warm front, and introducing mechanical disturbances derived from each. It appears that in the case of the Yilgarn Block, the elevated shield occupying the warm sector between the fronts has been worn down to its primordial Archaean roots. Traces of the diapiric granites may still survive in the form of the magnetic rings seen earlier in Figure 21, and repeated in Figure 26 in combination with the dykes from Figure 25. Once again, the intensity of occurrence increases toward the convergency of the "fronts".

A significant difference between a lithospheric subduction zone and an atmospheric cold front as depicted in Figure 24, is that the atmospheric cold front does not require any "conveyor belt" translation to and from an atmospheric spreading ridge, as in the case of a lithospheric subduction zone. It is simply an atmospheric fault or rift which moves progressively on, nudging, underthrusting, and elevating the trough that lies ahead of it. The cold front is the more aggressive arm of the "y" structure. Its counterpart, the warm front, is a more passive feature, acting like a slip plane over which the warm infilling trough of the "y" structure is thrust up and over the flanking cooler air mass lying ahead of it.

An important point to be emphasised in the comparison between the lithospheric Adelaide Geosyncline and an atmospheric trough is that they are initially trough-like structures which then become elevated, a process readily understandable in terms of the cross sections in Figure 24. The systems of lineaments already seen to be transecting the Adelaide Geosyncline help to establish another comparison. All airmen have observed remarkably persistent atmospheric lineaments visible in local cloud patterns, but it remained for satellite imagery to provide the synoptic view which reveals that atmospheric troughs are characterised by enormous transcurrent lineaments that are functions of linear thermal thresholds operating in patterns of cloud tectonics. An inspection of Figure 22 reveals prominent WNW trends, as well as a NNW and a NNE break within the trough sector. The application of the atmospheric analogy to geological observations leads to a contingent conclusion that the earth's early molten crust, behaving like a fluid encircling the rotating globe, developed structures and discontinuities like those in today's fluid atmosphere, and these have been preserved in the crust by consolidation to form the foundations of subsequent stratigraphic edifices. A few years ago it would have seemed unrealistic to compare the flow pattern of a molten crust to that of an atmosphere, because it would have been argued that the pattern of atmospheric circulation is determined by the sun, and that the sun could not possibly be regarded as affecting a molten (fluid) crust in the same way. However, from more recent interplanetary reconnaissance by

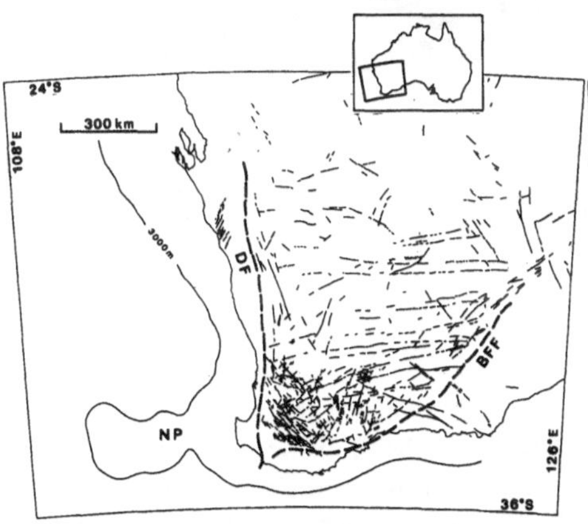

Figure 25. The Western Australian Yilgarn Block interpreted as a lithospheric cold front -warm front system. The linear network within the block represents the pattern of magnetic dykes from Tucker et al. 1986.

Figure 26. A combination of Figure 25 with the rings from Figure 21 suggest symptoms of an incipient occluded front.

Voyager 2, including observations especially of the anomalously-tilted planet Uranus, Ingersoll (1987) has reported that although the sun supplies the energy that drives the atmospheric circulation of a planet, it does not determine the circulation pattern. "Instead", writes Ingersoll, "the pattern is dominated by the effects of a planet's rotation". It is the Coriolis force from rotation, and not the sun, that causes planetary winds to blow in an east-west direction, and cloud bands to be concentric with the poles. It seems, therefore, that arising from a common cause, the pattern developed in a fluid atmosphere may realistically be compared with the pattern of a molten (fluid) crust. A final point worthy of note is that atmospheric patterns are seen to develop widely separated replications of identical forms and shapes, which, if superimposed, or "drifted" into contiguity, would fit one another not only in figure but in material contexts as well. But they were never together in the first place, and are not evidence of the drifting apart of cloud "continents" from an original super cloud mass.

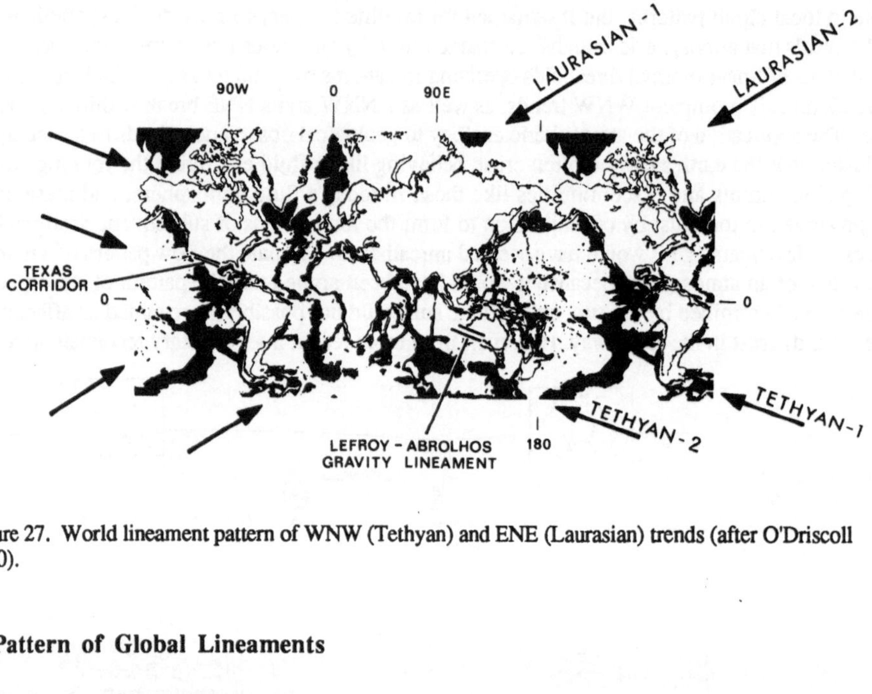

Figure 27. World lineament pattern of WNW (Tethyan) and ENE (Laurasian) trends (after O'Driscoll 1980).

A Pattern of Global Lineaments

Students of world maps drawn on a Mercator projection are familiar with the appearance of two crossing systems of global lineaments, one WNW causing sinistral dislocations, and the other ENE causing dextral dislocations. A map of continental outlines combined with a mid-level contour slice of sea-floor bathymetry, as illustrated in Figure 27, intensifies the prominence of these two world lineament systems. They are named in the figure as Tethyan (WNW) and Laurasian (ENE) systems respectively, and have been described in detail by O'Driscoll (1980,1986). Representatives of these two systems appear repeatedly in patterns of basement tectonics. Figure 27 shows two examples of the continental members of the Tethyan system, viz., the Lefroy-Abrolhos WNW gravity lineament in Western Australia, and the WNW Texas Corridor in the U.S.A. The former is the same as the gravity lineament G4, shown in Figure 1, and the parallel line to the north of it in Figure 27 is the axis of the WNW G9 system in Figure 1. In the same figure, gravity lineaments G1 and G7 correspond to the Laurasian system.

There is no problem reconciling the present positions of these global lineaments with the concept of Gondwanaland prior to its break-up. For instance, in South America, what would have been the Tethyan trend in pre-break up time is now rotated into a NNW orientation clearly visible as a continental lineament in South American geological ingredient data. The same comment applies to North America if it is considered to have drifted westerly from Euro- Africa. In both instances the pre-drift Tethyan lineament survives in a new NNW orientation, and a new post-drift Tethyan lineament has been developed in its place, as for example, the Texas Corridor, shown in Figure 27.

The fact that the global lineaments in Figure 27 are straight lines on a standard Mercator map identifies them as loxodromic curves, and therefore as spirals on the globe (O'Driscoll 1980). This is illustrated in Figure 28, in which the Tethyan lineaments, T1 and T2, from Figure 27, have been transcribed onto a globe. Intermediate parallel members of the system have been recognised in the U.S.A., and are indicated in Figure 28. The black dots on them denote major metalliferous and petroleum deposits identified by O'Driscoll (1986).

The linear trends in Figure 27 have been noted by other authors, but not plotted on projections which reveal their spiral nature. For example, Holmgren et al. (1975) depict WNW global rhegmatic shear systems which correspond in position to the Tethyan trends in Figure 27. Their map gives an equatorial view plotted on a projection with curved co-ordinates, which reduces the viewer's impression of the consistency of the shear systems. In Figure 29, I have transcribed their WNW shear systems onto a polar projection, which then reveals their real spiral nature, and their correspondence to the Tethyan spirals depicted in Figure 28.

Figure 28. Terrestrial globe with WNW Tethyan trends, T1 and T2, revealed as global spirals centred on the axis of rotation (after O'Driscoll 1982).

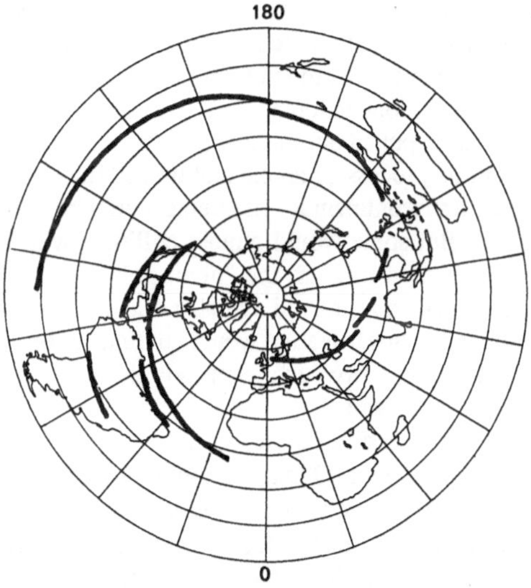

Figure 29. Rhegmatic world lineaments from Holmgren et al. (1975) transcribed to a polar projection to show their spiral nature.

The differential rotation between the equator and the poles of the sun and several planets led Gilliland (1973) to experiment with a rotating fluid-filled rigid sphere, rotating from west to east. This demonstrated that the fluid in the equatorial belt of the sphere had a higher angular velocity than in the polar regions, resulting from non-uniform east-west shearing in the fluid. The fluid also developed a flow pattern expressible in terms of opposing spiral forms trending WNW in the northern hemisphere, and ENE in the southern hemisphere. Global spirals are also seen in illustrations by Rance (1967), dealing with the theoretical brittle failure pattern of the earth's crust under the influence of both right-hand and left-hand torsion about its axis of rotation, i.e., of an oscillatory process. Between latitudes 60° north and south, Rance's theoretical pattern is dominated by two crossing shear-fracture trends, WNW and ENE, extending into both hemispheres. They are similar to the two global trends in Figure 27, especially to the Tethyan WNW trend, and can be similarly translated into global spirals centred on the poles. Rance has used the major physiographic lineaments in the Pacific Ocean to illustrate the consistent crustal expressions of his theoretical pattern.

An Expanding Earth Venture

Proponents of an expanding earth hypothesis might consider some possible consequences of global expansion if the process affected ocean floors rather than the continents. In that case, if a continent were to have moved radially outward, away from earth's centre, behaving as if it were fixed on the end of a lengthening earth radius passing through a fixed "pivot point" in the base of the continent, and if the continent has largely maintained its present size during the global expansion process, then it may be that the expanding ocean floor presently surrounding the continent has been pulled out from under it, in a manner similar to shallow subduction in reverse. It may then be that the present peri-continental sea floor patterns are registering the original sub-continental foundation patterns, exposing, as it were, the "foot prints" of the continental structures, as a carpet pulled out from under a table carries with it the indentations of the table-leg castors.

In this process of an expanding sea floor moving out from under a non-expanding continent, many transcontinental lineaments would be laterally displaced from their extensions in to the sea floor. But some would remain collinear with their original sea-floor extensions if their crustal tracks passed through the "pivot point" of the continental base, from which lateral expansion underneath had moved radially and horizontally outward. In that instance, the track of the continental linear feature would continue to lie along such a horizontal radius of lateral expansion. In this respect, two such features in the Australian continent suggest themselves. They are the belt of WNW lineament systems G4 and G9, seen in Figure 1, and the NNE lineament systems R12W and R12E, seen in Figure 16. At the present time, both these features are seen to extend on line into the sea floor (O'Driscoll 1980, Fig. 7; 1986, Fig. 4) Their intersection beneath the continent would correspond to a continental "pivot point" in southeastern Australia in the vicinity of the Australian Alps, that is, the point at the end of the radius from the earth's centre on which Australia has been fixed as it has moved centrifugally outward along it.

In making equations between sea-floor patterns and continental patterns under these postulated circumstances, critical factors include the assumed amount of global expansion, the position of the "pivot point", and the assumption of a non-rotating continent. Different amounts of expansion have been proposed by different authors. The trail example for the Australian continent shown in Figure 30 assumes an expansion in which the earth's radius has increased by one half. As discussed above, the "pivot point" is taken to be near the southeast corner of the continent, in the vicinity of the Australian alps.

For Figure 30, the expansion process has been reversed by shrinking the sea floor to its original dimensions. Thus, the plotted contour lines shown within the continental boundary are the present circum-Australian sea-floor contours (adapted from Campbell & O'Driscoll 1989) which have been slid back to their "shrunken" position beneath the continent where they would have been before expansion began.

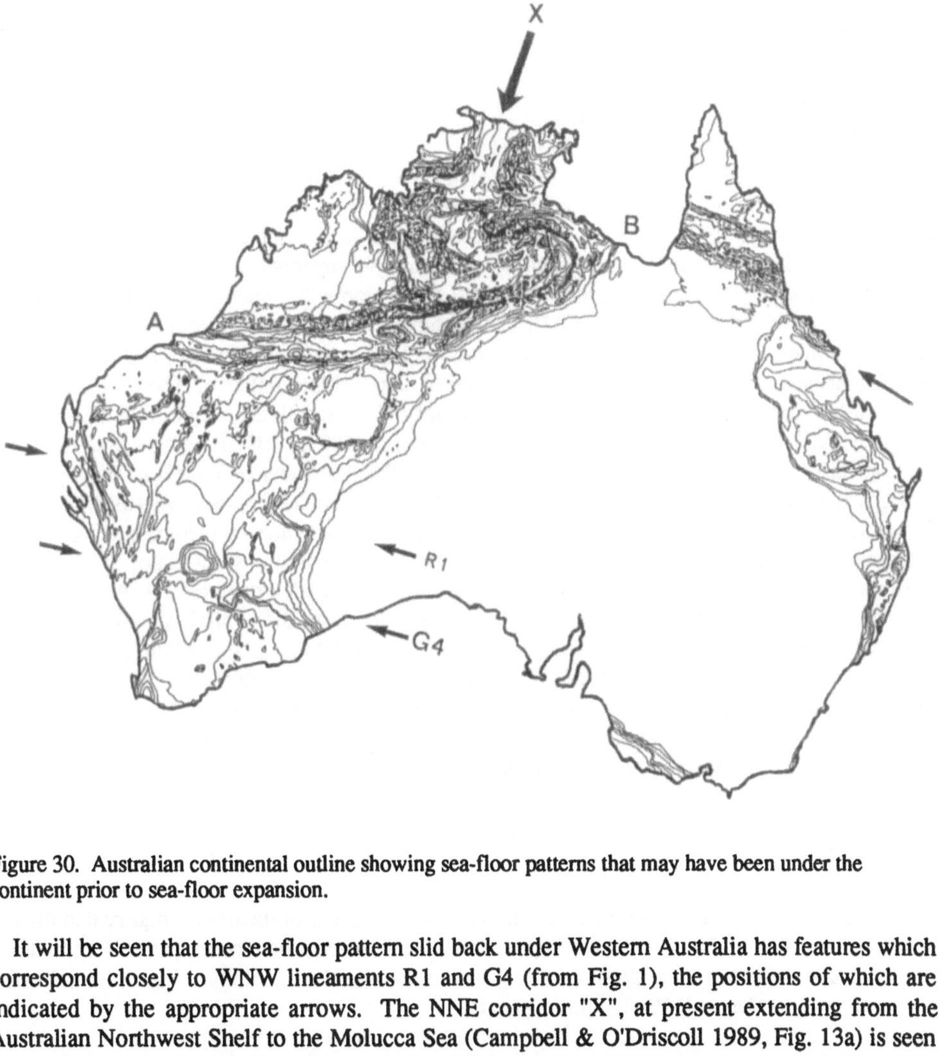

Figure 30. Australian continental outline showing sea-floor patterns that may have been under the continent prior to sea-floor expansion.

It will be seen that the sea-floor pattern slid back under Western Australia has features which correspond closely to WNW lineaments R1 and G4 (from Fig. 1), the positions of which are indicated by the appropriate arrows. The NNE corridor "X", at present extending from the Australian Northwest Shelf to the Molucca Sea (Campbell & O'Driscoll 1989, Fig. 13a) is seen returned to a pre-extension position corresponding to a major NNE lineament recorded by E.S. Hills in the Northern Territory of Australia (O'Driscoll 1989, Fig. 5). The broad arc "A-B" corresponds to a similar structural pattern visible in contour-slice treatment of the Australian Bouguer gravity pattern (not illustrated). In the northeastern quadrant of Figure 30, the WNW trends presently in the Coral Sea area have been slid back to their shrunken position under northeastern Queensland where they would have been before expansion. Here they show a strong correspondence to the present WNW systems of lineaments crossing through northern Queensland (O'Driscoll 1982).

Conclusions drawn from this treatment are very sensitive to changes in initial assumptions, and there are interesting variations which give remarkable geometric correlations. The discussion is not intended to be for or against an expanding earth, but simply an experiment with the thinking that can proceed from accepting the hypothesis.

Crustal Patterns and Core Patterns

New developments in studies of the shape of the earth's solid core by seismic (*catscan*) tomography may have interesting repercussions in concepts of basement tectonics. From results of these studies Horgan (1987) has reported that the core surface is not a smooth ellipsoid, but is marked by swells and depressions. From Horgan's map, it appears that when the positions of these "lumps and dents" are projected to the surface as a map with rectangular coordinates, they reveal a criss-cross pattern of WNW and ENE alignments. When his map is compared with Figure 27, it is seen that the core tomography trends conform closely to the directions of the Tethyan and Laurasian lineaments, i.e., they represent spirals traced on the ellipsoidal core. The bearings of the Tethyan and Laurasian trends in Figure 27 are 290° and 057° respectively. This slight asymmetry, clearly visible in the figure, has been explained (O'Driscoll 1980, p. 408) as being due to the left-lateral shear on the Tethyan system exceeding the right-lateral shear on the Laurasian system, causing a slight counter-clockwise transposition of the latter. The repetition of this asymmetry in the pattern of lumps and dents on the core raises the question whether the pattern may be one of accretionary growth, in which individual accretions are fitted together in spiral trains like the buds on a pine cone, i.e., according to the exponential laws of growth expressed by the Fibonacci sequence. In that case the frequency of major Tethyan and Laurasian spirals may be in a ratio expressed by numbers in the Fibonacci sequence 2, 3, 5, 8, 13, ..., the Tethyan spirals being fewer than the Laurasian spirals. For instance, it appears that a number of global features might fit into such a spiral network where the Tethyan/Laurasian frequency is 8/13. Whatever the correlation, it appears that trends on the core are repeated by parallel trends on the earth's surface — a conclusion that must have an important bearing on concepts of basement tectonics.

Conclusion

As a keynote speaker I have sounded out, in minor keys, a number of themes of inquiry that invite pursuit, but which are predicated by assumptions that are not everywhere compatible — a situation that is not uncommon among alternative theories in the world of geology. As I have said on another occasion (O'Driscoll 1986), crustal lineaments

> "are pre-eminently loci of anomalous geological occurrences and especially of changes in rock distribution, composition, metamorphism and alteration, and especially of en echelon repetitions of such effects, including ore deposits. Because they appear to have diverse

multi-metallic associations, depending on the type of petrologic environment found along them, and to be equally related to deposits of various ages, they may be identified as ancient fundamental features persisting through geologic time, and episodically reactivated by oscillatory movements.

Continental lineaments, some remarkably long and persistent, appear to be systematic, crossing continental margins and plate boundaries, and to be consistent with major trends visible in global patterns (O'Driscoll 1980, 1982). Many features of the observed lineament-ore relation are not easily reconciled with what is permitted by some current theories. Nevertheless, the relation is verifiable and observable in a great many diverse and widely separated ore occurrences and environments, (and) any theory that cannot accommodate such lineaments must itself be seriously re- examined"

There are many hidden unknowns which are not necessarily handed to us on a plate, and the elusive trails in the basement labyrinth remain as much a challenge as ever to our tectonic interpretive ingenuity.

Acknowledgements

My sincere thanks are due to Dr M.J. Rickard and the 9IBT committee for inviting me to contribute a keynote address, and for sponsoring my attendance at the Conference; and to Dr R Woodall for permission to reproduce data on the lineament-tectonics and morphology of the Andamooka Dome.

References

Bureau of Mineral Resources 1971. Australia Bouguer Anomalies (scale 1:2,534400), preliminary edition. Department of National Development, Canberra.

Bureau of Mineral Resources 1975. Australia Bouguer Anomalies (scale 1:5 000 000), preliminary edition. Department of National Development, Canberra.

Campbell, I.B., & O'Driscoll, E.S.T. 1989. Lineament-hydrocarbon associations in the Cooper and Eromanga Basins. In, O'Neil, B.J. (ed.) The Cooper and Eromanga Basins, Australia. Proceedings Petroleum Exploration Society Australia, Society Petroleum Engineers & Australasian Society Exploration Geophysicists, Adelaide., p. 295-313.

Gilliland, W.N. 1973. Zonal rotation and global tectonics. American Association Petroleum Geologist Bulletin **57**, 210-214.

Gostin, V.A., Haines, P.W., Jenkins, R.J.F., Compston, W., & Williams, I.S. 1986. Impact ejecta horizon within late Precambrian shales, Adelaide Geosyncline, South Australia. Science **233**, 198-200.

Greenhalgh, S.A., Singh, R., & Parham, R.T. 1986. Earthquakes in South Australia. Transactions of Royal Society of South Australia **110**,145-154.

Hills, E.S. 1956a. A contribution to the morphotectonics of Australia. Journal of the Geological Society of Australia **3**, 1-15.

Hills, E.S. 1956b. The tectonic style of Australia. In, Geotektonische Symposium zu Ehren von Hans Stille. Deutsche Geologische Gesellschaft, Stuttgart p. 336-346.

Hills, E.S. 1961. Morphotectonics and geomorphological sciences with special reference to Australia. Quarterly Journal Geological Society London **465**, 77-89.

Holmgren, D.A., Moody, J.D., & Emmerich, H.H. 1975. The structural settings for giant oil and gas fields. Proceedings 9th World Petroleum Congress **2**, 45-54. Applied Science Publishers, London.

Horgan, J. 1987. Core questions. Scientific American, **256** (2), p 43-44.

Ingersoll, A.P. 1987. Uranus. Scientific American **256**, 30-37.

O'Driscoll, E.S.T. 1980. The double helix in global tectonics. Tectonophysics **63**, 397-417.

O'Driscoll, E.S.T. 1981. Structural corridors in Landsat lineament interpretation. Mineralium Deposita **16**, 85-101.

O'Driscoll, E.S.T. 1982. Patterns of discovery — the challenge for innovative thinking. 1981 PESA Australian Distinguished Lecture. Petroleum Exploration Society Australia Journal **1**, 1-31.

O'Driscoll, E.S.T. 1983a. Deep tectonic foundations of the Eromanga Basin. The Australian Petroleum Exploration Association Journal, **23**, 5-17.

O'Driscoll, E.S.T. 1983b. Broken Hill at the cross roads. In, Broken Hill Conference 1983. Conference Series No. 12, The Australasian Institute of Mining and Metallurgy, Melbourne, p. 29-47.

O'Driscoll, E.S.T. 1985. The application of lineament tectonics in the discovery of the Olympic Dam Cu-Au-U deposit at Roxby Downs, South Australia. Global Tectonics and Metallogeny **3**, 43-57.

O'Driscoll, E.S.T. 1986. Observations of the lineament-ore relation. Philosophical Transactions of the Royal Society of London **A317**, 195-218.

O'Driscoll, E.S.T. 1989. Edwin Hills and the lineament-ore relationship. In, Le Maitre, R.W. (ed.) Pathways in Geology — Essays in honour of Edwin Sherbon Hills, Blackwell Scientific Publications, Melbourne, p. 247-267.

O'Driscoll, E.S.T. 1990. Lineament tectonics of Australian ore deposits. In, Hughes F.E. (ed.) Geology of the Mineral Deposits of Australia and Papua New Guinea. The Australasian Institute of Mining and Metallurgy: Melbourne, p. 33-41.

Preiss, W.V. 1987. The Adelaide Geosyncline — Late Proterozoic stratigraphy, sedimentation, palaeontology and tectonics. Bulletin of the Geological Survey of South Australia **53**.

Rance, H. 1967. Major lineaments and tortional deformation of the earth. Journal Geophysical Research **72**, 2213-2217.

Roberts, D.E., & Hudson, G.R.T. 1983. The Olympic Dam copper-uranium-gold deposit, Roxby Downs, South Australia. Economic Geology **78**, 799-822.

Rutland, R.W.R., Parker, A.J., Pitt, G.M., Preiss, W.V., & Murrell, B. 1981. In, Hunter, D.R. (ed) Precambrian of the Southern Hemisphere **3**, Elsevier, p. 309-360.

Tucker, D.H., Boyd, D.M., & Anfiloff, V. 1986. Magnetic dykes of Australia. Preliminary Edition Scale 1:5,000,000. Bureau of Mineral Resources, Canberra.

Williams, G.E. 1986. The Acraman impact structure: source of ejecta in late Precambrian shales, South Australia. Science **233**, 200-203.

Woodall, R. 1984. Success in mineral exploration: confidence in science and ore-deposit models. Geoscience Canada **11**, 127-133.

Woodall, R. 1985. Limited vision: a personal experience of mining geology and scientific mineral exploration. Australian Journal Earth Science **32**, 231-237.

Woodall, R. 1990. Gold in Australia. In, Hughes, F.E. (ed.) Geology of the Mineral Deposits of Australia and Papua New Guinea The Australasian Institute of Mining and Metallurgy: Melbourne, p. 45-67.

PART II

OTHER REGIONS

ANALYTICAL STUDY OF GEOLOGIC STRUCTURE IN THE COVER SEDIMENTS BY VIRTUAL BASEMENT DISPLACEMENT METHOD

K. KODAMA
Geological Survey of Japan
3-3-1, Higashi Tsukuba
305, Japan

ABSTRACT. Geologic structures of the cover sediments in the Niigata oil and gas basin, central Japan, are analyzed using the numerical simulation system (VBD method). It has been proved that these structures are formed through the accumulating process of the incremental deformation due to basement tectonics.

Introduction

Some geologic structures or sedimentary facies are analysed as the geologic records of basement tectonics. For example, the movement or tectonic process of the basement fault is discovered by the analysis of the abrupt change in lithology or thickness variation of the cover sediments (Beck 1987; Todd et al. 1989). Analysis of growth faults or growth folds also provides information about tectonics of the crustal movement (Vejaek & Andersen 1987; Badgley et al. 1989; Kodama et al. 1990). In this way, some structures within the sedimentary basin, except those such as gravity slides, gravity compaction or diapirs, are considered to be formed as a direct reflection of movemens in the basement. Therefore it is important to analyze quantitatively the interaction of the movement and tectonic process in the basement with the mechanism of growth of geological structures in the cover sediments.

The deformation of cover sediments in response to the movement of the basement has been analyzed for idealized simple geological structures (Sanford 1959; Kodama et al. 1976; Koopman 1987; Withjack et al. 1990), but actual structures of the basement which have been deformed through long geologic histories have not been discussed.

In this paper I introduce a new method of structural analysis, named 'the Virtual Basement Displacement (VBD) method' (Kodama et al. 1985), and present an example of analysis in the Niigata oil and gas basin, central Japan. This is followed by a discussion on how simple incremental deformation of the basement forms complicated structures in the cover sediments.

Outline of the virtual displacement method

Several oil and gas deposits have been found in Miocene volcanic-rock reservoirs at depths of 4,000-6,000m in the Nagaoka area, in the southern part of the Niigata oil and gas basin (Figs. 1 & 2). The rocks are mainly rhyolite and basalt of the Miocene Green Tuff Formation (Komatsu et al. 1984, Figs.2 & 5).

M. J. Rickard et al. (eds.), Basement Tectonics 9, 151–160.

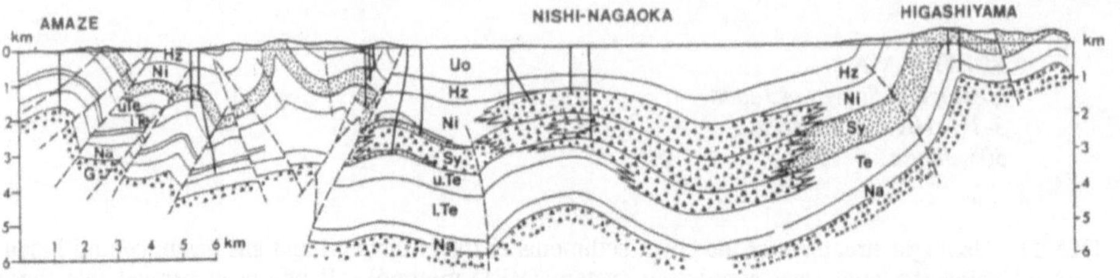

Figure 1. Geological cross section of the Nagaoka area on the line shown on Fig. 2. Uo:Uonuma Formation, Hz:Haizume Formation, Ni:Nishiyama Formation, Sy: Shiiya Formation, Te: Teradomari Formation, Na: Nanatani Formation, G: Green Tuff (after Aiba 1982)

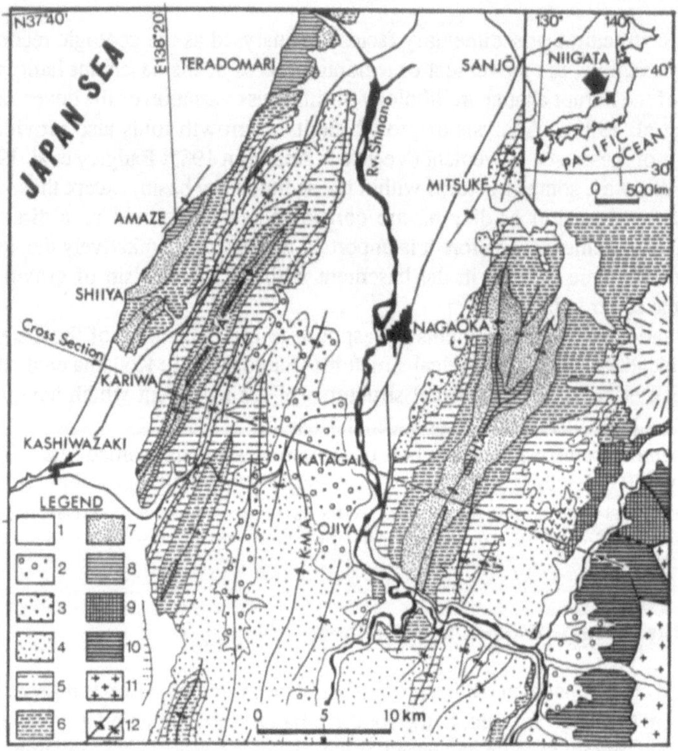

Figure 2. Geological map of the Nagaoka area. 1:Alluvial deposits, 2: Terrace deposits, 3: Yashiroda Formation, 4: Uonuma Formation, 5: Haizume Formation, 6: Nishiyama Formation, 7: Shiiya Formation, 8: Teradomari Formation, 9: Nanatani (Tsugawa) Formation, 10: Paleozoic, 11: Granitic rocks (Mesozoic), 12: Anticline and syncline, H.A. Higashiyama Anticline. (after Niigata Pref. 1989)

It has been partly proved that there are large-scale faults and complex horsts and grabens in the formations below 4,000m. Such structures in the deeper parts are very different from the shallower parts with gently folded structures. This is called the "duality of the oil and gas bearing structures" (Aiba 1982, Fig. 1). Therefore the conventional method of extrapolating the shallow geologic structures is not applicable to the exploration of deep-seated reservoirs, because it is well known that with such a method, anticlines and synclines overlap and deformation becomes flatter downwards. The method should be limited in use for the structures which were formed simultaneously with those of shallow parts.

The VBD method was developed to resolve these fundamental problems. This is one of the deterministic modeling methods to simulate deformation and fracturing of the strata in a sedimentary basin. Rock mechanics and deformation theory are the governing equations, and the Finite Element Method (FEM) is applied.

The VBD method is based on the following principal assumption — the geologic structures of the cover sediments in the basin are formed as a result of the deformation of the basement. Actually we have limited knowledge of the deformation of the basement and sometimes do not even know the depth or shape of the basement. In contrast, we have much information concerning the geological structures of the shallow parts which were formed as the result of the basement deformation. We introduce the concept of the inversion problem and devised a method for inferring hypothetical deformation of the basement (Fig. 3).

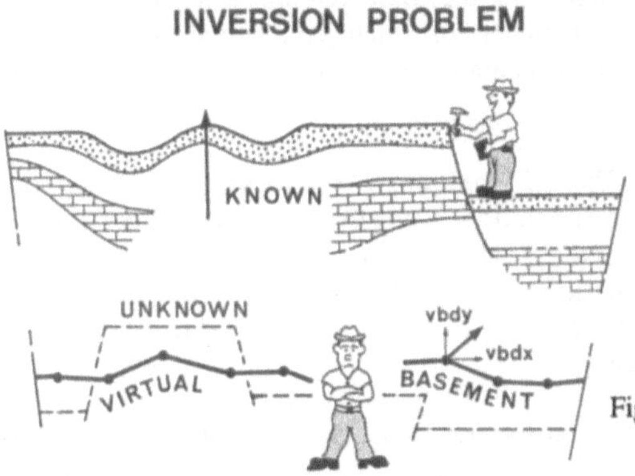

Figure 3. The concept of the VBD method.

The analysis using the VBD method is carried out as follows:-

Definition of the Tectonic System: In the case of a two dimensional (plane strain) problem, x, y and z coordinate axes are defined horizontal, vertical and perpendicular to the cross section respectively. The virtual basement is assumed at an appropriate depth below the geological unit to be discussed. The surface of the virtual basement does not correspond to the surface of the real basement. This surface always defines the lower deformation boundary of the model.

Definition of Tectonic Stage: The geological processes of the area are divided into several tectonic stages usually corresponding to the sedimentary stages of the formations or members. The boundary of each unit should be a chronological surface.

Estimation of Surface Deformation as a Control Datum: Thickness variation of each formation unit is measured on the present geological cross section. The present thickness should be decompacted into original thickness (tho), if possible. Eroded thickness at an unconformity is estimated. Paleodepth (dp) is estimated from paleo-environmental data and applied to the depth of the surface of each sedimentary layer.

The deformation of the surface of the basin is defined as the control datum (cntl) of deformation for each tectonic stage. This is calculated by the following equation — cntl = tho + dp - dp' — where (tho) is the original thickness, (dp) and (dp') are the paleodepth of the surface of sediments for a tectonic stage and the previous tectonic stage (Fig. 4).

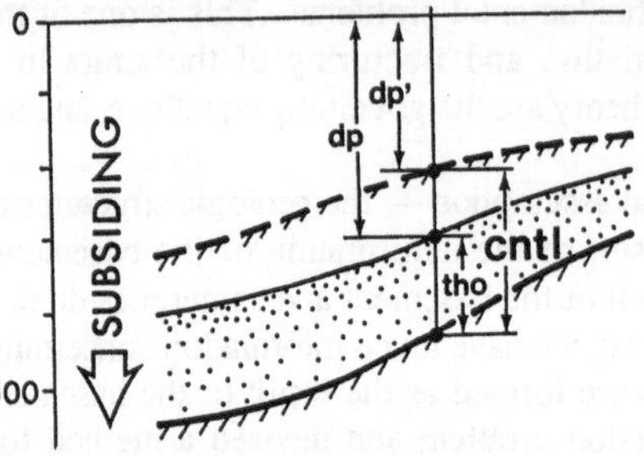

Figure 4. Definition of the deformation control datum (cntl).

Definition of Finite Element Mesh: Before simulating the first tectonic stage, finite-element meshes are defined in the lowest layer, the lower boundary of which is the virtual basement. The new finite-element meshes of the younger layers, which are stacked on the deformed older layer stage by stage, are defined automatically.

Input of Lithology and Physical Properties of Rock Deformation: The rocks behave as elastic-plastic material for the discontinuous faulting or fracturing. The effects of viscous flow or strain rate are not included here. The characteristics of the elastic-plastic rock materials are described by Young's modulus and Poisson's ratio for the elastic part, and yield strength and two parameters of strain hardening for plastic deformation (Yamada 1972). The values for these parameters are taken from triaxial deformation tests of actual rock samples.

Input of VBD Vector Components: In the case of the two dimensional problem, vertical and horizontal components of virtual displacement (vbdx,vbdy) are applied to every node along the virtual basement (Fig. 3). The value is first estimated and then adjusted during the correction process.

Execution of FEM.: As the deformation during one tectonic stage maybe very large, the calculation of F.E.M. is carried out step by step until the total deformation reaches a control datum. Only one or two elements are yielded at each step and their physical parameters are converted from the elastic state to plastic ones (Fig. 5).

The condition of yielding is evaluated by the von Mises criterion expressed by

$$eqs=[0.5\{(sx-sy)^2+(sy-sz)^2+(sz-sx)^2+6(txy^2+tyz^2+tzx^2)\}]^{0.5}$$

where sx, sy, sz, txy, tyz, tzx are components of a stress tensor on a point in a finite element. Eqs is called the equivalent stress, which corresponds to yield strength under the uniaxial test (Yamada 1972). After the finite- element mesh was deformed, the stiffness matrix is recalculated for the

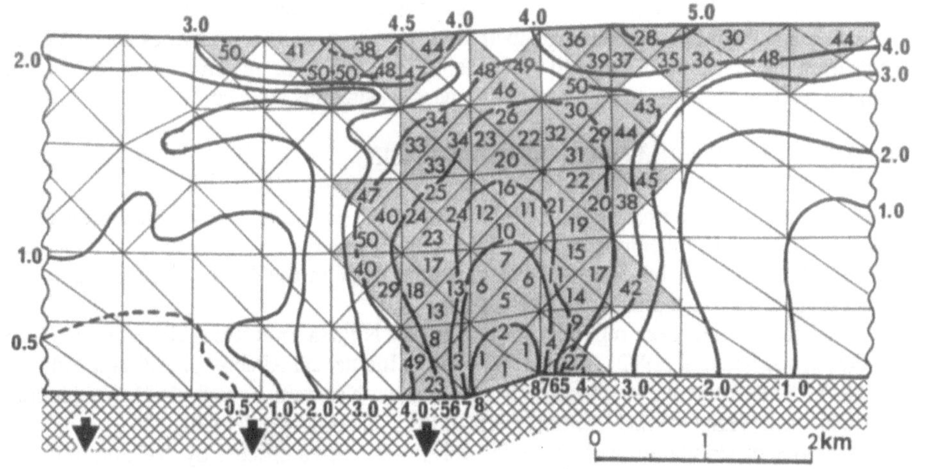

Figure 5. An example of propagation of the yielding during a tectonic stage. Shaded elements have been yielded in order of the step number drawn in the element. The contours show magnitude of eqs (unit: 10^2kg/cm^2). Note the concentration of eqs near the boundary of the elastic and plastic areas (Kodama 1976).

new geometry and new physical parameters at each step. Propagation of the fractured zone affects the stress state in the surrounding elements (Fig. 5). Physical parameters are reset in initial elastic state at the start of each tectonic stage.

Correction of VBD Vector: When the difference between the calculated deformation and the deformation control datum is large, the values of VBD vector components are adjusted. This process is interacted until the residual is below a nominated tolerance to obtain the optimized basement displacement.

Output of Incremental Deformation: The results of deformation under the optimized basement displacement are plotted for each tectonic stage. Deformation or inclination of the strata is shown by a mesh deformation diagram.

The magnitude of plastic strain is represented by the following equation;

$$eqe=\{(ex-ey)^2+(ey-ez)^2+(ez-ex)^2+gxy^2+gyz^2+gzx^2\}^{0.5}$$

where ex, ey, ez, gxy, gyz and gzx are components of the plastic strain tensor at a point within the yielding element. In the case of two-dimensional (plane strain) problem, ez, gyz and *gzx* are equal to zero. Plastic strain is residual after subtraction of the elastic-strain component from the total-strain component, which corresponds to the permanent deformation recorded as the geological structures.

The conjugate shear fracture is visually shown by two bars crossing orthogonally and diagonal to the principal-stress axes at 45°. One bar shows the sense of clockwise shear movement along a fault, and the other shows the counter-clockwise sense in the cross section. The magnitude of plastic strain (eqe) is expressed by the length (0 to 10%) of bars and the number (one for 10%) of bars.

Output of Cumulative Deformation: The analyses of incremental deformation are repeated for all tectonic stages and their deformations are superposed stage by stage. This provides the cumulative structures from the first tectonic stage to the present. The process of reactivation of faulting and the timing of the formation of structures can be discussed using this diagram.

AN EXAMPLE IN THE NIIGATA OIL AND GAS BASIN

The Nagaoka area, in the southern part of the Niigata oil and gas basin, was selected as the model field (Fig. 2). In this area oil companies have been actively conducting exploration with targets at 4,000m to 6,000m depths, but they did not penetrate through the Tertiary sediments to the basement except in some marginal areas of the basin.

Subsurface geology has been reported in detail by the Niigata Prefectural Government (1989, Fig. 1), Ogusa and Kikuchi (1982), Aiba (1982-Fig. 2), Inoma and Akabori (1982) and Komatsu et al. (1984). The Tertiary system consists of Tsugawa-Nanatani (Green Tuff), Teradomari, Shiiya, Nishi-yama, Haizume and Uonuma Formations in ascending order (Fig. 5).

The conditions of simulation were obtained for the formative process by division into four tectonic stages; Shiiya, Nishiyama, Haizume and post Uonuma stages, because reservoir structures have been formed after deposition of the Teradomari Formation which is the main source rock in this area (Fig. 6).

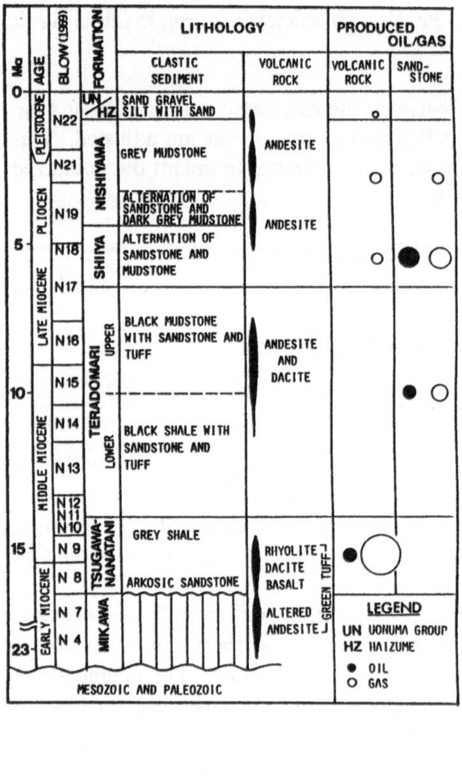

Figure 6. Generalized stratigraphic column in the Niigata area (Komatsu et al. 1984). At the top of the list of formations UN is Uonuma Formation and HZ is Haizume Formation (refer to Fig. 6).

The Virtual Basement was assumed to exist 1000m below the bottom of the Teradomari Formation. This surface does not correspond to the actual surface of the geological basement of the basin, but in part is probably in the Green Tuff Formation and in part in the pre-Tertiary basement. The Green Tuff Formation and the pre-Tertiary units are considered to behave in a similar manner during deformation.

The vertical component of the displacement vectors of the virtual basement was applied at 79 node points which are set every 500m along the Virtual Basement. The horizontal component was omitted in this case.

The data pertaining to the physical parameters of rock deformation are taken from Hoshino et al. (1972), Hattori et al. (1984) and others for the sedimentary and volcanic rocks of Niigata Sedimentary Basin.

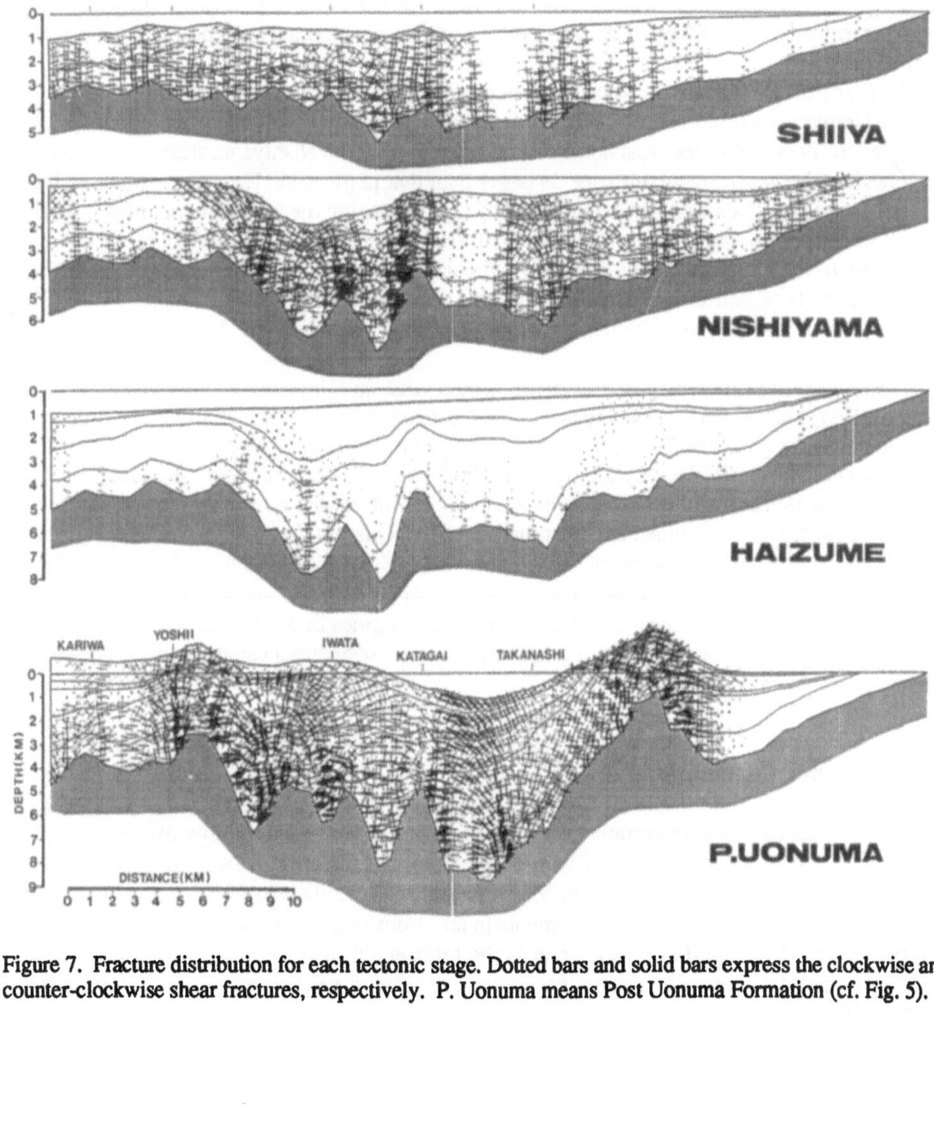

Figure 7. Fracture distribution for each tectonic stage. Dotted bars and solid bars express the clockwise and counter-clockwise shear fractures, respectively. P. Uonuma means Post Uonuma Formation (cf. Fig. 5).

The VBD analysis shows the following results:

The Formation of Fault Zones During Tectonic Stage: Conjugate fault planes formed by incremental displacements during each tectonic stage are shown by small cross bars in Figure 7. Zones of concentrated faults which continue from deeper to shallower parts are developed. One of the conjugate faults within the zone has a steep dip whereas the other is nearly horizontal. The former is parallel to or forms in echelon with the direction of elongation of the fault zone. Thus the sense of displacement along the fault zone is identified. The conjugate fault planes with vertical dip in the deeper parts gradually change their dip upwards, and the concentrated strain is also dispersed in the shallower parts.

Normal faults with dips of about 45° are developed near the surface layer in the upheaval areas, and groups of reverse faults are developed in areas of subsidence. The scale of the groups of normal and reverse faults becomes larger in the shallower parts and largest at the surface. They develop downwards and connect with the vertical fault zone which developed upwards from the deeper parts.

Of the four tectonic stages, significant development of the faults is observed at the Nishiyama and post Uonuma tectonic stages. The fault zones extending vertically from the deeper to shallower parts occurred to the west and east of Katagai during the Nishiyama stage. The western fault zone, i.e., the west-down fault zone, is larger than that in the east. Later there was a change and there were large-scale east-down faults east of Katagai during the post Uonuma stage. In this zone a horst structure with 2-3 km displacement was formed between these faults. Thus, the western limb of the horst, that is the anticline in the upper horizon, was developed mainly during the Nishiyama stage and the eastern limb mainly during the post Uonuma stage. A similar formative process is observed in the horst and anticlinal structure west of the Iwata area.

The Development of Sedimentary Basins at the Surface and the Formation of the Deep Structure: During the Shiiya stage, general subsidence of the basin area occurred and small deeper structures were formed. During the Nishiyama stage there were more significant differential movements and distinct parts of the basin developed. The largest subsidence occurred west of Katagai. The western limb of the Katagai horst was developed mainly during this stage. During the Haizume stage, gentle subsidence continued and no structure of particular characteristics was formed in the deep zones. The major formation of the eastern limb of the Katagai horst accompanied the subsidence east of Katagai, after the Uonuma stage. The Higashiyama and Yoshii horsts and anticlines were also formed during this stage. Horsts or anticlines in the deeper zone are generally found beneath localities where the dip or thickness of the younger strata varies abruptly, marking a boundary of a subsiding area.

Discussion and Conclusions

The deeper structures are very complex and different from those of the shallow part of this area, and together they are called "the duality of the oil bearing structures (Aiba 1982). It has been generally considered that the horst and graben structures of the units below Nishiyama Formation were formed by vertical differential movements in an extensional stress field during the stage of basin subsidence (Miocene-Pliocene), whereas the folds and thrusts of the shallower units were formed by horizontal shortening during the Quaternary (Uemura & Shimohata 1972; Aiba 1982; Komatsu 1990).

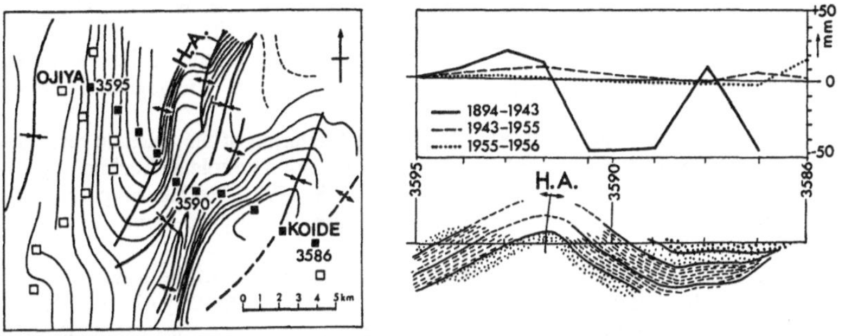

Figure 8. Secular crustal movement across the Higashiyama Anticline (H.A.).

On the contrary, the results of this simulation study show that the deeper geologic units are likely to have a more complex structure reflecting movements at various stages while the younger shallower units have simple structures reflecting the deformation of short time-spans. The deformation of the virtual basement is rather similar to block movements bounded by basement faults but the main movements on one of the faults can be at a different time from the main movements on the other fault.

Figure 8 shows the secular crustal movement measured by the levelling survey along the road across the axis of the Higashiyama anticline (Kodama et al. 1974). The crustal movement is a block-like and is not parallel to the anticlinal structure. The eastern wing of this anticline seems to be still growing. South of Tokyo, in the Boso area, Shibuya & Shinada (1986) and Kodama et al. (1990) analysed the syntectonic process of growth of the anticlines, the wings of which were deformed step by step during the process of shifting of the deposition centre of the formation. This process is also considered to be directly controlled by basement movements. It is widely agreed that the folds were developed in the region where thick sediments were deposited and succeeding general upheavals occurred. Such tectonics are explained by the inversion movements of the basement (Vejbaek & Andersen 1987; Todd 1989). The interaction between the cover sediments and the basement seems to be more closely related than previously thought. The analytical study of the processes of basement movement seem to be a most important field, and the VBD method should be useful in other regions.

Acknowledgement

The author wishes to express his appreciation to Dr M.J. Rickard of the Australian National University for his kind support and encouragement to complete this paper. He also appreciates the useful discussion on the model of Dr. Y. Suzuki of the Geological Survey of Japan and Dr. T. Mitsunashi, a former Professor of Shimane University.

References

Aiba, J. 1982. Duality of the oil bearing structures in Akita and Niigata sedimentary basins. In, Fujita, Y., Kobayashi, I., Uemura, T., Takahama, N., Shimazu M., & Aoki, S. (eds) Island-arc Disturbance, Monograph 24, Association. Geological Collab. Japan, p. 299-308.*

Badgley, M.E., Proce, J.D., & Backshall, L.C. 1989. Inversion, reactivated faults and related structures: seismic examples from the southern North Sea. In, Cooper, M.A., & Williams, G.D. (eds) Inversion Tectonics, Geological Society of London, Special Publication **44**, 201-219.

Beck, R.A., Vondra, C.F., Filkins, J.E., & Olander, J.D. 1988. Syntectonic sedimentation and Laramide basement thrusting, Cordilleran foreland; timing of deformation. Geological Society of America Memoir **171**, 465-487.

Hattori, M., Hoshino, K., & Inami, K. 1984. Relation of fracturing of the volcanic rocks to mechanical properties under high pressure. Journal of Japanese Association Petroleum Technology **49**, 20-28.*

Hoshino, K., Koide, H., Inami, K., Iwamura, S., & Mitsui, S. 1972. Mechanical properties of Japanese Tertiary sedimentary rocks under high confining pressure. Geological Survey of Japan Report **244**, 206p.

Inoma, A., & Akabori,Y. 1982. The exploration of the deeper zone of the Katagai gas field, Niigata Prefecture, Japan. Journal of Japanese Association Petroleum Technology **47**, 99-106.

Jaeger, J.C. 1962. Elasticity fracture and flow, with engineering and geological applications, 2nd. ed., Methuen & Co., London, 207p.

Kodama, K., Honda, S., Fujita, H., Nitta, K., & Suzuki, Y. 1976. Numerical analysis of faulting in the course of block deformation—(1) Effect of width of the basement block on faulting near the earth's surface. Geological Survey of Japan, Bulletin **27**, 123-134.*

Kodama, K., Suzuki,Y., Miyashita, M., & Soma, N. 1974. Studies on the relation between the recent crustal deformation and geologic structures in Niigata Tertiary Basin. Geological Survey of Japan Report **250-2**, 37-51.*

Kodama, K., Long Xue-Ming, Suzuki, Y. 1985. Structural analysis of deep-seated volcanic rock reservoirs by tectonic simulation. United Nations ESCAP, CCOP Technical Bulletin **1**, 61-79.

Kodama, K., Otake, N., Tomita, H., Ono, S., Nakajima, Y., Kii, K., & Suzuki, Y. 1990. On the stratigraphic subdivision of the Amatsu formation and process of folding in the central part of the Boso Peninsula. Geological Society of Japan, Memoir **34**, 105-115.*

Komatsu, N. 1990. Folds of Niigata oil fields and their folding process. Geological Society of Japan, Memoir **34**, 149-154.*

Komatsu, N., Fujita, Y., & Sato, O. 1984. Cenozoic volcanic rocks as potential hydrocarbon reservoirs. In, Proceedings of 11th W.P.C. (1983) **2**, 411-420, John Wiley & Sons Ltd., Chichester.

Koopman, A., Speksnijder, A., & Horsfield, W.T. 1987. Sandbox model studies of inversion tectonics. Tectonophysics **137**, 379-388.

Ogusa, K., & Kikuchi, Y. 1982. On the exploratory work of the deeper part in the vicinity of the Sekihara-Koshijihara, Niigata Prefecture, Japan. Journal of Japanese Association Petroleum Technology **49**, 11-19.*

Niigata Prefectural Government 1989. Geological map of Niigata Prefecture (1:200,000) 2nd. ed.

Sanford, A.R. 1959. Analytical and experimental study of simple geologic structures. Geological Society of America Bulletin **70**, 19-52.

Shibuya, T, & Shinada, S. 1986. The formation of the Sakuna anticline in the southern part of the Boso Peninsula. Journal, Geological Society of Japan **92**, 1-13.*

Todd, S.P. 1989. The role of the Dingle Bay Lineament in the evolution of the Old Red Sandstone in southwest Ireland. In, Arthurton, R.S., Gutteridge, P., & Nolan, S.C. (eds) The role of Tectonics on Devonian and Carboniferous Sedimentation in the British Isles. Occasional Publication of the Yorkshire Geological Society **6**, 35-54.

Uemura, T., & Shimohata, I. 1972. Neutral surface of a fold and its bearing on folding. 24th International Geolgical Congress, Sect. 3, 599-603.

Vejbaek, O.V., & Andersen, C. 1987. Cretaceous-Early Tertiary inversion tectonism in the Danish Central Trough. In, Ziegler, P.A. (ed.) Compressional Intra-Plate Deformations in the Alpine foreland, Tectonophysics, **137**, 221-238.

Withjack, M.O., Olson, J., & Peterson, E. 1990. Experimental models of extensional forced folds. American Association of Petroleum Geologists Bulletin **74**, 1038-1054.

Yamada, Y. 1972. Plasticity, visco-elasticity. Baifukan, Tokyo, 240p

(*in Japanese with English abstract).

GEOCHEMICAL EVOLUTION AND BASEMENT TECTONISM OF THE ARABIAN-NUBIAN DOME

H.O. SINDI
Department of Earth and Planetary Sciences
Harvard University
20 Oxford Street
Cambridge Massachusetts 02138
U.S.A.

ABSTRACT. Geological and geochemical investigations including major, trace and rare-earth elements have been carried out on selected areas from the Arabian Shield. The results are compared with those from similar terrains elsewhere. This study includes accounts of age determinations, types of tectonism and zones of metamorphism.

The oldest rock in the eastern flank of the Arabian-Nubian Dome is the Khamis Mushayt fundamental gneiss in the southern region of the Arabian Shield which has a distinct REE pattern and a pre Pan-African age older than 1800 Ma. The age of the oldest continental rock in this sub-plate is older than 2500 Ma, comparable to similar ages in Egypt.

The major components in the Arabian Shield are plutonic rocks that range from felsic to basic in composition, with strongly alkaline or primitive tholeiitic series to mature calcic rocks occurring in a systematic trend. The rocks define at least five different subduction zones with a growth rate of about 20% of the modern global rate. These subduction zones are also indicated by the shallow crustal lithologies, by poly-metallic mineralization, and by oxygen-isotope and heat-flow data and by geophysical and geochemical behaviour. These rocks were formed in ensimatic island-arc and continental marginal-arc environments.

Most of the granitic rocks in the Arabian sub-plate have geochemical compositions and genetic relationships which suggest that they include both S-type and I-type granites. The youngest post-accretion granitic bodies have an age of about 450 Ma. and were formed by massive crustal fusion modified by fractional processes. The plutonic rocks are both mesozonal to epizonal. The crustal thickness of the Arabian sub-plate is estimated to be about 45 km.

Introduction

In the Precambrian there were at least five subduction zones in the Arabian-Nubian Dome. A reconstruction of the complete dome before the opening of the Red Sea is presented in Fig. 1. Several stages of magmatic evolution have affected the Arabian Shield. The oldest stage is dominated by an intermediate plutonic series of tholeiitic to calc-alkaline affinities followed by island-arc sequences and lastly by the widespread post-orogenic phase. The true continental crust in the Arabian peninsula is about 45 km thick (Knopoff & Fouda 1975).

The oldest units in the Arabian-Nubian Dome as a whole are present in Egypt and consist of a mafic and highly serpentinized ultramafic igneous sequence with chilled margins and ocean-floor and ophiolite features. These units have an age older than 2.3 Ga. (Dixon 1979).

The Precambrian basement in Arabia consists of similar groups in addition to immature sediments, tuffaceous wackes, quartzite, jasper, banded iron lenses, Proterozoic metavolcanics,

M. J. Rickard et al. (eds.), Basement Tectonics 9, 161–168.

plutonics, metaplutonics, unconsolidated metaclastics and metasediments (e.g. marble, skarn, and calcsilicate bodies) plus granite pebbly and cobbly mudstones (Sindi 1976, 1982). Volcanic successions are chemically mature with calc-alkaline to tholeiitic affinities. These rocks are formed in oceanic island arcs. Zones of ultramafic, basic and ophiolite complexes are tectonically emplaced oceanic crust. Younger plutons consist of granitic, granodioritic, dioritic, gabbroic, tonalitic and trondhjemitic rocks with ages ranging from 1300 Ma. to 550 Ma. in age.

Geophysical Studies

Several geophysical studies have been carried out on the Arabian-Nubian Dome including the Red Sea Graben and its extension to the pull-apart Dead Sea basin in the north. Some of these data, with discussion and interpretations will be presented in Sindi (1991 in press, & 1992 in press).

The seismic data are compatible with gradual attenuation of the continental margin of the Arabian plate and a discontinuous transition to the Red Sea oceanic crust (Milkereit & Fluh 1985; Mechie et al. 1986). The micro-seismicity data indicate a high level of activity in the southwestern part of the Arabian Peninsula and show that Red Sea extension is still active.

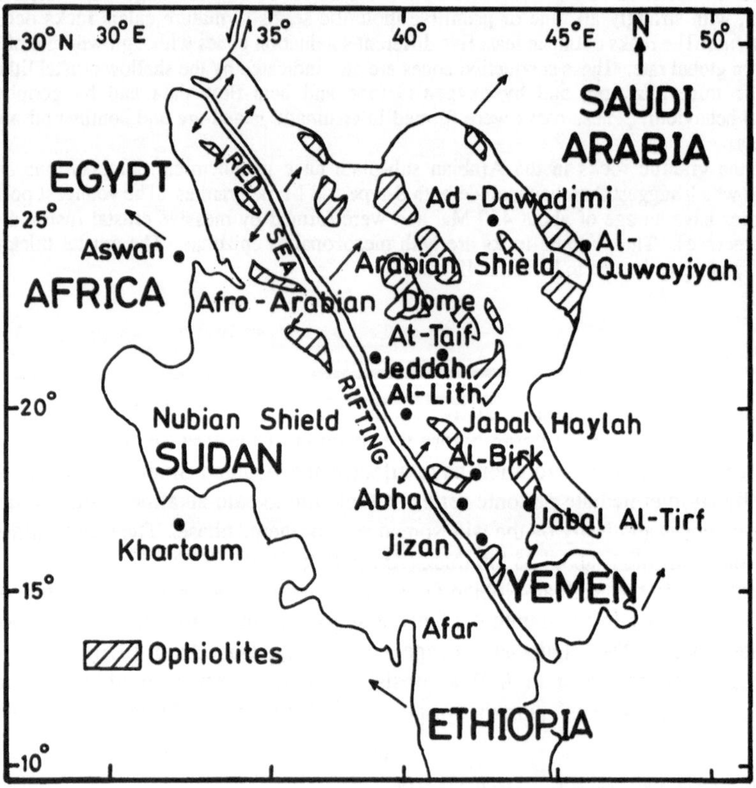

Figure 1. Reconstruction of the Arabian-Nubian Dome showing the locations of the ophiolite complexes in relation to the Red Sea axial rifting and transitional region

The gravity anomalies have a steep gradient in the eastern part of the coastal plain (Gettings 1977; Mechie et al. 1986).

The magnetic data for southwest Arabia have low amplitudes and indicate linear magnetic anomalies parallel to the trend of the axial trough of the Red Sea (Gettings 1977).

Petrology and Petrochemistry

In the African Nubian plate, the hawaiite-mugearite in the Wadi Natash area in the Eastern Desert of Egypt (24° 30 N, 34° 10 E) has been fractionated from alkali-olivine basalt that in turn was formed by olivine fractionation from a primary picrite. The olivine (F_{085}) is altered to serpentine and iddingsite (Hubbard et al. 1987).

In the Arabian plate, the Jabal Abu-Shidad at 39° 32' N, 20° 59' E consists of rhyolite and trachyte (Fig. 1).

The Al-Lith area to the south of Jeddah contains several formations (e.g. the Tertiary hawaiite Sita volcanics and their feeder zone the Wadi Ad-Damm dyke complex). It has rocks transitional from sodic-alkaline to subalkaline. The plutons in the area are generally composed of quartzdiorite, diorite, monzodiorite, anorthosite and leucogabbro. Fresh non-layered Tertiary olivine gabbro occurs near Jabal Sita with a whole-rock K-Ar age of about 27 Ma. Harrat Hadan has ferruginous and pisolitic laterite, and ferruginous and oolitic mudstone, while the Usfan and Shumaysi formations near Jeddah (predating the Sita formation) are intruded by sparse basalt and plagioclase-megacryst-bearing hawaiite dykes (Pallister 1987). The Harrat Al-Birk basalts (18° 20' N, 41° 50' E) near Jizan (Tihamat Asir) consist mainly of alkali-olivine basalt.

A complex in the Jabal At-Tirf area consists of different plutons with ages from 24 to 20 Ma. (Coleman et al. 1979). A huge lopolithic layered gabbroic body, and dyke swarms, intrude the detachment terrain, and steeply tilted strata of Cambrian age. This large mafic pluton is intruded by a 23.3-20.6 Ma. granophyric mass with a thermally altered contact zone some 20-30 m wide. Some other layered gabbroic plutons occurring in this region consist of gabbro, amphibolitized gabbro, pyroxene gabbro, olivine gabbro, leucogabbro and anorthosite.

The At-Taif and Al-Jibub area consists of several calc-alkaline granitic plutons of both S-type and I-type, layered gabbroic bodies, anorthosite, dioritic intrusions and volcanic sequences in addition to several dykes of different ages and assorted composition (Sindi 1976, 1981).

The Tihamat Asir magmatic complex consists of different plutonic bodies, granophyres and several generations of a N-S unaltered dyke swarm with assorted compositions and eastward dips. This magmatic complex is the eastern boundary of the ophiolitic and Red Sea oceanic crust.

The Abha and Khamis Mushayt in the southern area consist mainly of the oldest fundamental gneiss, metamorphic series, gabbroic and dioritic bodies and some acidic intrusions. The northern part "i.e. Jabal Haylah near Wadi Hali" of the Al-Birk alkali-olivine plateau basalts (18° 20' N, 41° 50' E) rests on the continental crust. The southern part of these flood basalts "i.e. Jabal Akwah", lies on the Miocene oceanic crust and is directly related to the separation of the Arabian plate from the African plate and the sea-floor spreading of the Red Sea (Sindi 1991, in press). In these rocks, olivine grains, F_{091}, are cut by strain bands with trains of fluid inclusions, while pyroxenes occur moulded against, and in some crystals enclosed by olivine crystals and show undulatory extinction in the polygonal mosaic texture. By applying the Carmichael et al. (1977) P-T model to the Al-Birk megacrysts and lava, it can be suggested that the temperature was >1350°C and the pressure was 16 kb. Layered hypersthene-normative gabbroic rocks occur in these areas.

Mineralogy

The Al-Birk alkali-olivine basalts have highly altered plagioclase An_{50-71}, alkali feldspar $Or_{60}Ab_{30}An_{10}$, orthopyroxene, clinopyroxene with reaction rims, olivine F_{091}, opaque minerals including ilmenite and a minor amount of elongated Cr-spinel. The plagicoclase has irregular zones of dusty inclusions. The chemistry of these minerals has been studied by Ghent et al. (1980) using the electron microprobe. They have shown that chemical zoning in the spinel crystals and both pyroxenes is related to the reaction texture which suggests a rapid crystallization from a melt.

Selected minerals from the At-Tirf layered gabbro have been analysed using the Hitachi electron microprobe S450 SEM. Ten feldspar crystals, 17 clinopyroxene grains and 20 amphibole specimens have been studied and the results are presented in Table 1.

Selected minerals from the Median alkali granite have also been analysed. Some of the amphibole laths have a blue colour, whereas others have a brown colour, and a few have colourless laths. Twenty-five amphibole samples and 15 pyroxene crystals have been analysed (Table 1).

The basic dykes from the southern region are equigranular to intergranular with cumulate, interseptal and subophitic texture. They are composed of calcic-plagioclase, olivine, clinopyroxene and ilmenite, with minor amount of quartz and iron sulfides. These high-alumina specimens are both quartz-normative and nepheline-normative. Five biotite crystals from the Asir biotite granite have been chemically analysed (Table 1).

TABLE 1 MICROPROBE CHEMICAL ANALYSES (NORMALIZED VALUES) OF SELECTED MINERALS FROM SAMPLES COLLECTED FROM THE ARABIAN SHIELD.

Area	Midian		Asir	Jabal At-Tirf		
Rock name	Alkaki granite		Biotite granite	Layered gabbro		
No. of samples	25	15	5	10	17	20
Mineral	Amphibole	pyroxene	Biotite	Feldspar	Clinopyroxene	Amphibole
SiO_2	50.12	52.91	37.56	50.31	51.46	42.89
TiO_2	1.63	1.29	3.43	-	0.90	3.33
Al_2O_3	1.18	0.18	19.11	30.34	2.90	10.96
FeO	33.88	30.75	20.82	0.53	9.75	12.75
MnO	0.67	0.21	0.08	-	0.35	0.15
MgO	0.76	0.06	8.76	-	16.33	14.78
CaO	2.95	2.51	0.01	15.77	17.78	11.76
Na_2O	6.78	11.98	0.16	2.94	0.39	2.65
K_2O	2.01	0.01	9.99	0.11	0.01	0.71
Cr_2O_3	0.01	0.01	0.04	-	0.12	0.01
NiO	0.01	0.09	0.04	-	0.01	0.01

Geochemistry

Pillowed basalts and cogenetic gabbros from the Arabian shield have a tholeiitic composition with low K_2O, high Mg, Ni, Cr and Ca/Al and are depleted in the chondrite-normalized LREE. The low value of the initial $^{87}Sr/^{86}Sr$ (0.7031 - 0.7047) for the Pan-African granites suggests an oceanic-island and oceanic-mantle source region (Coleman et al. 1983). This value is attributed to isotopic heterogeneity of the mantle and to assimilation of the lower crust.

Most of the alkali-olivine basalts have Ni about 250 ppm (Donnelly et al. 1980), so the low Ni means fractionation of olivine at high water pressure which tends to increase the stability field of olivine. Olivine is more stable in water-saturated melts (Kushiro 1972).

The rhyolites from the Sita formation and Ad-Damm dyke complex have Eu-depletion and strong HREE values.

Fifty samples from the Jabal At-Tirf layered gabbros have been analysed (Table 2). These samples are low in TiO_2, Fe_2O_3, K_2O, Ba, Sr, V and Zr and high in SiO_2, MnO, Cr, Ni and Zn compared with layered gabbros from the Al-Jibub area.

Ten samples from the granitic batholith in the Al-Amar region, near Al-Quwayiyah zone have also been analysed along with 5 samples from the Al-Ehen monzogranite at the western side of the Margin fault to the west of Al-Quwayiyah (Table 2).

Analyses of 8 samples from the Asir Alkali granite in the southern province are also presented in Table 2.

Twenty samples from the median alkali granite in the northern sector of the Arabian Shield are low in Al_2O_3, MgO, CaO and Sr, and high in TiO_2, iron oxides, Ce, La, Nb, Y, Zn, Zr, and Nd (Table 2). These analyses fit well with the results presented by Ramsay et al. (1986).

TABLE 2. AVERAGE CHEMICAL ANALYSES OF MAJOR OXIDES (WT%) AND TRACE ELEMENT CONCENTRATIONS (PPM) OF SELECTED ROCKS FROM THE ARABIAN SHIELD

A - Major oxides (wt%)

Area	Midian	Al-Amar	Al-Ehen	Asir	Jabal At-Tirf
Rock name	Alkali granite	Granite	Monzogranite	Alkali Granite	Layered gabbro
No. of analyses	20	10	5	8	30
SiO_2	72.85	71.75	70.51	75.19	48.32
TiO_2	0.33	0.21	0.24	0.13	0.91
Al_2O_3	12.59	14.48	14.63	12.70	16.95
Fe_2O_3	2.54	1.09	1.47	1.41	2.81
FeO	1.96	0.81	0.89	0.52	7.19
MnO	0.07	0.04	0.05	0.03	0.28
MgO	0.11	0.58	0.80	0.13	7.88
CaO	0.66	1.99	2.56	0.58	11.95
Na_2O	3.82	4.25	3.98	3.97	2.12
K_2O	4.54	2.57	3.65	4.68	0.21
$T.H_2O$	0.41	1.85	0.92	0.57	1.31
P_2O_5	0.03	0.01	0.09	0.01	0.07
Others	0.09	0.37	0.21	0.08	-

B - Trace element concentrations (ppm)

Ba	214	986	413	118	40
Ce	149	52	41	79	40
Co	4	-	5	6	38
Cr	-	-	-	-	372
Cu	8	-	16	8	10
La	54	23	26	38	18
Li	15	-	39	12	15
Mo	8	-	3	1	5
Nb	34	-	10	48	20
Ni	10	-	9	8	250
Pb	17	-	7	5	15
Rb	101	95	231	165	5
Sr	58	463	159	30	248
V	10	35	18	10	20
Y	83	9	73	55	20
Zn	89	-	65	70	100
Zr	781	-	116	240	30
Th	16	-	-	-	-
Nd	85	15	21	-	-
Sc	2	-	-	2	-
Sm	1.1	2.5			
Eu	0.28	0.78			
Dy	1.1	0.77			
Er	0.5	1.8			
Tm	1.1	0.74			
Yb	0.68	0.56			
Lu	0.04	0.10			

Geochronology

The andesitic volcanism and some plutonism increase in age from the northern to the southern section of the Arabian-Nubian Dome.

The syntectonic quartz dioritic and granodioritic intrusions in southeastern Egypt are about 750 m.y. old (Dixon 1979). The slightly metamorphosed volcanic sequence of the Dokhan area in northeastern Egypt has a whole-rock Rb/Sr age of about 602 +/- 13 Ma. (Stern 1979). The alkali-olivine basalt and hawaiite mugearite that occur in the Wadi Natash area at the Eastern Desert of Egypt have an age of about 90 Ma. (Hubbard et al. 1987). These rocks may be related to early eclogitic mantle tectonic activity.

Jabal Abu-Shidad in the Arabian Shield has a K/Ar age of about 20 Ma while the Ad-Damm dyke complex have an age about 50 Ma. with the exception of the Harrat Ad-Damm which is dated at 11.3 Ma (Pallister 1987).

The At-Taif granitic rocks have a Rb-Sr age of 520-618 Ma, while the K-Ar age is about 608 Ma (Sindi 1976). Other age determinations for the Arabian Shield have been given by Stoeser (1986).

Structures

The central part of the Afro-Arabian Dome has been refolded, sheared and thrust faulted, generally parallel to the Red Sea feature, reflecting the extensive crustal shortening in this direction.

The E striking Bitlis/Zagros suture that started in the middle Miocene, slowed the northward drift of Africa/Arabia and marks the boundary between the Arabian and Anatolian plates.

The important features in the Arabian-Nubian Dome are the fault systems which run parallel to the Red Sea. In the Jizan area, there are several generations of NW-SE trending antithetic low-angle normal faults that dip E to NE. Some of these faults have accommodated uplift of the escarpment with some clockwise block rotation. Some dextral transfer faults and a few sinistral strike-slip faults represent reactivated Precambrian Pan-African structures (Greenwood et al. 1980).

The earthquake that happened in north-central Iran on 21 June, 1990 is a good recent example of the movements of the Arabian plate against the Eurasian plate. It illustrates the effect of the stress on the Anatolian and the Bitlis/Zagros sutures as a result of the widening of the Red Sea.

References

Carmichael, I.S.E., Nicholls, J., Spera, F.J., Wood, B.J., & Nelson, S.A. 1977. High-temperature properties of silicate liquids: applications to the equilibration and ascent of basic magma. Royal Society London, Philosophical Transactions A. **286**, 373-431.

Coleman, R.G., Hadley, D.G., Fleck, R.J., Hedge, C.T., & Donato, M.M. 1979. The Miocene Tihama Asir ophiolite and its bearing on the opening of the Red Sea. In, A.M.S. Al-Shanti (Convener), Evolution and mineralization of the Arabian-Nubian Shield, Proceedings of a symposium held at King Abdulaziz University, Jeddah, Saudi Arabia. I.A.G. Bulletin **3**, 173-186, Pergamon Press, Oxford, U.K.

Coleman, R.G., Gregory, R.T., & Brown, G.F. 1983. Cenozoic volcanic rocks of Saudi Arabia. U.S.G.S. — Saudi Arabian project Open-File Report 83-788, Department of Geology and Mineral Resources, Jeddah, Saudi Arabia, 82p.

Dixon, T.H. 1979. The evolution of continental crust in the late Precambrian Egyptian Shield. University of California, San Diego, **231** p.

Donnelly, T.W., & Rogers, J.J.W. 1980. The northeastern Caribbean compared with worldwide island-arc assemblages. Bulletin Volcanologique **43**, 347-382.

Engel, A.E.J., Dixon, T.H., & Stern, R.J. 1980. Late Precambrian evolution of Afro-Arabian crust from ocean arc to craton. Geological Society American Bulletin **91**, 699-706.

Gettings, M.E. 1977. Delineation of the continental margin in the southern Red Sea region from new gravity evidence. U.S.G.S. — Saudi Arabian project Open-File report, Department of Geology and Mineral Resources Bulletin 22, Red Sea Research 1970-1975, Kl-Kll, Jeddah, Saudi Arabia.

Ghent, E.D., Coleman, R.G., & Hadley, D.G. 1980. Ultramafic inclusions and host alkali olivine basalts of the southern coastal plain of the Red Sea, Saudi Arabia. American Journal of Science **280-A**, 499-527.

Greenwood, W.R., Anderson, R.E., Fleck, R.J., & Roberts, R.J. 1980. Precambrain geologic history and plate tectonic evolution of the Arabian Shield. U.S.G.S. Saudi Arabian project Open-File report, D.G.M.R. Bull. 24, Jeddah, Saudi Arabia, 35p.

Hempton, M.R. 1987. Constraints on Arabian plate motion and extensional history of the Red Sea. Tectonics **6**, 687-705.

Hubbard, H.B., Wood, L.F., & Rogers, J.J.W. 1987. Possible hydration anomaly in the upper mantle prior to Red Sea rifting: Evidence from petrologic modeling of the Wadi Natash alkali basalt sequence of eastern Egypt. Geological Society America Bulletin **98**, 92-98.

Knopoff, L., & Fouda, A.A. 1975. Upper mantle structure under the Arabian Peninsula. Tectonophysics **26**, 121-134.

Kushiro, I. 1972. Effect of water on the composition of magmas formed at high pressures. Journal of Petrology **13**, 311-334.

Mechie, J., Prodehl, C., & Koptschalitsch, G. 1986. Ray-path interpretation of the crustal structure beneath Saudi Arabia. Tectonophysics **131**, 333-352.

Milkereit, B., & Fluh, E.R. 1985. Saudi Arabian refraction profile: crustal structure of the Red Sea-Arabian Shield transition. Tectonophysics **111**, 283-298.

Pallister, J.S. 1987. Magmatic history of Red Sea rifting: Perspective from the central Saudi Arabian coastal plain. Geological Society America Bulletin **98**, 400-417.

Ramsay, C.R., Odell, J., & Drysdall, A.R. 1986. Felsic plutonic rocks of the Midyan region, Kingdom of Saudi Arabia — II. Pilot study in chemical classification of Arabian granitoids. Journal of African Earth Sciences **4**, 79-85.

Sindi, H.O. 1976. The Geology and Geochemistry of the At-Taif area, Saudi Arabia. M. Phil. thesis. Leeds Univ. 304p.

Sindi, H.O. 1981. The Petrology and geochemistry of the plutonic intrusives of the Al-Jibub area, Kingdom of Saudi Arabia. Ph.D thesis, The University of London (Queen Mary College), 422p.

Sindi, H.O. 1991 in press. The geochemical-geophysical aspects of tectonism in the Arabian Shield. In, Workshop on Geophysics and its tectonic implications in the Arabian Peninsula and the Red Sea region, 25-31 October, 1986, Department of Geology, Faculty of Science, Sana'a University, Sana'a Yemen Arab Republic. Bulletin of the Faculty of Science (Special volume), Sana'a University, Yemen.

Sindi, H.O. 1992 in press. The geology and geochemistry of the Red Sea, Saudi Arabia, and its relation to the Pacific region. Bulletin of the fifth conference and exhibition of the Circum-Pacific council for Energy and mineral resources July 28 - August 3 1990, Honolulu.

Sindi, H.O., & French, J.W. 1983. The geology and geochemistry of the metamorphic rocks of the Al-Jibub area, Kingdom of Saudi Arabia. In, Proceedings of the first Jordanian geological conference (JGC), Geology of Jordan, University of Jordan, Amman, Jordan, 1982, Jordanian Geologists Association (JGA), Hitten printing Press, Amman, Jordan, 352-382.

Stern, R.J. 1979. Late Precambrian ensimatic volcanism in the central eastern Desert of Egypt. Ph.D thesis, University of California, San Diego, 210p.

Stoeser, D.B. 1986. Distribution and tectonic setting of plutonic rocks of the Arabian Shield. Journal of African Earth Sciences **4**, 21-46.

BASEMENT TECTONICS OF SAUDI ARABIA AS RELATED TO OIL FIELD STRUCTURES

H.S. EDGELL
King Fahd University of Petroleum & Minerals
KFUPM Box 940
Dhahran 31261
Saudi Arabia

ABSTRACT. All the oil fields of Saudi Arabia are of the structural type and they all lie in the northeastern part of the country, including the Saudi offshore portion of the Persian Gulf.

These oil field structures are mostly produced by extensional block faulting in the crystalline Precambrian basement along the predominantly N-S Arabian Trend which constitutes the 'old grain' of Arabia. This type of basement horst, which has been periodically reactivated, underlies the world's largest oil field, Ghawar, and other major oil fields, such as Khurais, Mazalij and Abu Jifan. The basement horst beneath Ghawar Anticline has been suggested by Aramco (1959), from a positive Bouguer gravity anomaly which practically mirrors the field, and more recently, in greater detail, by Barnes (1987).

All Saudi Arabian offshore oil fields, and some near coastal fields, such as Abu Hadriya, Abqaiq and Dammam, are also produced by basement faulting which has cut the saliferous, Upper Precambrian Hormuz Series, triggering deep-seated salt diapirism. Consequently, Saudi Arabian offshore and coastal fields are denoted by distinctive negative gravity anomalies. Some of these oil fields are circular, such as Dammam, Abu Hadriyah and Karan, whereas others are elongated, due to salt wall diapirism, as in the case of Khafji, Kurayn, Jana and Jurayd. The latter oil fields all follow a NE to NNE trend believed to be due to left-lateral strike-slip faulting in the basement, as seen in Kuh-e Namak on the Iranian side of the Gulf. One large offshore oil field at Manifa trends NW - SE, paralleling the Persian Gulf and is due to right-lateral strike-slip faulting in the basement along the well-known Erythraean Trend (von Wissman et al. 1942). An undeveloped offshore oil field at Hasbah trends E-W, due to basement faulting in this direction. Some major, elongated offshore oil fields with negative gravity anomalies trend almost N-S along the Arabian Trend, including Berri and Qatif, as well as the nearby Bahrain and Dukhan fields. All the known oil fields of Saudi Arabia and its offshore are thus related to four major directions of basement faulting, namely N-S, NE-SW, NW-SE and E-W. The major fault trend is the N-S Arabian Trend which has produced repeated basement horsts and grabens due to extensional tectonics. In the Saudi offshore portion of the Persian Gulf, NE-SW, left-lateral strike-slip faulting of the basement is also important in forming oil-field structures.

A major difference between Saudi Arabian onshore and offshore oil fields is that the former show strong positive gravity anomalies due to block uplift of basement, whereas the latter have pronounced negative anomalies due to deep-seated salt diapirism induced by faulting in the crystalline basement.

Detailed analysis of the potential field of gravity in the oil-field areas of Saudi Arabia, both offshore and onshore, substantiates these major basement-fault directions and shows the marked coincidence of oil-field outlines with large basement-induced gravity anomalies. The regmatic shear pattern of basement faulting is also clearly shown by mapping the second derivative of the potential field of gravity.

M. J. Rickard et al. (eds.), Basement Tectonics 9, 169–193.

Introduction

Anticlinal or domal structures in the sedimentary sequence of the northeastern Arabian Platform and its offshore extension, contain all the known oil and gas fields of Saudi Arabia (Ministry of Petroleum 1984). These currently comprise some fifty six oil fields and four gas fields (Fig. 1), all of which owe their origin to deep-seated tectonic movements in the Precambrian crystalline basement (Edgell 1987). The thickness of the sedimentary sequence varies from 4500m to 13,700m by a gradual thickening towards the northeast.

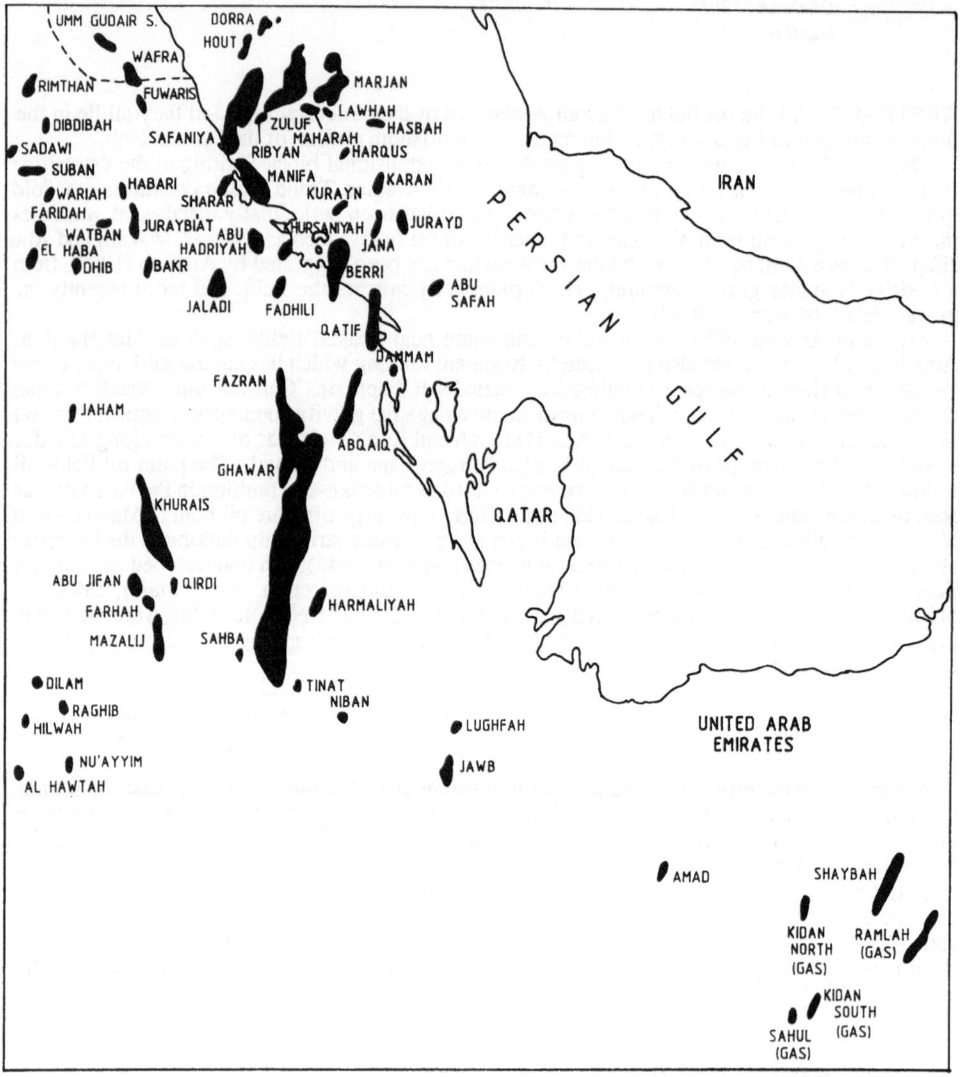

Figure 1. Oil and Gas Fields of Saudi Arabia.

Underlying these sediments in most of onshore northeastern Saudi Arabia, there is a faulted Precambrian basement, with alternating horsts and grabens directed along the N-S Arabian Trend. This is the 'old grain' of the Arabian Peninsula formed by repeated, E-W, extensional tectonism (Henson 1951). As a result, almost all the onshore oil fields of northeastern Saudi Arabia exhibit distinctive positive gravity anomalies due to the presence of denser uplifted basement beneath them. As in the case of Ghawar, the world's largest oil field, the presence of an underlying, N-S directed, horst block was first suggested by Aramco (1959). Further refined computerized study of the potential field of gravity, including running difference gravity (Barnes 1987), shows a gravity high almost exactly underlying Ghawar Oil Field. Study of the second derivative of the potential field of gravity shows the location at depth of basement faults bordering this underlying horst, as well as some oblique, cross-cutting faults, which are apparently left-lateral, strike-slip faults in the basement. Many other oil fields to the west and south of Ghawar also show a striking coincidence of positive gravity anomalies with their known extent, as in the cases of Mazalij, Abu Jifan, Tinat, Khurais, Bakr and Faridah (Barnes 1987). They are all underlain by uplifted basement blocks and their crestal stratigraphic sequences indicate repeated rejuvenation of their underlying basement uplifts.

In offshore northeastern Saudi Arabia and adjacent coastal areas, more than 20 oil fields occur, all marked by distinctive negative gravity anomalies (Edgell 1987). This is because basement faulting has also penetrated the Upper Precambrian halite beds of the Hormuz Series, thus triggering deep-seated salt diapirism and producing domal oil-field structures. Some of these oil fields, such as Karan, Abu Sa'afah, Abu Hadriyah and Dammam, have the typical circular shape of diapiric oil structures. Most offshore oil-field structures are elongated, doubly plunging anticlines, such as Safaniyah, Khafji, Kurayn, Jana and Jurayd. They are elongated along a general NE to NNE direction (Aualitic Trend of von Wissmann et al. 1942) and are due to left-lateral, strike-slip faulting in the basement, which has also penetrated the salt beds of the Hormuz Series causing diapiric salt-wall structures at depth (Player 1969), which have uplifted the overlying strata in elongated anticlines.

Uplift of an elongated oil-field structure, in the Manifa Field, parallels the coast of northeastern Saudi Arabia and also exhibits a strong negative gravity anomaly. It is interpreted as originating from basement faulting along a NW trend (the Erythraean Trend of von Wissmann et al. 1942) which has also cut the Upper Precambrian salt beds of the Hormuz Series, allowing the lighter salt to move upwards as a broad salt wall, thus pushing up overlying Phanerozoic strata into a NW-SE elongated, doubly plunging, anticline. Only one, undeveloped, oil field at Hasbah/Farsi (Mina et al. 1967), near the Saudi/Iranian offshore border, is clearly developed along an E-W structural alignment (i.e. the Tethyan Trend, Henson 1951). This is the Hasbah structure, which is also an elongated doubly plunging anticline with a negative gravity anomaly (Player 1969), due to basement faulting and rupture of the Hormuz Series salt beds, causing deep-seated diapirism and consequent structural growth. A close examination of the gravity pattern, especially of the second derivative of the potential field of gravity, shows quite a number of basement elements aligned along the Tethyan Trend.

Four trends of basement faulting control the oil fields of Saudi Arabia (Henson 1951), most prominently the N-S Arabian Trend, as well as the N-E Aualitic Trend and the N-W Erythraean Trend in offshore diapiric fields. Least conspicuous is the E-W Tethyan Trend.

Geological Background

The oil-field areas of Saudi Arabia are all located in the northeast part of the Kingdom, both onshore and offshore, in a tectonostratigraphic province referred to as the Arabian Platform, or Unstable Shelf (Henson 1951). It is composed of a sedimentary sequence, which is virtually subhorizontal, except for a few basement-induced and diapiric folds (Greig 1958), comprising the Arabian Platform, and thickening northeast from 4,500m to 13,700m. To the southwest of the Arabian Platform, the NE-dipping, scarp-forming, Mesozoic and Paleozoic strata form the Interior Homocline, which rests nonconformably on the northeast edge of the Arabian Shield. Crystalline Precambrian igneous and metamorphic rocks, similar to those exposed in the Arabian Shield, underlie both the Interior Homocline and the Arabian Platform, where a few deep wells have cored crystalline basement.

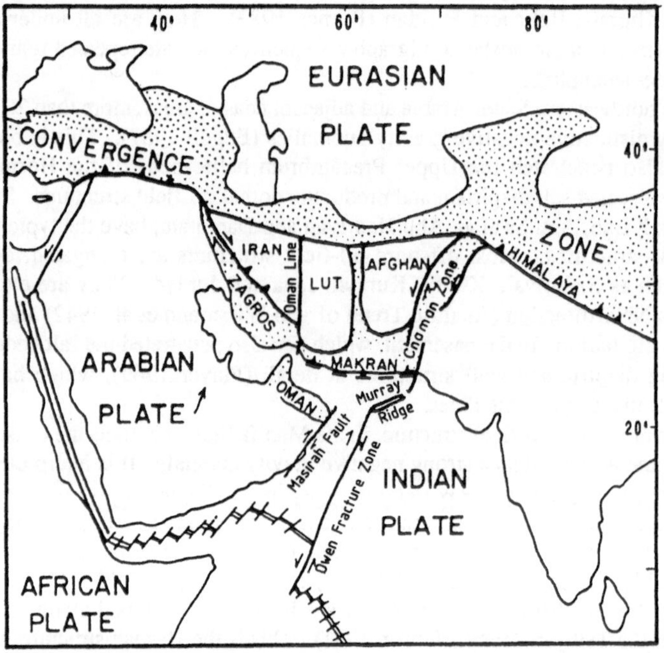

Figure 2. Plate Tectonic Setting of Arabia.

On the northeast side of the Persian Gulf, and especially in southern Iran and northeast Iraq, the Zagros Fold Belt (Falcon 1967, 1969), with row upon row of NE-SE trending anticlines forms a separate tectonostratigraphic province, to the northeast of the Arabian Platform. The Main Zagros Reverse Fault (Berberian 1976, 1981), sometimes called the Zagros Thrust, limits the Zagros Fold Belt along its northeast side and separates it from the Iranian Plate.

All these tectonic units (Fig. 2), from the opening Red Sea, across the Arabian Shield to the Interior Homocline, Arabian Platform, Zagros Fold Belt, Zagros Thrust, Iranian Plate, Convergence Zone, and Eurasian Plate, form part of a dynamic, plate tectonic system (Jackson et al. 1981). The separation of the Arabian Plate from Africa along the Red Sea, which is widening at 1.2 cm per year in a NNE direction (NASA personal communication), has propelled the Arabian Plate towards Iran, buckling, folding and faulting the Zagros Ranges until their sediments are subducted beneath the Iranian Plate along the Main Zagros Reverse Fault. There is every evidence that the Arabian Plate extends beneath southern Iran, as pointed out by Falcon (1967), Morris (1977) and Jackson et al. (1981).

Within the Arabian Platform (Fig. 3) there are very few fold structures paralleling the Zagros Fold Belt, although both structural zones are subparallel and elongated in a NW-SE direction. The few weak folds of the Iranides are seen in an indistinct ridge, joining Dammam-Bahrain and northern Qatar, and in some NW-SE folds in the Gulf of Bahrain and at Al 'Uqayr (Kassler 1973). Another NW-SE trending anticlinal fold along the northeastern coast of Saudi Arabia forms the Manifa Oil Field. However, the major fold structures in the Arabian Platform trend almost N-S along the Arabian Trend and are due to extensional faulting of the crystalline Precambrian basement, giving rise to an alternating system of horsts and grabens (Edgell 1987), which have pushed up overlying Phanerozoic strata. This has produced major elongated, doubly plunging, anticlinal oil fields, such as Ghawar (140 km long), Khurais Oil Field (75 km long), Qatif Oil Field (55 km long), and numerous other, N-S directed, oil fields of basement origin, such as Abu Jifan, Mazalij, Harmaliyah, Bakr, Juraybi'at, Wari'ah, El Haba, Faridah, Rimthan, Dibdibah, and Jaham. The recently discovered Permo-Carboniferous sand-reservoir fields at Al Hawtah, Dilam, Hilwah, Ragibh and Nu'ayyim also lie along the Arabian Trend, and form an exception in that they lie within the Interior Homocline, where Jurassic and Cretaceous strata are exposed. Outcrops in the Arabian Platform, on the other hand, are predominantly Upper Tertiary (Neogene) with minor areas of Lower Tertiary strata (U.S. Geological Survey 1963).

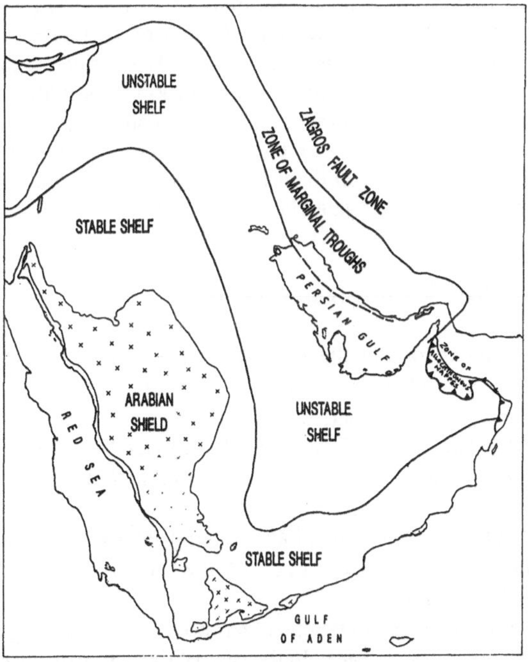

Figure 3. Major Tectonic Zones of Arabia (modified from Henson 1951).

In coastal parts of northeastern Saudi Arabia, and in the Saudi offshore portion of the Gulf, different structural trends and structural types exist, but are all basement related. A NE to NNE trend is clearly seen in major offshore oil field structures, such as Safaniya, Khafji, Marjan, Kurayn, Jana and Jurayd, as well as onshore in oil fields, such as Abqaiq and Khursaniyah. These fields are all related to faulting of the basement along a NE trend as seen in Kuh-e Namak (Talbot and Jarvis 1984). The underlying basement faults are considered to be left-lateral, strike-slip faults which have also cut the thick saliferous Precambrian Hormuz Series providing release for the lighter halite. As a consequence, deep-seated diapirism of the salt-wall type has pushed up overlying Phanerozoic strata producing elongated doubly plunging anticlines in the Saudi offshore and coast and forming major oil fields. In some cases, intersecting NE and NW basement faults have provided a focal point for salt diapirism and typical circular salt domes have been formed, as at Abu Hadriyah, Abu Sa'afah, Karan and Dammam.

Salt diapirism is also associated with some of the trends of oil-field structures, where deep-seated diapirism has been caused by basement faults. These include the old N-S Arabian Trend, as seen in the Berri Oil Field, the NW-SE Erythraen Trend as seen in the Manifa Oil Field, and the E-W Tethyan Trend as seen in the undeveloped Hasbah Oil Field.

Surface strata are mainly subhorizontal in the oil-field areas of northeastern Saudi Arabia, with only very gentle dips (<1°) on the flanks of anticlines. However, structural closure increases with depth in the Saudi oil field anticlines by two methods. Firstly, repeated rejuvenation of the major onshore fields has occurred by repeated intervals of basement uplift. These fields are marked by thinning, or truncation, over the fold axis, accompanied by a number of unconformities (Edgell 1987). The oil-field structures of coastal and offshore Saudi Arabia represent a second type and exhibit gradual growth as the deep-seated salt diapirs have slowly pushed up overlying strata. These diapiric structures are usually due to salt growth from Mid-Cretaceous to Recent and, in some cases, have started growth as far back as Late Jurassic. They originate from basement faults cutting the salt beds of the Hormuz Series, which are up to 2 km thick in the Gulf (Rönnlund and Koyi 1988; Player 1969). Intersecting basement faults may produce typical, circular, salt-dome structures, whereas single faults allow salt-wall diapirism to dome up more elongated structures. In general, oil fields with a deep-seated salt diapiric origin are either domes, such as Dammam, Abu Hadriyah, and Khursaniyah, or brachyanticlines, such as Berri, Manifa, Jana and Kurayn.

Nature of Basement in Northeastern Saudi Arabia

The evidence for the type of rocks forming the Precambrian basement under the sedimentary cover in the oil-field areas of northeastern Saudi Arabia is only direct in a few instances. This applies in the case of some deep wells which have been drilled to basement as at Jaham No. 2, El Haba No. 2, Haradh No. 51, Abu Jifan No. 23, Al Hawtah No. 1, Ain Dar No. 196 and Qirdi No. 5. All these wells have encountered crystalline igneous or metamorphic rocks, similar to those found in the Arabian Shield.

As can be extrapolated, these Upper Precambrian crystalline rocks dip beneath the sediments of the Interior Homocline and the Arabian Platform, so that they are found at depths of from 4,500 m to 13,700 m in the Saudi oil-field areas.

Much of the evidence regarding depth to basement has been deduced from geophysical surveys, primarily gravity and magnetic investigations. In the case of the newly discovered oil fields at Al Hawtah, Dilam, Raghib and Nu'ayyim, there is also direct seismic evidence of the depth to basement.

	System/Series	Stage	Formation	Generalized Lithological Description	Thickness Type section
		Quaternary	Surficial	Gravel, sand & silt	
CENOZOIC	MIOCENE and PLIOCENE	Pliocene	Kharj	Limestone, lacustrine, gypsum and gravel	28m
			Hofuf	Sandstone, marly & sandy limestone; subordinate calcareous sandstone. Gravel in lower part.	95m
		Miocene	Dam	Marl & shale, minor sandstone chalky limestone & coquina	91m
			Hadrukh	Sandstone, calcareous silty; sandy limestone; local chert	84m
	EOCENE	Lutetian	Damman	Limestone, dolomite, marl-shale	33m
		Ypresian	Rus	Marl, chalky limestone, & gypsum; common chert & geodes Mainly anhydrite in subsurface.	56m
	PALEOCENE	Thanetian Montian (?)	Umm er Radhuma	Limestone, dolomitic limestone & dolomite	243m
MESOZOIC	CRETACEOUS	Maastrichtian -Campanian	Aruma	Limestone; some dolomite & shale. Lower sands in NW & S.	142m
		Cenomanian -Albian	Wasia (Sakaka)	Sandstone; subordinate shale, rare dolomite lenses	42m
		Aptian	Shu' aiba	Limestone, calcarenite, dolomite	±100m
		Barrenian	Riyadh	Sandstone, subordinate shale	425m
		Hauterivian	Buwaib	Calcarenitic limestone, upper fine sandstone, minor marl	18m
		Valanginian	Yamama	Calcarenitic & aphanitic limestone & calcarenite	45m
		Berriasian	Sulaiy	Limestone, chalky aphanitic & calcareous limestone	170m
	JURASSIC	Tithonian	Hith	Anhydrite	90m
			Arab	Calcarenite, dolomite with anhydrite interbeds	124m
		Kimmeridgian	Jubaila	Limestone, aphanitic & calcarenitic. Some sand in S.	118m
			Hanifa	Limestone, aphanitic & calcarenitic	113m
		Oxfordian Callovian	Tuwaiq Mountain	Limestone, aphanitic, minor calcarenite. Upper corals	203m
		Bathonian Bajocian	Dhrusa	Limestone, aphanitic + shale Mainly sandstone in N and S	375m
		Toarcian	Marrat	Shale, dense limestone + sand	103m
	TRIASSIC	Upper	Minjur	Sandstone; some shale	315m
		Middle	Jilh	Sandstone, limestone, shale	326m
		Lower	Sudair	Shale, brick-red & green	116m
PALEOZ	PERMIAN	Upper	Khuff	Limestone, dolomite & shale	171m
		Lower	'Unayzah	Sandstone & fluvioglacials	33m
	CARBONIFEROUS	Lower	Berwath	Sandstone, minor shale (in N)	696m
	DEVONIAN	Lower	Jauf	Limestone, shale, some sand	299m
	SILURO-ORDOVICIAN		Tabuk	Sandstone & shale	1,072m
			Saq	Sandstone, reddish, quartzose	+600m
	UPPER PROTEROZOIC		Hormuz Series	Halite, interbedded with dolomite & shale	0m to 2,500m

Precambrian crystalline basement.

____ = unconformity

Table 1. Stratigraphic Succession of the Saudi Arabian Oil Field Areas.

Sedimentary Cover

The sedimentary succession in the oil-fields area of northeastern Saudi Arabia can be generally described as a Lower Paleozoic clastic supergroup, unconformably overlain by a Permian to Lower Tertiary carbonate supergroup, which is in turn disconformably overlain by an Upper Tertiary (Neogene) clastic group.

In descending stratigraphic sequence, the typical units of sedimentary cover overlying the Precambrian crystalline basement in Saudi oil-field areas (Powers et al. 1966; Powers 1968) are as shown in Table 1.

These stratigraphic units comprise the normal stratigraphic sequence in most of the oil-field areas of Saudi Arabia. However, there are lateral variations in lithology and a considerable thickening of most of these units towards the northeast. In addition, the Upper Proterozoic salt beds of the Hormuz Series are found in coastal northeastern Saudi Arabia (Greig 1958) and in the Saudi Arabian offshore (Fig. 4) but do not occur in major onshore fields, such as Ghawar and Khurais, where the Saq Sandstone rests directly on Precambrian crystalline basement.

A striking feature of sedimentation in the Arabian Platform is the very extensive lateral persistence of many formations over distances of up to several thousand kilometres. The lateral continuity of these blanket lithosomes, such as the Aruma, Wasia, Arab and Khuff formations (Edgell 1977, 1987), is noticeable in a NW-SE direction paralleling the Gulf, but across the Gulf, as seen in Iran, almost all these units are replaced by different lithostratigraphic units (James and Wynd 1965). Many unconformities, or disconformities, occur within the stratigraphic succession of the Saudi oil-fields areas. One of the most pronounced of these is the sub-Unayzah (or sub-Haushi) unconformity (Powers et al. 1966; Al-Laboun 1988), which is of regional extent, affecting not only the Arabian Platform but also the Mobile Belt of northeast Iraq and southern

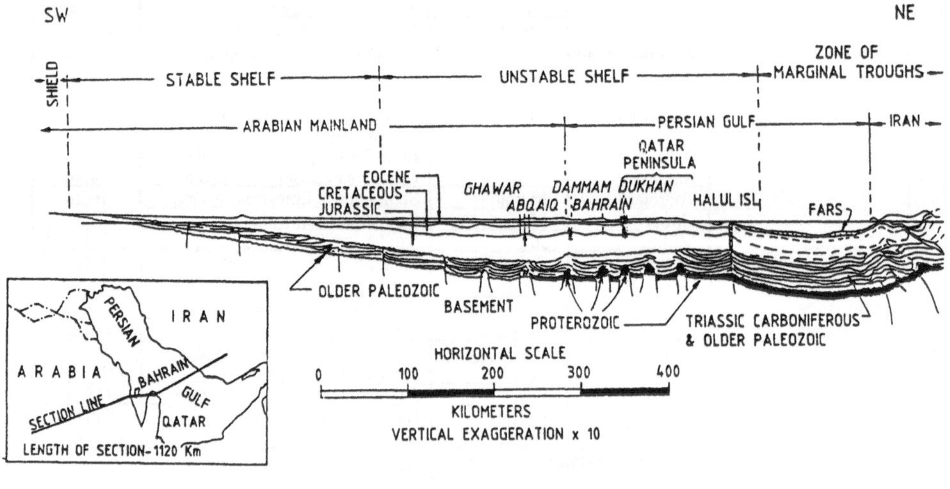

Figure 4. Geological Cross Section of the Persian Gulf Basin with halokinetic Proterozoic Hormuz Series.

Iran. A second most significant unconformity, or nonconformity, exists between the crystalline Precambrian basement rocks and the overlying sedimentary sequence. There is evidence to suggest that the basement was already faulted along the Arabian Trend in the Late Precambrian, so that the Proterozoic Hormuz Series sediments were deposited in a series of N-S troughs, between uplifts or headlands. Repeated rejuvenation of these uplifts, or basement horsts, particularly along the N-S Arabian Trend, is reflected in the pattern of deposition throughout the Phanerozoic. It is seen in many formations, such as the Arab Formation, which has oolitic facies in the Arab C and D member over these highs (Steineke et al. 1958), and in the Rus Formation, which is a dolomite over basement highs and a thicker anhydrite-marl sequence in the intervening troughs.

Structure of Basement in Saudi Oil-Field Areas

Most of the deductions concerning the structure of basement in the Saudi Arabian oil-field areas have been based on interpretations of gravity data (Aramco 1959; Barnes 1987). In areas near the Arabian Shield, as with the new oil fields at Al Hawtah, Dilam, Raghib and Nu'ayyim, the shallow basement can be interpreted by seismic profiles.

The dominant structure of crystalline Precambrian basement is a series of N-S trending basement horsts with intervening graben structures. These are due to E-W extensional tectonics and the pattern is so regular that Khatieb and Norman (1982) have suggested a series of basement-induced lineaments with a systematic spacing of about 40 km. It is improbable that basement faulting is so precisely regular, and many of the basement uplifts are not simple horsts but complex horst blocks with step faults on either side, as is the case with Ghawar. Similarly, the troughs in the basement are quite broad and also appear to be complex, as with the Central Basin to the west of Ghawar. These horst and graben structures do not trend exactly N-S but mainly in a N17° E direction, which Henson (1951) noted as the main alignment of the Arabian Trend. The structural relief of basement horst blocks is interpreted as being up to 4.5 km, with regard to adjacent basement troughs. It is not implied that such horsts were originally so elevated, or were suddenly uplifted by such a magnitude. Instead, it is considered that repeated reactivation along the original faults bordering the basement horst blocks has accounted for most of the structural growth, as can indeed be demonstrated by thinning of overlying formations onto the axes of the uplifts (Sugden 1962) and by truncation of strata in their axial areas.

The most pronounced of these basement uplifts has produced the En Nala Axis (Steineke et al. 1958), extending 280 km from Haradh through the greater Ghawar Anticline to Fazran, Fadhili and near Khursaniyah. From west to east across the Arabian Platform, there is a succession of basement uplifts (Fig. 5) and intervening basins (Edgell 1987) including:

1) Hail-Rutbah Arch
2) Widyan Basin
3) Khufaisah-Mubayhis Uplift
4) Jaham-Ma'aqala-Faridah-Wariah Uplift
5) Dibdibah Trough
6) Burgan-Wafra-Juraybi'at-Khurais Uplift
7) Central Basin
8) En Nala Axis Uplift
9) Gulf of Salwa Depression
10) Qatar Arch
11) Duwayhin Depression

12) Matti Uplift
13) Nahaidin Anticlinal Uplift
14) Mushash Anticlinal Uplift
15) Bu Hasa-Kidan Anticlinal Trend

The predominance of SW trending, left-lateral basement shears, already well documented in southern Iran (Fürst 1970), has given rise to a number of NE trending anticlinal structures in the Saudi Arabian offshore and the adjacent coastal areas. Such structures have distinct negative gravity anomalies caused by the basement faults, cutting through the salt beds of the Proterozoic Hormuz Series and triggering salt diapirism, which has updomed overlying strata (Player 1969). Some examples of oil-field structures along this NE-SW basement shear trend are Abqaiq, Kurayan, Jana, Jurayd, Safaniya and Khafji. This trend has been interpreted by Hancock and Bevan (1987) as due to extension, which does not fit the regional tectonic pattern.

Basement faulting parallel to the shoreline of northeastern Saudi Arabia has also transected the Hormuz Salt Series causing diapiric updoming, as in the Manifa and Abu Hadriyah oil fields. It is thought that basement faults along this NW trend are right-lateral, strike-slip faults, due to offsetting of the median gravity low in the Safaniya Anticline, and also parallel to a right-lateral component on the Main Zagros Reverse Fault (Berberian 1976, 1981). An E-W basement fault trend is less clearly seen in the oil fields of offshore Saudi Arabia and appears as alignments of gravity highs along this direction, both onshore and offshore. Only one undeveloped oil field, at

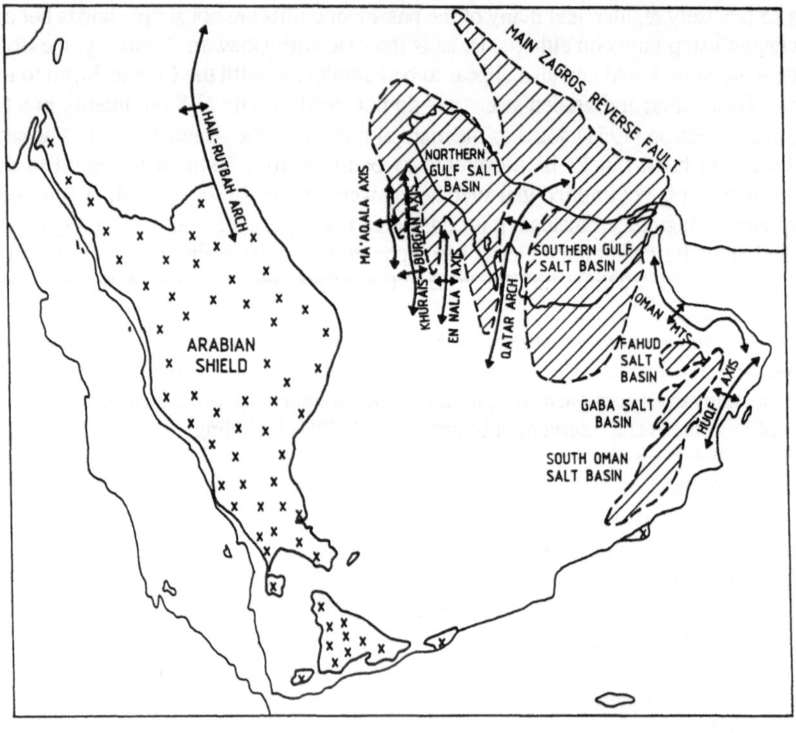

Figure 5. Late Precambrian Salt Basins of the Persian Gulf and Arabian Peninsula.

Hasbah, shows this E-W Tethyan Trend, and is attributed to basement faulting cutting the Hormuz Series salt beds and causing salt wall diapirism, which has domed up overlying strata forming an elongated anticline. This E-W basement fault is clearly seen in the Central Arabian Graben System (Al-Kadhi and Hancock 1980; Hancock and Al-Kadhi 1985; Hancock and Bevan 1987), which continues eastward to truncate the southern end of the En Nala Anticlinal Axis (Fig. 11).

Intersection of basement faults in offshore and coastal northeastern Saudi Arabia has permitted salt diapirism from the Hormuz Series along pipe-like pathways causing circular, salt-dome oil fields, as seen at Abu Sa'afah, Abu Hadriyah, Dammam and Karan. In the case of the Abu Sa'afah Oil Field, the second derivative of the potential field of gravity indicates deep basement faults intersecting along NW-SE and N-S directions. The Dammam Dome contains a typical keystone graben (Fayyad 1966) trending NW and also lies on a basement fault trend.

An indirect method of interpreting deep salt diapirism in the formation of oil-field anticlines has

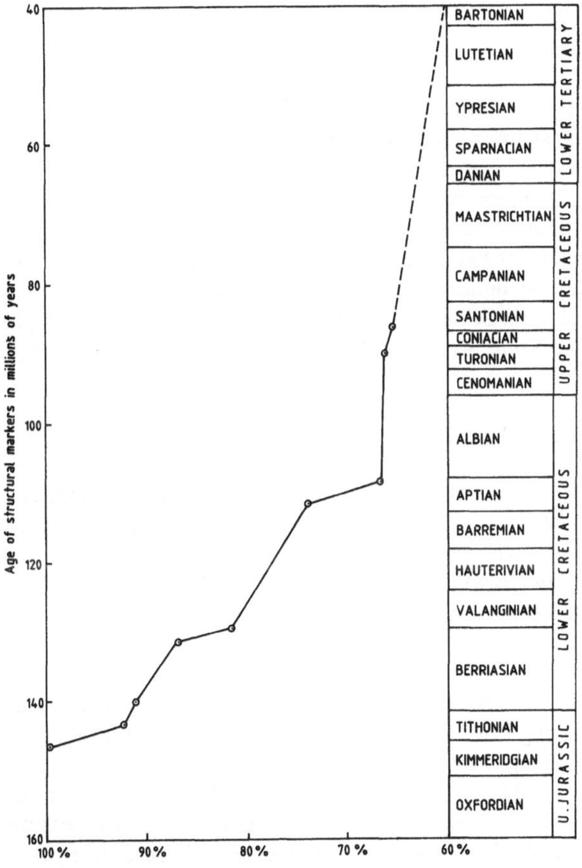

Figure 6. Continuous Structural Growth of the Jawb Anticline from Late Jurassic to Tertiary due to Halokinesis (modified from Benzakour 1986).

been proven by Sugden (1962) for the Dukhan Oil Field by comparing thicknesses of stratal units in flank and crest wells. This shows a steady growth of the Dukhan Anticline with time from Mid Cretaceous to Miocene, which is consistent with slow growth of a basement-induced salt-wall diapir and not with an irregular basement uplift. A similar steady growth curve (Fig. 6) has been shown for the Jawb Oil Field (Benzakour 1986), proving that it is deep-seated salt diapir induced by basement faulting along the Arabian Trend. Abu Hadriyah and Khursaniyah are known to show similar steady growth curves, as also Dukhan Oil Field (Sugden 1962).

As stated by North (1985) 'some of the great productive oil regions of the world are characterized by traps due either to basement horst-block uplift, halokinesis (deep-seated salt movement) or a combination of the two. In the salt basin of the Persian Gulf, the great domes of Burgan, Bahrain, and Dukhan form obvious surface features'. All these salt-dome oil fields lie in the Northern Gulf Salt Basin (Edgell 1987; Husseini 1989) of the Arabian Platform, adjacent to northeastern Saudi Arabia.

Geophysical Evidence of Basement Tectonics

Most of the data on geophysics of the basement underlying the Arabian Platform in the Saudi oil-field areas is confidential and restricted by major oil companies, such as Saudi Aramco and the Arabian Oil Company.

A small amount of information has been published, such as on Ghawar (Aramco 1959), and more recently on the onshore northeastern Saudi Arabian oil-field areas by Barnes (1987). A report by Player (1969) contains some generalized gravity information on the Persian Gulf, as also in Kassler (1973). In addition, limited information has been released in open meetings, such as the recent SPE-Dhahran Geological Society Meeting (26th-27th May 1990). Geophysical evidence of basement tectonics in the Saudi Arabian oil-fields areas is from seismic, magnetic or gravity methods.

SEISMIC EVIDENCE OF BASEMENT TECTONICS

Until recently, it has been difficult, if not impossible, to carry seismic reflection right through the overlying strata to detect basement structure. Saudi Aramco have now improved seismic shooting and recording techniques, as well as processing, so that they can now obtain data as deep as six seconds two way return. In a NNW-SSE seismic profile across the western Rub' al Khali from Ash Shuqqan to 'Uruq al Mawarid, Dyer et al. (1990) have shown the deep, extensional block faulting of Precambrian, crystalline basement, underlying the Phanerozoic strata in this area. Similar basement has been shown in the eastern Rub' al Khali (Aigner 1989; Aley and Nash 1985). These block faults trend NNE and have been repeatedly reactivated, so that most of them extend through the overlying Lower Paleozoic, while a few even cut Lower Cretaceous strata. This extensional block faulting, with grabens flanked by step-faulted horsts, seems typical of the basement in Saudi Arabia and is seen again in the Dilam, Al Hawtah, Raghib, and Nu'ayyim oil fields area, where these fields are basically underlain by sub-'Unayzah Unconformity block uplifts along the N-S Arabian Trend. Faults bordering these block uplifts are clearly defined by seismic profiles, together with rollover to the west in the overlying Permo-Carboniferous strata, induced by further rejuvenation of block uplift.

MAGNETIC EVIDENCE OF BASEMENT TECTONICS

Data from airborne magnetometer traverses now covers almost all of Saudi Arabia. In the oil-fields areas of the Kingdom, such data shows the predominant N-S Arabian Trend in the Ghawar, Mazalij and Abu Jifan fields. It also shows the basement highs which underlie the newly discovered fields at Al Hawtah, Dilam, Hilwah, Raghib and Nu'ayyim. The map of aeromagnetic anomalies for northeastern Saudi Arabia is most useful in indicating variations in the magnetic attraction of basement due to different lithological types in these crystalline Precambrian rocks. In particular, subcircular plutonic intrusions in the basement are distinct.

A broad, N trending belt of magnetic highs, about 100 km wide, lies between about longitude 45°E and longitude 46°30'E. It is considered by Stewart (1990) to be part of a major Late Proterozoic island arc in the basement rocks and may represent the substratal continuation of the Al Amar-Al 'Idsas area, where ophiolites and ultrabasic rocks mark on old plate margin (Stacey and Hedge 1983, Sustras 1980, Davies 1984).

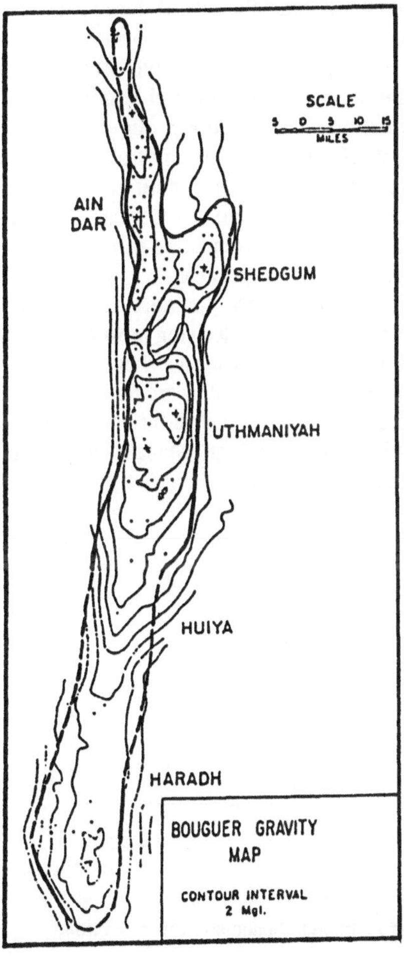

Figure 7. Bouquer Gravity Map of Ghawar Anticline and Comparison with Field Outline (after Aramco 1959).

GRAVITY EVIDENCE OF BASEMENT TECTONICS

The most definitive evidence of the structure of basement in the Saudi Arabian oil-fields areas has been provided by detailed measurements of the potential field of gravity and the enhancement of this data into various types of map. The earliest of these maps, published by Aramco Staff (1959), was a Bouguer Gravity Map of the greater Ghawar Anticline contoured at 2 milligal intervals (Fig. 7). This map shows the outline of the field distinctly coinciding with a positive gravity anomaly, which is clearly comparable with a structure contour map on the top Arab-D Member for the same oil field. From this survey, it was deduced that the Ghawar Oil Field was underlain by an upfaulted basement block, or horst (Fig. 8), which has updomed overlying strata.

A considerably more detailed analysis of gravity data for onshore oil-field areas of northeastern Saudi Arabia has been presented by Barnes (1987), based on observations at one kilometre intervals, later resampled and interpolated into a grid spacing of 0.5 km. A Bouguer Gravity map prepared at 1 milligal intervals already shows the broad outlines of the Ghawar and Khurais oil fields (Fig. 9). Image enhancement processes that approximate the second derivative were used to reduce the dominant, northerly Arabian Trend and to show subtle cross trends. One of the most effective of these was the running difference method, in which the value of a grid point was compared with the weighted average for a square region of 11 x 11 grid points. Local gravity maxima thus appear as dark areas and the correlation with the known outline of Ghawar and Khurais oil field is striking, (Fig. 10). In addition, many basement fault trends can be deduced from the running difference map either by optical filtering (Ronchi grating) or by performing Fourier filtering. Some of the major basement trends seen in this gravity map are N 5° E, N 20°E, N 70°E, N 25°W, and N 80°W. Of course, the N and NNE trends are most striking, but the presence of basement faults along the other trends is clearly recognizable.

The running difference method and the second derivative of the potential field of gravity both produce maps which clearly define the structure of basement. Underlying most onshore Saudi oil fields there are distinct positive gravity anomalies which almost mirror the shape of these anticlinal structures. The reasons for these underlying positive gravity anomalies are basement horst uplifts which have been repeatedly reactivated. Onshore oil fields which exhibit underlying basement uplift include the giant Ghawar Field, as well as Khurais, Abu Jifan, Mazalij, Jurabiy'at, Jaham, El Haba, Wariah, Rimthan, Dibdibah, Bakr, Fazran and Harmaliyah.

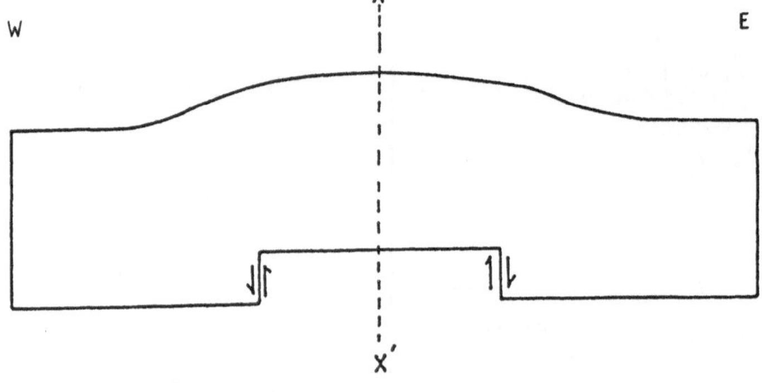

Figure 8. Uplifted Basement Block Updoming Ghawar Anticline; Schematic Cross-Sectional Representation (after Aramco 1959).

In the coastal area of northeastern Saudi Arabia and in the Saudi offshore, there are a number of circular oil-field structures and brachyanticlines (Fig. 1) which cannot be explained, except as salt domes induced by intersecting basement faults transecting the Hormuz Series salt beds and releasing diapiric salt. Greig (1958) realised that basement faults provided a mechanism for deep salt diapirism, as shown in his NE-SW structural section across the Persian Gulf (Fig. 4). The subcircular Dammam Dome is clearly shown as a salt dome, while its shape, and the NW-SE keystone graben across its center, are clear evidence that it is a salt dome (Fayyad 1966). It is also a distinct negative gravity anomaly, as is also shown in the nearby Bahrain Dome and Dukhan Anticline (Edgell, 1987).

Abu Hadriyah Oil Field, in coastal northeastern Saudi Arabia, also shows a very similar, subcircular shape to the Dammam Dome and is believed to be another diapiric structure induced by intersecting basement faulting which has allowed deep-seated upward movement of the lighter salt beds of the Upper Precambrian Hormuz Series. Khursaniyah and Fadhili oil fields are similarly domal and are difficult to explain by uplift of basement blocks. They are also considered to be due to deep-seated salt diapirism. Fuwaris and Wafra fields, in the Saudi-Kuwait Neutral Zone, also show a similar structure and are clearly salt structures, as is Burgan Oil Field (Fox 1956, North 1985).

In the Saudi offshore area of the Persian Gulf, there are a number of subcircular oil fields which almost certainly have a diapiric origin by upward movement of Hormuz Series halite, triggered by intersecting basement faults. Amongst these are Karan, Harqus, Lawhah, Abu Sa'afah, Ribyan and Maharah, as well as the undeveloped Arabiyah structure.

There are also a number of brachyanticlines in the offshore area with a slightly elongated form, similar to the Bahrain Dome. These are all thought to be due to halokinesis, probably as deep-seated salt-wall structures produced by basement faults. Offshore Saudi oil fields with this type of brachyanticlinal structure include Jana, Jurayd, Kurayn, Safaniya, Khafji, Marjan/Fereidoon (Iran), Manifa and Zuluf, as well as Berri with its bulbous north end, which is mostly offshore.

In contrast to the larger onshore oil fields of Saudi Arabia, those in the offshore area and near the coast are almost exclusively basement fault-induced, deep-seated, diapiric salt structures, and can be shown as gravity negative features by examining the gravity residuals after removing the regional gravity gradient. These fields all fall within the Northern Gulf Salt Basin (Edgell 1987; Husseini 1989). This northern Proterozoic salt basin of the Hormuz Series (Fig. 5) also extends onshore, especially in Kuwait, southernmost Iraq, coastal northwestern Saudi Arabia and extends regionally into southern Iran, between Kuh-e Namak and Bushire. It also connects through Fars Province (Iran) with the Southern Gulf Salt Basin, which covers southern Iran, the U.A.E. and most of the southern Persian Gulf as far east as the Straight of Hormuz. The persistence of old Precambrian basement uplifts, such as the Qatar Arch, the En Nala Axis and the Kuwait-Wafra-Khurais Axis are often not realized. It can be shown that the Northern Gulf Salt Basin extended as an embayment in the Late Proterozoic to the south of Bahrain, along the trend of the present Gulf of Salwa, reaching as far as 23° N in the Jawb and Lughfah oil fields. A study of the Jawb brachyanticline has been made by Benzakour (1986) on the basis of detailed well-log correlation. Using the classical method of Sugden (1962) and comparing flank and crestal wells, Benzakour (1986) has demonstrated conclusively that the Jawb Structure exhibited continuous steady growth from Mid Cretaceous to the present (Fig. 6), as is characteristic of the Dukhan salt-wall diapiric structure (Player 1969).

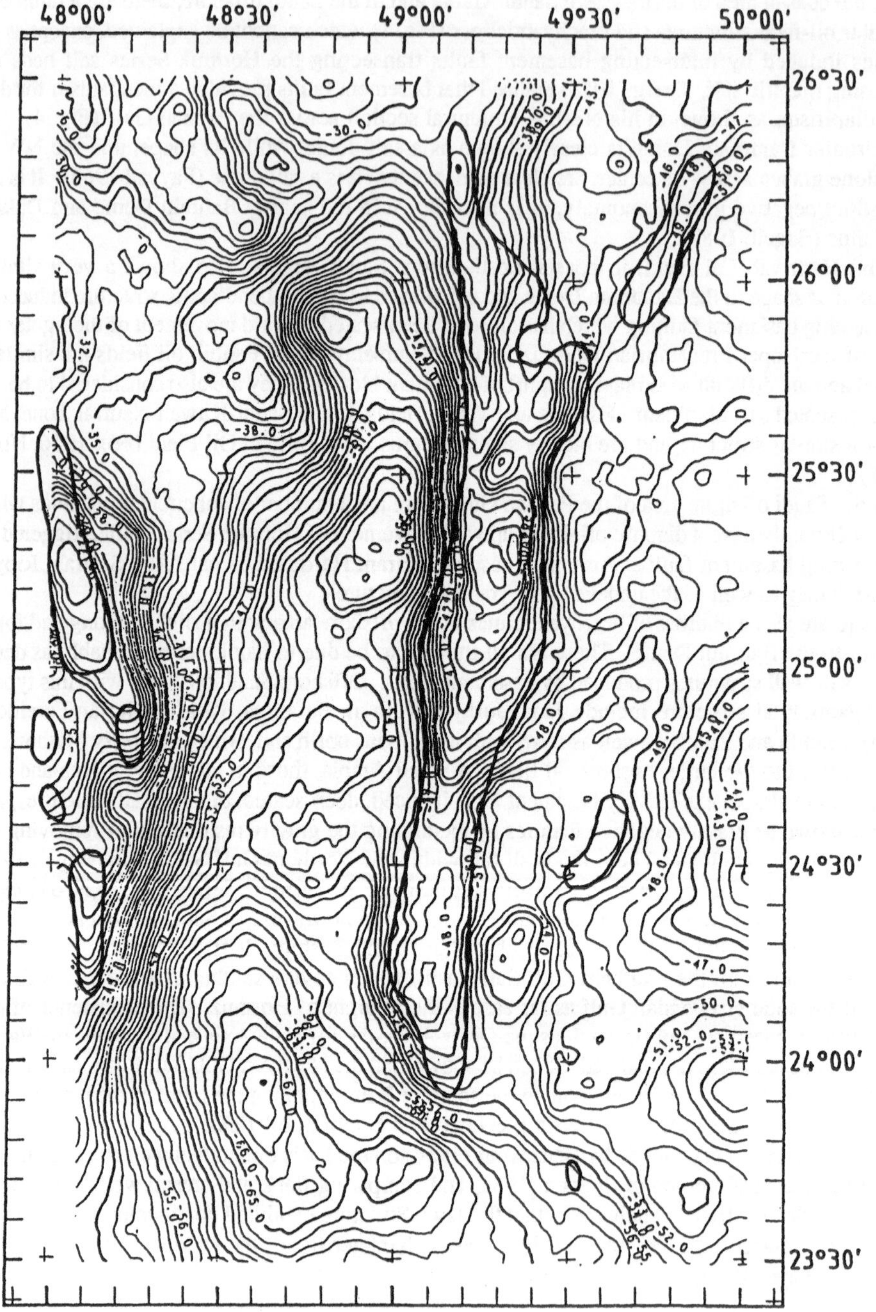

Figure 9. Bouguer Gravity Map of Main Saudi Onshore Oil Fields (1 milligal intervals). Oil fields outlined by solid lines, modified from Barnes 1987.

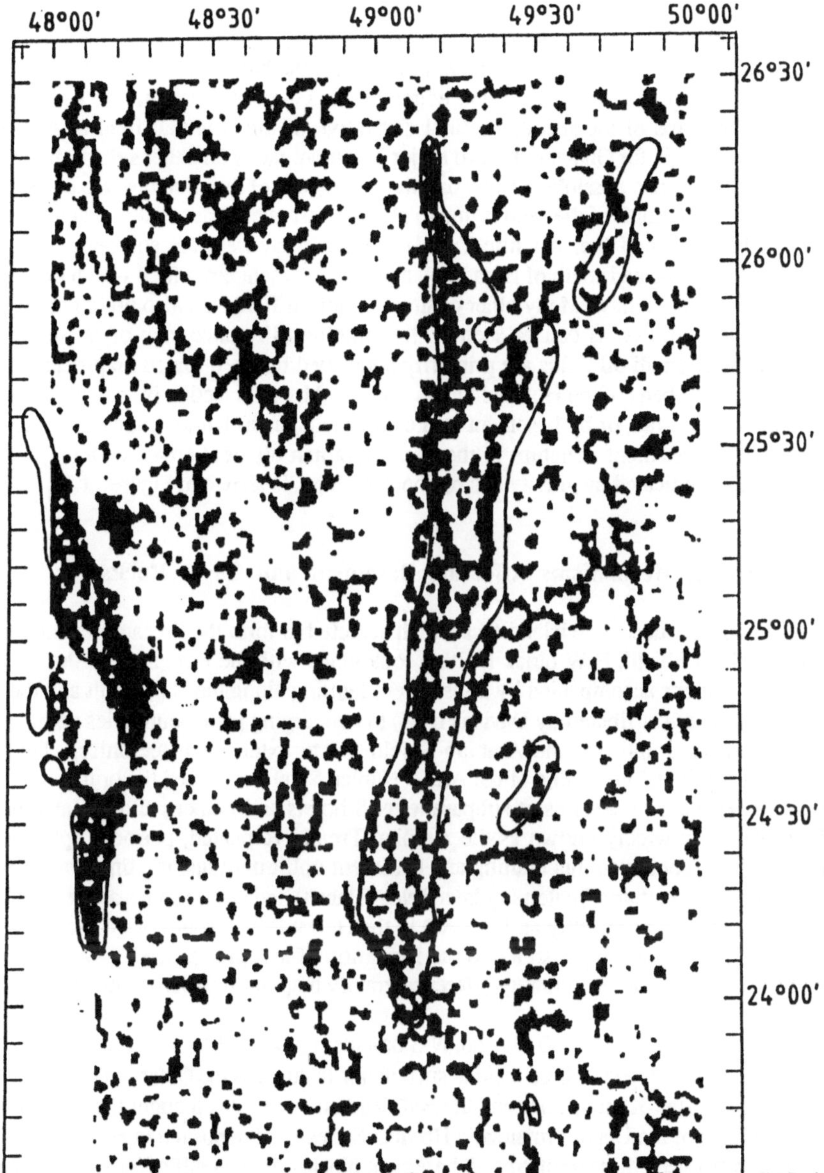

Figure 10. Gravity Map-Running Difference (digitized) of Main Saudi Onshore Oil Fields. Oil fields outlined by solid lines, modified from Barnes 1987.

Oil-Field Structures and Basement Tectonics

Major Saudi Arabian oil-field structures, such as the Ghawar Anticline, show a very clear relationship to a complex underlying basement uplift along the N-S Arabian Trend. On running difference gravity maps (Barnes 1987), the edges of the Ghawar Anticline are quite sharply delineated. When maps of the second vertical derivative of the potential field of gravity are prepared on a larger scale for Ghawar, the two major basement faults on either side of the anticline are clearly shown. It is also possible to show that the divergence of the structure into the Shedgum and Ain Dar axes at the north of the field is due to oblique basement cross-faulting along the NNE left-lateral, strike-slip trend (i.e. Aualitic Trend). In a similar way, the southern edge of the Ghawar Anticline is truncated south of Haradh by a major basement cross fault which trends NW-SE (i.e. the Erythraean Trend). Many other onshore Saudi oil fields are also underlain by fault-controlled basement uplifts, as can be seen on the running difference map of Barnes (1987). Offshore and coastal Saudi oil fields are primarily controlled by deep-seated diapirism from the Upper Proterozoic salt beds of the Hormuz Series, triggered by intersecting basement faults, as in Dammam and Abu Sa'afah, or by basement strike-slip faults, as in Kurayn, Jana and Jurayd. A provisional map of basement structure in the Saudi Arabian oil field areas, with emphasis on basement fault lines indicated by gravity and regional geology, is shown in Figure 11.

Structure Forming Mechanisms Affecting Basement and Cover Rocks

After the formation of successive island arcs which accreted to form the Precambrian crystalline rocks of the Arabian Shield, E-W block faulting took place with the formation of the collapsed Central Arabian grabens accompanied by transcurrent faulting along the Najd Fault and the Wadi Fatima Fault. These are attributed by Davies (1984) to converging plate boundaries up to 700 Ma ago. Late Proterozoic uplift of the Arabian Shield led to extensional tectonics. This E-W extension, due to the initial updoming of the Arabo-Nubian Shield, stretched the brittle crystalline, Precambrian rocks, causing a series of subparallel N-S horsts and grabens and initiated the 'old grain of Arabia', now widely known as the Arabian Trend. Repeated, periodic uplifts of the Arabian Shield have continued this faulting and basement uplift until historic times, as shown by the N-S trending Late Tertiary-Quaternary lava fields (Harrat) and cinder cones along fractures in the shield, with eruption as recently as 1250 AD in the Harrat Rahat, south of Al Madinah. This uparching of the shield probably also led to the formation of less pronounced, E-W fault trends, which can be seen clearly on the running difference gravity map of northern onshore Saudi Arabia (Barnes 1987).

By the Late Proterozoic, about 600 Ma ago, N-S basement block uplifts with intervening grabens had already developed. These include such major features as the Qatar Arch, the En Nala Axis, the Burgan-Wafra, -Khurais Axis and the Hail-Rutbah Arch. It is thought that evaporites, dolomites and shales of the Upper Proterozoic Hormuz Series were deposited in embayments along the troughs between these uplifted N-S basement blocks. In the Gulf of Salwa area, and onshore to at least 23°N latitude, it can be shown that Hormuz Series evaporites extend between the eastern edge of the En Nala Axis and the western side of the Qatar Arch, just east of Dukhan Anticline. It is thought that similar, though smaller, embayments of the Hormuz Series exist in the

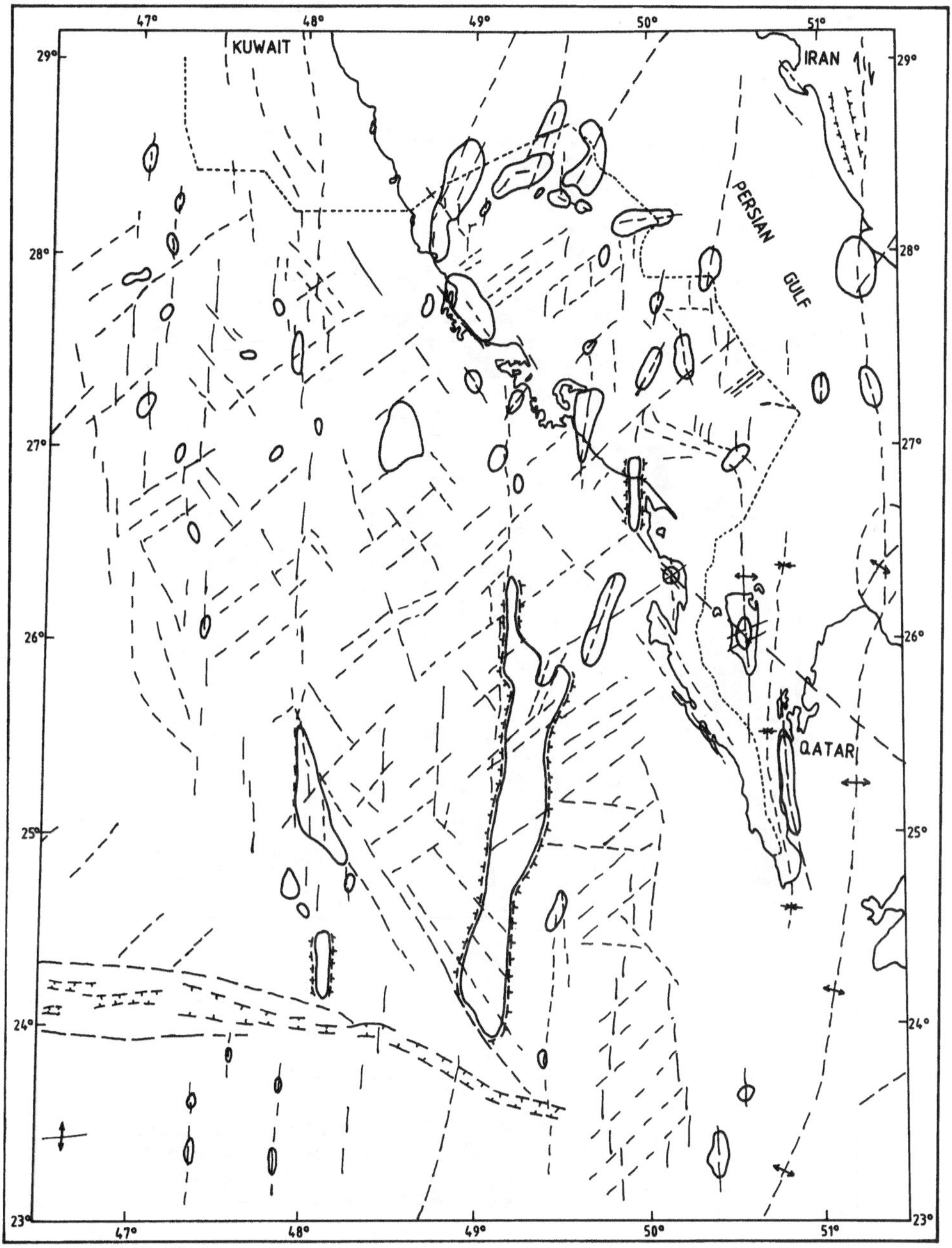

Figure 11. Basement Fault Structures in the Saudi Arabian Oil Fields and Adjacent Areas. Oil fields outlined by solid lines.

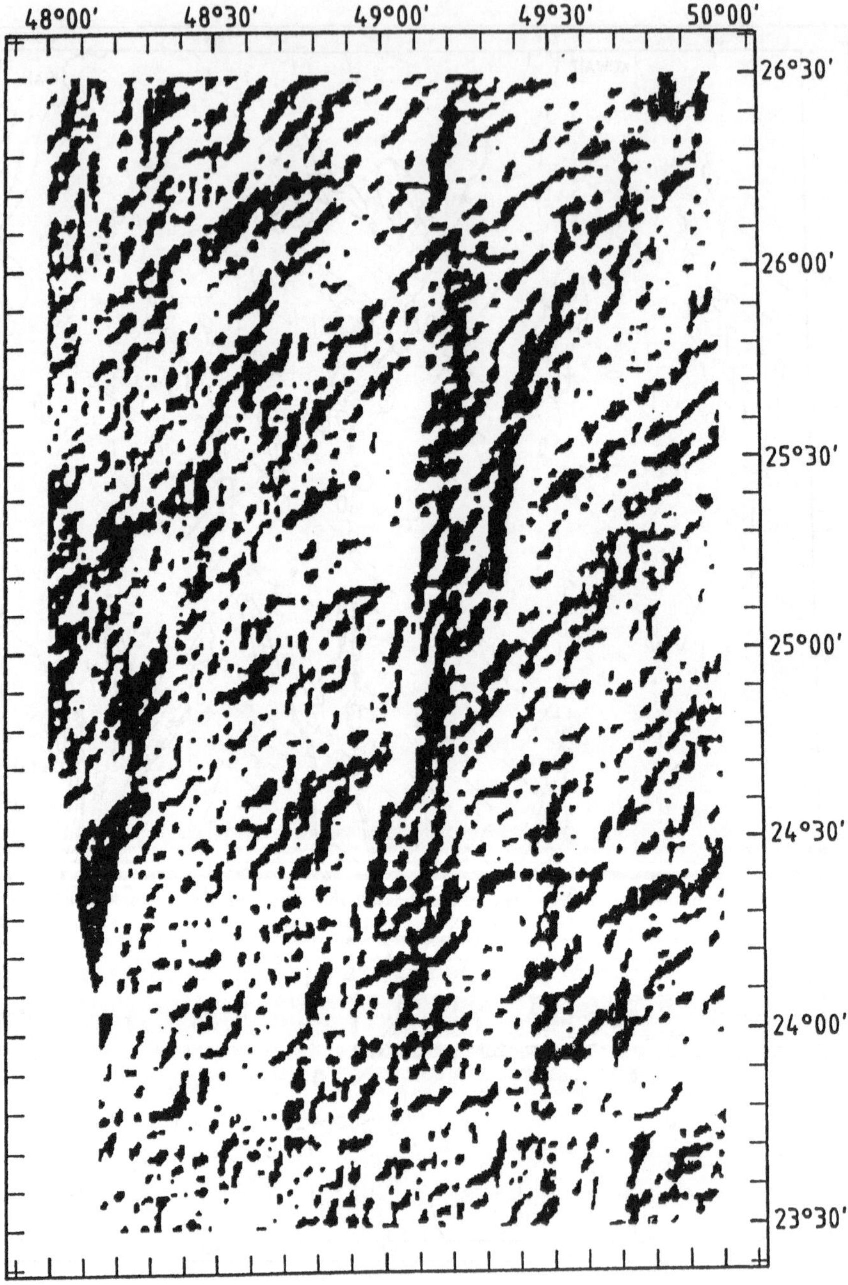

Figure 12. Computer Enhancement of Gravity Residuals in Azimuths from 15° to 75°, Indicating NE Basement Fault Trends (modified from Barnes 1987).

northern part of the Central Basin and in the Dibdibah Trough.

The significance of the extent of the Hormuz Series in the Northern Gulf Salt Basin is that it underlies all offshore Saudi Arabian oil fields, and many onshore ones. The Upper Proterozoic Hormuz Series consists largely of bedded halite and these salt beds are cut by many basement faults, triggering salt diapirism as the major structure-forming mechanism.

During the Mesozoic, and especially in the Triassic and Late Cretaceous, and even probably during the Permian, the N-S uplifts and basins of the Arabian Trend were intermittently reactivated, as seen in the Ghawar Anticline, Dibdibah Trough and Qatar Arch. There is even evidence in the Ma'aqala Arch, and by truncation of the Lower Eocene, Rus Formation in Ghawar, that tectonism along the Arabian Trend extended to at least Mid Tertiary times. The prominence of the Hofuf Formation, and its topographic expression over the Ghawar Anticline, indicates that intermittent reactivation of these N-S trends probably continued until the end of the Pliocene. With

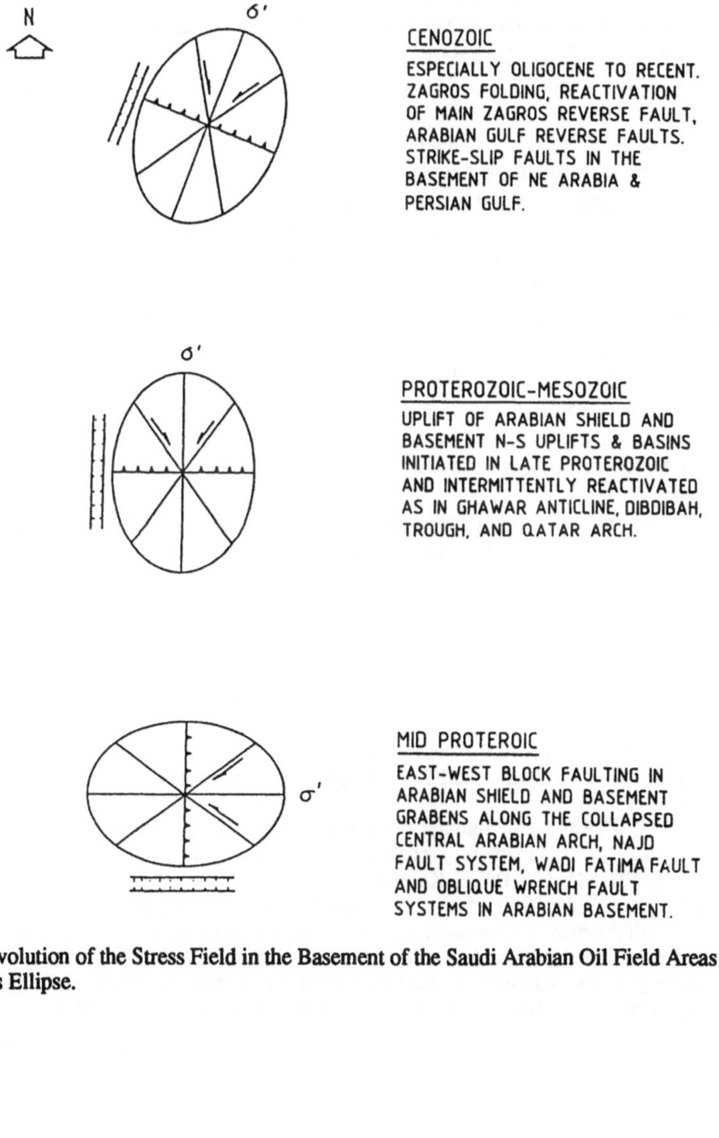
N
σ'
CENOZOIC
ESPECIALLY OLIGOCENE TO RECENT. ZAGROS FOLDING, REACTIVATION OF MAIN ZAGROS REVERSE FAULT, ARABIAN GULF REVERSE FAULTS. STRIKE-SLIP FAULTS IN THE BASEMENT OF NE ARABIA & PERSIAN GULF.
σ'
PROTEROZOIC-MESOZOIC
UPLIFT OF ARABIAN SHIELD AND BASEMENT N-S UPLIFTS & BASINS INITIATED IN LATE PROTEROZOIC AND INTERMITTENTLY REACTIVATED AS IN GHAWAR ANTICLINE, DIBDIBAH, TROUGH, AND QATAR ARCH.
σ'
MID PROTEROIC
EAST-WEST BLOCK FAULTING IN ARABIAN SHIELD AND BASEMENT GRABENS ALONG THE COLLAPSED CENTRAL ARABIAN ARCH, NAJD FAULT SYSTEM, WADI FATIMA FAULT AND OBLIQUE WRENCH FAULT SYSTEMS IN ARABIAN BASEMENT.

Figure 13. Evolution of the Stress Field in the Basement of the Saudi Arabian Oil Field Areas as Indicated by the Stress Ellipse.

the NNE drift of the Arabian Plate, and its separation from the African Plate from Oligocene onward (McKenzie 1972), a new set of stress conditions prevailed in basement beneath the Saudi oil fields. In this movement, the Arabian Plate was not only pushed north-northeast but was also stressed in a shear couple formed by the left-lateral, Aqaba- Dead Sea Fault (Freund et al. 1970) and the right-lateral Masirah Fault and Owen Fracture Zone (Jackson et al. 1981).

With the greatest principal-stress axis directed NNE towards the Zagros Ranges of southern Iran, two major vertical shears were produced at about N 36° E and N 20° W. The NW direction of vertical shear stress has produced right-lateral, strike-slip faults in the basement parallel to the shoreline of northeastern Saudi Arabia, and roughly parallel to the Najd Fault Zone (Agar 1985) and the axis of the Red Sea. This basement fault trend offsets the axial trace of the Safaniya Anticline in a right-lateral fashion, as shown from running difference gravity.

It is also thought to underlie the Manifa Oil Field, which is a broad, doubly plunging anticline with a NW-SE trend. The Manifa Anticline is also considered to be caused by basement wrench faulting, cutting the Hormuz Series salt beds and giving rise to deep-seated diapirism which has pushed up overlying strata. It is said that the Manifa structure lies near a regional gravity high, but when the regional gradient of gravity is removed it appears as a broad, negative gravity feature.

A whole series of left-lateral, strike-slip faults were also produced in the basement along the N 36° E trend (i.e Aualitic Trend), as can be clearly seen in southern Iran, where rows of Hormuz Series salt-piercement structures are aligned along this trend (Fürst 1970). These transcurrent basement faults extend southwestward beneath the Persian Gulf and even under the main oil fields of Saudi Arabia, as shown by computer enhancement of gravity residuals in azimuths from 15° to 75° (Fig. 12) by Barnes (1987). In many of the Saudi offshore oil fields, this series of left-lateral basement faults is also evident and has extended up into the Hormuz Series salt beds, causing salt-wall type halokinesis and pushing up overlying strata to form doubly-plunging oil-field anticlines. Some of the Saudi oil fields which have this type of origin are Kurayan, Khafji, Jana, Jurayd and Safaniya. Most probably, the large Abqaiq Field is also of this type, and by removal of the outermost structure contours (Levorsen 1954) it can be seen that the majority of closure on the Arab-D (>1000 ft) forms a distinctive, NE trending brachyanticline, which is difficult to explain except as a deep-seated salt diapir induced by this NE-SW left-lateral, basement wrench-fault trend.

An interpretation of the evolution of the stress field in northeastern Saudi Arabia and the Persian Gulf from Late Proterozoic to Late Cenozoic is shown in Fig. 13 by the changing stress ellipse.

References

Agar, R.A. 1985. The Najd fault system revisited; a two-way strike-slip orogen. Journal of Structural Geology 9, 41-48.

Aigner, T., & van Vliet, A. 1990. Hierarchy of Depositional Cycles in Epeiric Carbonate Basin-Fills: An Example from Oman (Rub' al Khali Basin Margin). International Geological Congress, 28th, Abstracts 1, 23-24.

Aley, A.A., & Nash, D.F. 1985. A Summary of the Geology and Oil Habitat of the Eastern Flank Hydrocarbon Province of South Oman. Proceedings, Seminar on Source and Habitat of Petroleum in Arab Countries, p. 521-541, O.A.P.E.C, Kuwait.

Al Kadhi, A., & Hancock, P.L. 1980. Structure of the Durma-Nisah segment of the central Arabian graben system. Saudi Arabian Directorate General of Mineral Resources, Bulletin **16**, 40p.

Al-Khatieb, S.O. 1981. Detection of surface evidence of sub-surface structures by interpretation methods of Landsat imagery in the sedimentary rocks of central Saudi Arabia. Ph.D. thesis filed as Saudi Arabian Deputy Ministry for Mineral Resources Open-File Report AC-OF-01-1, 417p.

Al-Khatieb, S.O., & Norman, J.W. 1982. A possibly extensive crustal failure system of economic interest. Journal Petroleum Geology **4**, 319-327.

Al-Laboun, A.A. 1988. The distribution of Carboniferous-Permian siliclastic rocks in the greater Arabian Basin. Geological Society America Bulletin **100**, 362-373.

Al-Sawari, A.M. 1980. Tertiary faulting beneath Wadi Al-Batin (Kuwait). Bulletin Geological Society America **29**, 610-618.

Al-Shanti, A.M., & Mitchell, A.H.G.I. 1976. Late Precambrian subduction and collision in the Al-Amar-Idsas region. Arabian Shield. Journal Geological Society of London **140**, 867-876.

Arabian American Oil Company 1959. Ghawar Oil Field, Saudi Arabia. American Association Petroleum Geologists, Bulletin **43**, 434-454.

Barnes, B.B. 1987. Recent Developments in Geophysical Exploration for Hydrocarbons. In, Short Course on Hydrocarbon Exploration, KFUPM Press, Dhahran.12p.

Benzakour, T.M. 1986. Petroleum Geology of the Jawb Oil Field on the Basis of Well Log Analysis. King Fahd University of Petroleum and Minerals, Thesis, 134p.

Berberian, M. 1976. Contribution to the Seismotectonics of Iran (part 2). Geological Survey of Iran, Report **39**, 516p.

Berberian, M. 1981. Active Faulting and Tectonics of Iran. In, Gupta, H.K., & Delaney, F.M. (eds) Zagros-Hindu Kush-Himalaya Geodynamic Evolution. Geodynamics Series **3**, American Geophysical Union, Washington, p. 33-69.

Brown, G. 1972. Tectonic Map of the Arabian Peninsula. Ministry of Petroleum and Mineral Resources. Directorate General of Mineral Resources, Saudi Arabia, Map AP-2.

Brown, G.F., & Coleman, R.G. 1972. The tectonic framework of the Arabian Peninsula. 24th International Geological Congress, Montreal, 1972, Proceedings, Section 3, p.300-305.

Davies, F.B. 1984. Strain Analysis of Wrench Faults and Collision Tectonics of the Arabian-Nubian Shield. Journal of Geology **82**, 37-53.

Doornkamp, J.C., Brunsden, D., & Jones, D.K.C. (eds) 1980. Geology, Geomorphology, and Pedology of Bahrain. Geobooks, Norwich, U.K., 443p.

Dunnington, H.V. 1967. Stratigraphical Distribution of Oilfields in the Iraq-Iran-Arabia Basin. Journal Institute Petroleum **53**, 129-161.

Dyer, R.A., Husseini, M.I., & Saudi Aramco Exploration Staff 1990. The Western Rub' al Khali Infracambrian Graben System, Saudi Arabia. SPE/Dhahran Geological Society Symposium (unpublished).

Edgell, H.S. 1977. The Permian System as an oil and gas reservoir in Iran, Iraq and Arabia. Second Iranian Geological Symposium, Teheran, p. 161-195.

Edgell, H.S. 1987. Regional Stratigraphic Relationships of Arabia in Exploration for Oil and Gas. In, Short Course on Hydrocarbon Exploration, KFUPM Press, Dhahran, p. 1-44.

Edgell, H.S. 1987. Structural Analysis of Hydrocarbon Accumulation in Saudi Arabia. In, Short Course on Hydrocarbon Exploration, KFUPM Press, Dhahran, 26p.

Edgell, H.S. 1987. Stratigraphic Control of Oil and Gas Accumulation in Saudi Arabia. In, Short Course on Hydrocarbon Exploration, KFUPM Press, Dhahran, 23p.

Edgell, H.S. 1987. Geology of Studied Areas. In, KFUPM Ground Water Resources Evaluation in Saudi Arabia, King Fahd University of Petroleum and Minerals, KFUPM Press, Dhahran, p. 24-77.

Falcon, N.L. 1967. The Geology of the north-east Margin of the Arabian Basement Shield. Advancement of Science **24**, 31-42.

Falcon, N.L. 1969. Problems of the relationship between surface structure and deep displacements illustrated by the Zagros Range. In, Spencer, A.M. (ed.) Time and Place in Orogeny, Special Publication Geological Society of London, **4**, 9-22.

Fayyad, A. 1966. Lithofacies of Eastern Arabia. American University of Beirut, M.S. Thesis (unpublished).

Fox, A.F. 1956. Oil Occurrences in Kuwait. 20th International Geological Congress, Mexico, Symposium sobre jacimientos de petroleo y gas **2**, 131-148.

Freund, R., Garfunkle, Z., Zak, I., Goldberg, M., Weisbrod, T., & Derin, B. 1970. The shear along the Dead Sea Rift, Philosophical Transactions Royal Society London, **267**, 107-130.

Frisch, W., & Al-Shanti, A.M. 1977. Ophiolite Belts and the collision of island arcs in the Arabian Shield. Tectonophysics **43**, 293-306.

Fürst, M. 1970. Stratigraphic und Werdegang der Ostlichen Zagrosketten (Iran). Erlanger. Geol. Abh., Heft **80**, p. 1-51.

Greig, D.A. 1958. Oil Horizons in the Middle East. In, Weeks, L.G. (ed.) Habitat of Oil A Symposium conducted by the American Association of Petroleum Geologists, Tulsa, U.S.A. p. 1182-1193.

Halbouty, M.T. (ed.). Geology of Giant Petroleum Fields. American Association Petroleum Geologists, Mem. No. **14**, 556 p.

Hancock, P.L., Al-Khatieb, S.O., & Al-Kadhi, A. 1981. Structural and photogeological evidence for the boundaries to an East Arabian block. Geological Magazine **118**, 533-538.

Hancock, P.L., Al-Kadhi, A., & Sha'at, N.A. 1984. Regional joint sets in the Arabian Platform as indicators of intraplate processes. Tectonics **3**, 27-43.

Hancock, P.L., & Al-Kadhi, A. 1985. Structure of the Qirdan segment of the Central Arabian Graben System, Kingdom of Saudi Arabia. Deputy Ministry for Mineral Resources, Professional Paper No.**PP-2**, 63-74.

Hancock, P.L., & Bevan, T.G. 1987. Brittle modes of foreland extension. In, Coward, M.P., Dewey, J.F., & Hancock, P.L. (eds) Continental Extension Tectonics, Geological Society London Special Publication No. **28**, 127-137.

Henson, F.R.S. 1951. Observations on the Geology and Petroleum Occurrences of the Middle East. 3rd World Petroleum Congress, The Hague, Proceedings, Section **1**, 118-140.

Husseini, M.I. 1988. The Arabian Infracambrian extensional system. Tectonophysics **148**, 93-103.

Husseini, M.I. 1989. Tectonic and Deposition Model of Late Precambrian-Cambrian Arabian and Adjoining Plates. American Association of Petroleum Geologists, Bulletin **73**, 1117-1131.

Jackson, J.A., Fitch, T.J., & McKenzie, D.P. 1981. Active thrusting and evolution of the Zagros fold belt. In, McClay, K.R., & Price, N.J. (eds) Thrust and Nappe Tectonics, Geological Society London, Special Publication No.**7**, 371-379.

Jackson, J.A. 1987. Active normal faulting and crustal extension. In, Coward, M.P., Dewey, J.F., & Hancock, P.L. (eds) Continental Extensional Tectonics, Geological Society of London, Special Publication No. **28**, 3-17.

James, G.A., & Wynd, J.G. 1965. Stratigraphic Nomenclature of Iranian Oil Consortium Agreement Area. American Association Petroleum Geologists, Bulletin **49**, 2182-2245.

Johnson, P.R., Scheibner, E., & Smith, A. 1987. Basement fragments, accreted tectonostratigraphic terranes, and overlap sequences; elements in the tectonic evolution of the Arabian Shield. Geodynamics Series, American Geophysical Union **17**, 323-343.

Kamen-Kaye, M. 1970. Geology and Productivity of Persian Gulf synclinorium. American Association Petroleum Geologists, Bulletin **54**, p. 2371-2394.

Kassler, P. 1973. The Structural and Geomorphic Evolution of the Persian Gulf. In, Purser, B.H. (ed.) The Persian Gulf, Springer Verlag, Berlin, p. 11-32.

Kent, P.E. 1970. The Salt Plugs of the Persian Gulf Region. Transactions Leicester Literary Philosophical Society **64**, 55-88.

McClay, K.R., & Ellis, P.G. 1987. Analogue models of extensional fault geometries. In, Coward, M.P., Dewey, J.F., & Hancock, P.L. (eds) Continental Extensional Tectonics, Geological Society of London, Special Publication, No.**28**, 109-125.

McKenzie, D. 1972. Active tectonics of the Mediterranean region. Geophysical Journal of the Royal Astronomical Society **30**, 109-185.

Millon, R. 1970. Structures géologiques révélees par l'interpretation du léve aeromagnetique du bouclier arabe. Bulletin B.R.G.M. (deuxieme serie) Section iv, No.**2**, 15-27.

Mina, P., Razaghnia, M.T., & Paran, Y. 1967. Geological and Geophysical Studies and Exploratory Drilling of the Iranian Continental Shelf, Persian Gulf. 7th World Petroleum Congress, Mexico, Proceedings P.D. 9, No.**19**, 179-222.

Ministry of Petroleum Resources, Deputy Ministry of Mineral Resources 1984. Geographic Map of the Arabian Peninsula, (showing oil field outlines). Map AP-5B-2; Scale 1:2,000,000; compiled by the United States Geological Survey and the Arabian American Oil Company.

Moore, J. McM. 1979. Tectonics of the Najd Transcurrent Fault System, Saudi Arabia. Journal Geological Society of London **136**, 441-454.

Moore, J. McM. 1982. Structure, stratigraphy and mineralization in the southern Arabian Shield — A study in satellite-image interpretation. Precambrian Research **16**, abstract, 53-54.

Moore, J. McM. 1983. Tectonic Fabric and Structural Control of Mineralization in the Southern Arabian Shield: A Compilation based on Satellite Imagery Interpretation. Deputy Ministry for Mineral Resources, Open File Report USGS-OF-03-105, 61p.

Morris, P. 1977. Basement structure as suggested by aeromagnetic surveys in S.W. Iran. Second Geological Symposium of Iran, Proceedings Iranian Petroleum Institute, Teheran.

Murris, J.J. 1980. Middle East; Stratigraphic Evolution and Oil Habitat. American Association Petroleum Geologists, Bulletin **64**, 597-618.

Nelson, P.H. 1968. Wafra Field-Kuwait-Saudi Arabia Neutral Zone. Second Regional Technical Symposium, Society of Petroleum Engineers, Saudi Arabia Section, Dhahran, 16 p.

North, F.K. 1985. Petroleum Geology. Allen and Unwin, Boston, 607 p.

O'Brien, C.A.E. 1957. Salt Diapirism in South Persia. Geologie en Mijnbouw, (N.W. Ser), **19** , 357-376.

Onyedim, G.C., & Norman, J.W. 1986. Some Appearances and Causes of Lineaments seen on Landsat Images. Journal Petroleum Geology **9**, 179-194.

Perrodon, A. 1983. Dynamics of Oil and Gas Accumulations. Bulletin Centre de Recherches Exploration-Production, Elf Aquitaine, Memoir **5**, 360 p.

Picard, L. 1970. On Afro-Arabian graben tectonics. Geologisches Rundschau **59**, 337-381.

Player, R.A. 1969. Salt Plug Study. Iranian Oil Operating Companies, Geological and Exploration Division, Report No. **1146**, (unpublished) 123 p.

Powers, R.W., Ramirez, L.F., Redmond, C.D., & Elberg, E.L. Jr. 1966. Geology of the Arabian Peninsula-Sedimentary Geology of Saudi Arabia. U.S. Geological Survey, Professional Paper 560-D, 147 p.

Powers, R.W. 1968. Saudi Arabia (excluding Arabian Shield) Asie. In, Lexique Stratigraphique International, **3**, Parsis CNRS, 177 p.

Rönnlund, P., & Koyi, H. 1988. Fry spacing of deformed and undeformed modeled and natural salt domes. Geology **16**, 465-468.

Sha'ath, N.A. 1986. The structure of the Majma' ah graben complex. Unpublished Ph.D. thesis, University Bristol.

Steineke, M., Bramkamp, R.A., & Sander, N.J. 1958. Stratigraphic Relations of Arabian Jurassic Oil. In, Weeks, L.J. (ed.) Habitat of Oil Tulsa, American Association of Petroleum Geologists Symposium, p. 1294-1329.

Steineke, M., Harriss, T.F., Parsons, K.R., & Berg, E.L. 1958. Geologic map of the Western Persian Gulf quadrangle, Kingdom of Saudi Arabia. United States Geological Survey Investigations, Map 1-208A, Scale 1:500,000.

Stewart, I.C.F., & Hargan, B.A. 1990. Enhancement and interpretation of gravity and magnetics data in the Arabian Peninsula. Society of Petroleum Engineers, Dhahran Geological Society Symposium (unpublished).

Stöcklin, J. 1968. Structural history and tectonics of Iran: a review. American Association Petroleum Geologists. Bulletin **52**, 1229-1258.

Sugden, W. 1962. Structural analysis and geometrical prediction for change of form with depth of some Arabian plains-type folds. American Association Petroleum Geologists, Bulletin **46**, 2213-2228.

Talbot, C.J., & Jarvis, R.J. 1984. Age, budget and dynamics of an active salt extension in Iran. Journal of Structural Geology **6**, 521-533.

Trusheim, F. 1974. Zur Tektogenese der Zagros-Ketten, Süd-Irans. Zeitschrift Deutschen geologischen Gesellschaft **125**, 119-150.

United States Geological Survey Staff and Arabian American Oil Company Staff 1963. Geologic Map of the Arabian Peninsula. United States Geological Survey Miscellaneous Geologic Investigations, Map 1-270A; Scale 1:2,000,000.

Waite, M.W., Dyer, R.A., & Simms, S.C. 1990. Seismic stratigraphy modeling of Central Arabia. Society of Petroleum Engineers, Dhahran Geological Society Symposium (unpublished).

Wissmann, H. von, Rathjens, C., & Kossmat, F. 1942. Beiträge zur Tektonik Arabiens. Geologisches Rundschau **33**, 221-353.

AN E-W TRANSECT SECTION THROUGH CENTRAL IRAQ.

S.A. ALSINAWI AND A.S. AL-BANNA
Department of Geology
University of Baghdad
Iraq

ABSTRACT. Geological and geophysical measurements give an idea about the crustal thickness beneath an E-W transect 620 km long through central Iraq. Because the basement of Iraq is hidden below younger rocks, several studies in neighbouring countries have been compiled by Buday and Jassim (1987) who deduced that the basement is metamorphosed and of Precambrian age. The Basement depth was determined from an aeromagnetic survey and is used in this study to calculate a density contrast of 0.1 gm/cc between the sedimentary cover and the basement. A gravity model for the crust in central Iraq has been built to fit the smoothed anomaly. The crustal thickness along the transect ranges between 31 km and 37 km with an average of 34 km. The geologic and tectonic data along the transect are compared with the available aeromagnetic and Bouguer anomalies and seismic-reflection data. The resulting model is compared with published empirical relations for crustal thickness and a close correlation is noted.

Introduction

This study is the first published work on crustal thickness in the middle of Iraq. It was carried out on an E-W Transect profile 620 km long (Fig. 1).

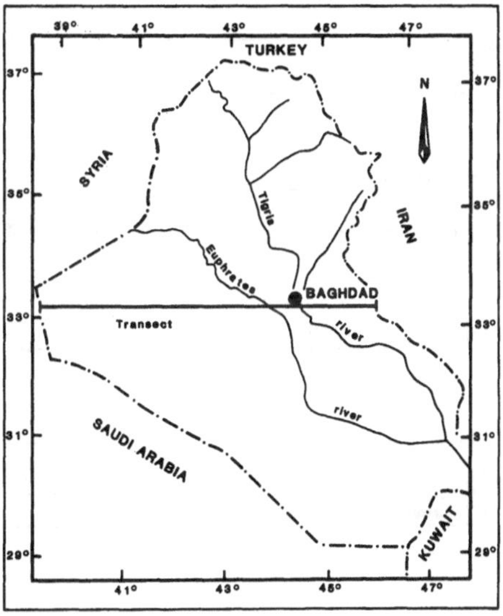

Figure 1. Location map of the studied transect.

M. J. Rickard et al. (eds.), Basement Tectonics 9, 195–200.

Iraq lies on the border area between the Arabian platform and the Iranian part of the Asian platform. The boundary between them is a geosyncline created during the Alpine Orogeny (Buday and Jaseem, 1987). The platform part within the territory of Iraq is divided into two basic units, the stable and the unstable shelves.

Structurally the transect passes through the stable shelf which is characterized by a reduced thickness of sedimentary cover and lack of folding. The Mesopotamian zone i.e. the unstable shelf, is characterized by great subsidence since the Mesozoic, by thick sedimentation culminating in the late Cenozoic, and by slight folding of sedimentary cover (Buday, 1980, Fig. 2). The oldest geological unit, of Permian age, is exposed on the western side of the transect near Rutba.

For our crustal study, gravity data were taken from the unified Bouguer map of Iraq with a scale of 1:1,000,000 and with a contour interval of 1.0 mgal. An aeromagnetic map at a scale of 1 to 1,000,000 was also used (C.C.G., 1973; SEGSMI, 1987).

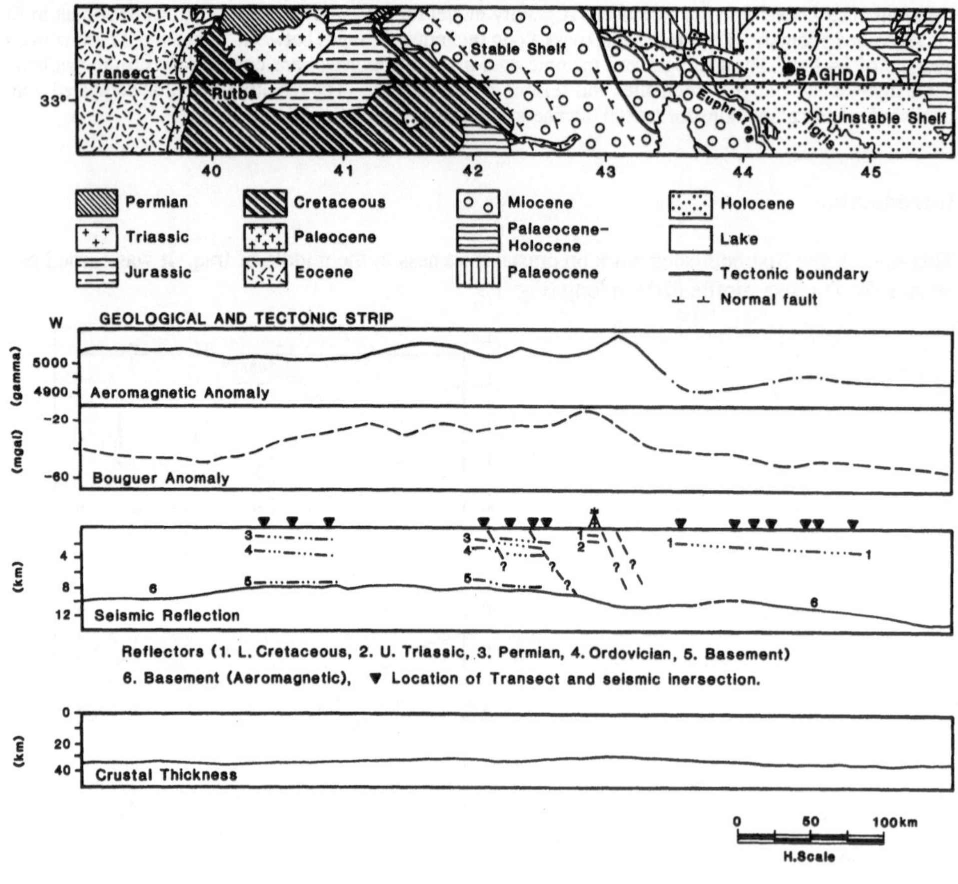

Figure 2. Geology and tectonic profiles: Aeromagnetic anomaly, Bouguer anomaly, Seismic reflection and crustal thickness (model) along the transect.

The Basement of Iraq

There is at present no direct information about the depth and composition of the basement complex. Aeromagnetic seismic, and gravity studies were therefore carried out to gain information about the depth of the basement. Aeromagnetic and seismic surveys provide more reliable results than gravity in most of Iraq.

Buday and Jaseem (1984) in their tectonic map used mainly the thickness of sedimentary cover determined from the aeromagnetic interpretation map for basement depths with modifications in some areas using seismic information.

Basement rocks are exposed in all countries surrounding Iraq, namely Saudi Arabia, Jordan, Turkey and Iran. Studies of basement rocks in these countries were integrated and compiled by Buday (1980) to give an idea about the basement composition of Iraq, he deduced that the basement of Iraq is metamorphic and mostly of Precambrian age. Table 1 summarizes several geophysical studies dealing with the basement in Iraq.

TABLE 1. Earlier geophysical studies of the basement of Iraq.

Author(s)	Area	Methods	Basement Depth km
Ditmar et al. 1971, 1972	most of Iraq	Gravity Magnetic	4-14
C.C.G., 1973	most of Iraq	Aeromagnetic	5-13
Abdel Dayam & Al-Din, 1975a, 1975b	West of Karbala	Gravity	8-10
Al-Khatib, 1976	Latitude 32-33 Longitude 42-47	Gravity	3.5-11.5
Al Sinawi & Hassan, 1977	Al-Fatha	Gravity	6
Al Sinawi & Al-Rawi, 1978	Badra	Gravity	9-12
Najam, 1979	Latitude 31-32 Longitude 42-47	Gravity	4.5-13.5
Razkalla, 1981	Debaga, Erbil	Gravity	3-9
Zaynal, 1981	Latitude 33-34 Longitude 41-46	Gravity	4.5-10.5
Mohammad, 1981	Hatra Uplift	Gravity	6-7
Al Sinawi & Al-Banna, 1983	Hatra Uplift	Gravity Magnetic	5-6.7
Shaswar, 1983	Anah	Gravity & Total Magnetic	8
Baban, 1983	Abu-Rassian	Gravity	5-6
Aziz, 1986	Western part of Iraq	Reflection Seismic	6-7.5

POLYNOMIAL REGRESSIONS AND REGIONAL GRAVITY

Polynomial regressions for the Bouguer anomalies of different orders were carried out along the transect. Only the first and the sixth are shown in Fig. 3.

The regional gravity of the transect was obtained using the Griffin method for radii ($5\sqrt{5}$ $10\sqrt{5}$ $15\sqrt{5}$) km. A comparison between them and the sixth order polynomial regression shows close concidence. All of them showed a wide positive anomaly in the middle of the transect. This might be due to thinning of the crust, or to the presence of a positive density contrast lying at the lower part of the crust from the boundary of the stable shelf towards the west.

INTERPRETATION OF GRAVITY

A sedimentary column about 7-12 km thick, increasing from west to east, which overlies the basement, affects the Bouguer anomalies along the transect. Interpretation of gravity depends mainly on the density contrast and the dimensions of the body. A homogeneous density for the crust is a usual assumption in crustal studies. Using the sedimentary thickness obtained from aeromagnetic interpretation and a density contrast of 0.12/cc between the sedimentary and basement rocks, the calculated sedimentary effect is eliminated from the measured gravity anomaly. The corrected gravity anomaly for the sedimentary column with additional smoothing of local anomalies, was compared with a model computed by the authors. A density contrast of 0.4 g/cc were used for the model between the crust and the mantle, and the thickness of the crust obtained from this modeling is given in Table 2.

DIFFERENT EMPIRICAL RELATIONS FOR CRUSTAL THICKNESS CALCULATIONS

Empirical relations deduced by various authors were used for calculations of crustal thicknesses along the transect and compared with the result of the model. The results are given in Table 2. The empirical relations used were as follows:

(1) $Z = 33 — 0.055\ Dg$ (Worzel and Shurbet 1955)
(2) $Z = 35\ (1— \tanh 0.0037)\ Dg$ (Demenitskaya 1958)
(3) $Z = 30 — 0.1\ Dg$ (Andreev, 1959)
(4) $Z = 32 — 0.08\ Dg$ (Woollard, 1959)
(5) $Z = 40.5 — (32.5\ \text{tahn}\ (Dg + 75)/275)$ (Woollard and Strange, 1962)

Conclusions

The comparison shows that the minimum crustal thickness is in the central part of the transect and is equal to 31.3 km. The maximum thickness is towards the east and equal to 37 km. Only equation 2 shows a different minimum and maximum, but it also gives the same difference between the thin and thick crust of about 4 km to 5 km.

It is necessary to carry out deep seismic-reflection profiles and studies on isostasy using refraction data to come closer to the actual dimensions and variations with the crust beneath Iraq.

Acknowledgement

The authors are grateful to the Establishment of Geological Survey and Minerals Investigation for their kind permission to use their geophysical and geological maps for this investigation.

TABLE 2. Estimates of Crustal Thickness on the traverse across central Iraq

A	B	1	2	3	4	5
100	36.4	35.83	41.67	35.15	36.12	37.72
150	35.0	35.03	39.79	33.70	34.96	36.00
200	33.3	34.51	38.56	32.75	34.20	34.89
250	33.2	34.65	38.88	33.0	34.4	35.18
300	32.1	34.65	38.88	33.0	34.4	35.18
350	31.5	34.26	37.98	32.3	33.84	34.35
400	31.7	34.65	38.88	33.0	34.4	35.18
450	34.6	35.5	40.89	34.55	35.64	37.01
500	36.0	35.89	41.80	35.25	36.16	37.84

Column A — Distance in km from western end of the traverse
Column B — Crustal thickness estimated by the writers
Columns 1-5 — Crustal thickness calculated using the empirical relations of other authors set out in the text. 1, Worzel and Shurbet, 1955; 2, Demenitskaya, 1958; 3, Andreer, 1959; 4, Wollard, 1959; 5, Woollard and Strange, 1962.

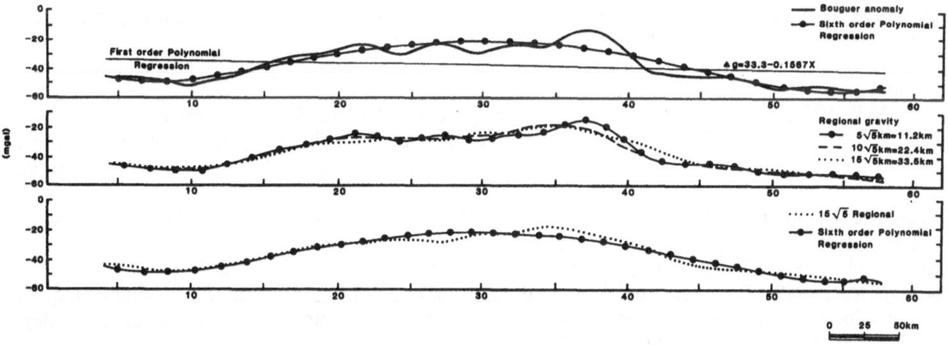

Figure 3. Comparison between regional anomalies (Griffin method) of different radii, with the polynomial regression of the gravity along the transect.

References

Abbas, M.J., & Masin, J. 1975. New geophysical aspects of the basement structure in western Iraq. Journal Geological Society Iraq. Special Issue **8**, 1-13.

Abbas, M.J., Al-Khadimi, J.A., & Fatah, A.S. 1984. Unifying gravity maps of Iraq. Department of Geophysics, Directorate General of Geological Survey and Mineral Investigations, unpublished report.

Abdel-Dayem, M 1977. A contribution to gravity interpretation of Al-Najaf area, Iraq. Bulletin College Science **18**, 139-148.

Abdel-Dayem, M., & Al-Din, T.S. 1975a. Two-dimentional filtering of gravity fields in the area west of Karbala, Iraq. Bulletin College Science **6**, 135-151.

Al-Khatib, H.H. 1977. Downward analytical continuation of the gravity field in central Iraq between latitude 32 and 33N', M.Sc. thesis, College Science Baghdad University (unpublished).

Al Sinawi, S.A., & Hassan, H.H. 1977. The gravity field at Al-Fatha area, central Iraq. Bulletin College Science **18**, 149-180.

Al Sinawi, S.A., & Al-Rawi, F.R. 1978. A gravity investigation of Badra area'. Iraqi Journal Science **19**, 87-100.

Al Sinawi, S.A., & Al-Banna, A.S. 1983. Regional gravity and magnetic interpretation of Hatra uplift and its surroundings, west central Iraq. Iraq Journal Science **24**, 49-63.

Al Sinawi, S.A., Rizkallah, R.I., & Al-Rawi, F.R. 1987. On the gravity field of Iraq, Part 1, the state of the Art. Journal Geological Society Iraq **20**, 17-37.

Aziz, B.K. 1986. Study of thickness distribution of the Paleozoic sediments and basement depth in parts of the Western desert based on seismic reflection data. M.Sc thesis, University of Baghdad unpublished.

Baban, E.N. 1983. Analysis of geophysical data available on Abu Rassain area. M.Sc. thesis, Mosul University (unpublished).

Bird, P. 1978. Finite element modeling of lithosphere deformation in Zagros collision orogeny. Tectonophysics **50**, 307-336.

Buday, T. 1980. Regional Geology of Iraq, Stratigraphy and Paleogeography. Directorate General of Geological Survey and Mineral Investigations, unpublished report, 445p.

Buday, T., & Jassim S.Z. 1984. Tectonic Map of Iraq. Directorate General of Geological Survey and Mineral Investigations, 1st Edition, National Library Catalogue Card 597/1984.

Choudhury, S.K. 1975. Gravity and crustal thickness in the Indo-Gangetic Plains and Himalayan Region, India. Geophysical Journal Royal astronomial Society **40**, 441-452.

Compagnie Generale de Geophysique 1973. Aeromagnetic Map of Iraq. Department of Geophysics, Directorate General of Geological Survey and Mineral Investigations (unpublished).

Ditmar, et al. 1971. Geological conditions and hydrocarbon prospects of the Republic of Iraq (northern and central parts). Technoexport, INOC Library, Baghdad vol. **1**.

Ditmar, et al. 1972. Geological conditions and hydrocarbon prospects of the Republic of Iraq (northern and central parts). INOC Library, Baghdad vol. **2**.

Dobrin, M.B. 1976. Introduction to Geophysical Prospecting. McGraw Hill, Inc, New York.

Jassim, S.Z., Hogopian, D.H., & Al-Hashimi, H.A. 1986. Geological Map of Iraq. Directorate General of Geological Survey and Mineral Investigations, 1st Edition, National Library Catalogue Card No. 1487/1986, Baghdad

Mohammed, A.A.M. 1981. Geophysical study to establish basement configuration in the Western Desert. M.Sc. thesis, University of Mosul (unpublished).

Monger, J.W.H. 1986. The global geoscience transects project. Episodes **9**, 217-222.

Najem, H.A.R. 1979. Basement structures in the area between latitudes 31 and 32, Southern Iraq, An application of downward continuation method on the gravimetric potential data. M.Sc. thesis, College Science, Baghdad University (unpublished).

Rezkalla, J.S. 1981. Geophysical gravimetric, investigation over Debaga Plain area, south-west Arbil. M.Sc. thesis, College Science, Baghdad University (unpublished).

Shaswar, O.K.A. 1983. Analytical study of the geophysical data over Anah Area. M.Sc. thesis, University of Mosul, (unpublished).

Zaynal, M.S. 1981. Downward analytical continuation of the gravity field in the Central Iraq between Latitudes 33-34 N, basement rock study. M.Sc. thesis, University of Baghdad (unpublished).

CONJUGATE BASEMENT RIFT ZONES IN KANSAS, MIDCONTINENT, USA

D.L. BAARS
Kansas Geological Survey
University of Kansas
1930 Constant Avenue, Campus West
Lawrence Kansas 66047
USA

ABSTRACT. The structure of the Precambrian basement of Kansas, Midcontinent USA, is dominated by conjugate NNE and NW trending wrench fault zones. NNE trending faults of the Midcontinent Rift System (MRS) extend from Lake Superior across Kansas and into north-central Oklahoma. The fault zone widens from about 100 km in northeast Kansas to more than 160 km in south-central Kansas in a series of horsetail splays. NNE-trending structures of the MRS are displaced by about 80 km of dextral offset by the NW trending strike-slip fault zone.

Apparently penecontemporaneous NW trending wrench faults of the Bourbon Arch-Central Kansas uplifts cross the state from southeast to northwest, offsetting MRS structures. The two conjugate wrench fault zones are complexly interrelated in central Kansas, where internal synthetic shears complicate axial horsts and grabens of the MRS. The Bourbon Arch is offset about 100 km by sinistral slip from the Central Kansas uplift along the MRS. The Humboldt fault zone at the eastern margin of the MRS was not offset significantly by NW-trending faults, suggesting that the present-day expression of the southward-weakening fault zone was created during Pennsylvanian (Upper Carboniferous) rejuvenation of the basement fabric. Stratigraphic relationships record a history of repeated reactivation in Paleozoic time that strongly affected petroleum entrapment, with an especially strong pulse of uplift during Pennsylvanian time.

These rift zones are segments of continental-scale basement lineaments that are fundamental to the structural fabric of the North American basement. The Bourbon Arch-Central Kansas structural lane lies sub-parallel to the Olympic-Wichita Lane that extends from southern Oklahoma to the northwest through the Paradox basin of eastern Utah, and the MRS lies sub-parallel to the Colorado Lineament which extends from the Grand Canyon in Arizona to the Lake Superior region. Thus, the basement of the western Midcontinent and Southern Rocky Mountains consists of large-scale fault zones that delineate sub-orthogonal basement blocks.

Introduction

The Midcontinent U.S.A., and the state of Kansas in particular, are generally believed to be geologically structureless like the scenery. Yet upon close examination of geophysical and subsurface data the basement is found to be complexly faulted. In spite of earlier interpretations to the contrary, the basement of Kansas is here interpreted to be broken by wrench fault zones of regional proportions. These fault zones readily fit in a continental-scale structural fabric that will be seen to delineate regional crustal blocks that together form the North American craton.

M. J. Rickard et al. (eds.), Basement Tectonics 9, 201–210.

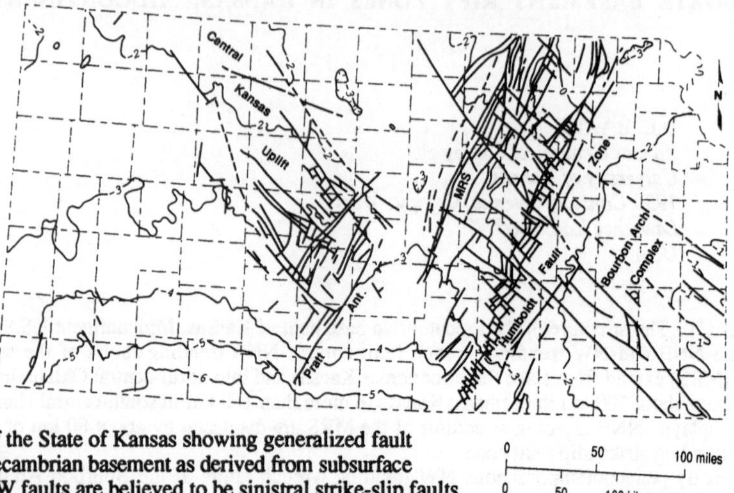

Figure 1. Map of the State of Kansas showing generalized fault patterns in the Precambrian basement as derived from subsurface studies. NNE-SSW faults are believed to be sinistral strike-slip faults and NW-SE structures are interpreted to be dextral strike-slip faults.

Basement Structure of Kansas

The Midcontinent Rift System (MRS) (sometimes referred to as the Central North American Rift System or Midcontinent Gravity High) has long been recognized from geophysical mapping to trend south-southwestward from Lake Superior into central Kansas in a snake-like curving pattern. Faults of this zone trend NNE-SSW across east-central Kansas, splaying outward toward the south (Fig. 1). Lying parallel to the MRS immediately to the east is a complexly faulted positive feature known as the Nemaha uplift, bounded on the east by the Humboldt fault zone (Berendsen & Blair 1986). In spite of its proximity and parallel trend, the Nemaha fault zone has generally been considered as a separate structure from the MRS, extending from southeastern Nebraska across eastern Kansas into north-central Oklahoma (Dolton & Finn 1989).

Another prominent trend of basement faults crosses the MRS-Nemaha fault zone at near right angles. Numerous NW-SE oriented faults extend from central Missouri into east-central Kansas, comprising what is often referred to as the Bourbon arch complex (Fig. 1). The Bourbon arch that affected lower to middle Paleozoic depositional patterns includes, and is bounded by, this zone of faults. The Bourbon arch complex appears to abut the Nemaha uplift. Another swarm of NW-SE faults in west-central Kansas marks a structually high platform known as the Central Kansas uplift (CKU) (Fig. 1). This faulted block appears to abut fault extensions of the MRS in central Kansas.

Subsurface mapping of lower Paleozoic rocks and the upper Precambrian surface reveals that the NW trending Bourbon arch fault zone complexly offsets faults of the Nemaha-MRS fault zone in east-central Kansas, breaking the basement into myriad suborthogonal fault blocks (Fig. 1) (Berendsen & Blair 1986). Similarly, the southeastern extension of the Central Kansas uplift is complexly offset by faults that appear to be southerly extensions of the MRS (Fig. 1). Geophysical maps, especially the second vertical derivative of gravity map (Fig. 2), confirm these strongly intersecting relationships.

Figure 2. Second vertical derivative magnetic map of Kansas (reduced to the pole) compiled by the Kansas Geological Survey. The MRS is strongly apparent in the north-central part of the State, but other basement structures with NNE and NW orientations are apparent. Compare with the basement structure map of Figure 1.

Berendsen and Blair (1986) interpreted the Nemaha-MRS fault system to exhibit sinistral strike-slip displacement along a regional wrench-fault zone, and this study confirms that interpretation. Many local offsetting relationships found on the Central Kansas uplift strongly suggest that dextral strike-slip movement has occurred along the NW trending fault zone as well.

In an effort to make order out of this chaos, an interpretive sketch map (Fig. 3) demonstrates the present interpretation that the two fault trends intersect in central Kansas, each set displacing the other in their respective senses. The NW-oriented Bourbon arch-Central Kansas trend is offset sinistrally by Nemaha-MRS faults, and the Nemaha-MRS fault zone is offset in a dextral sense by the Bourbon arch-Central Kansas faults. These zones of intersecting faults are further complicated by synthetic shears along the Nemaha uplift. That these faults displace one another suggests that movement of both sets was essentially penecontemporaneous. Thus, the intersection of the two major fault zones is interpreted as forming a conjugate set in central Kansas.

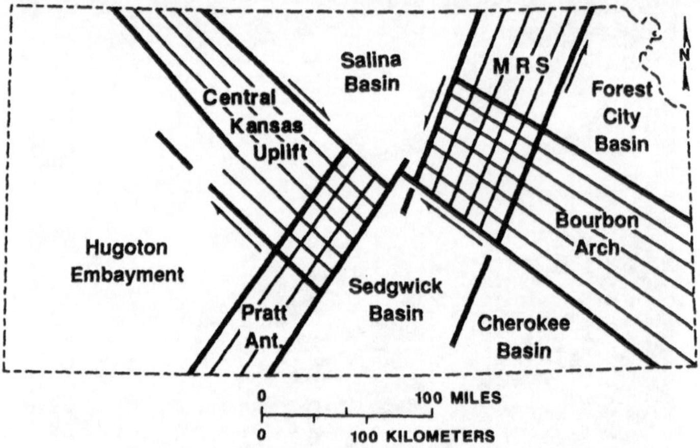

Figure 3. Grossly generalized cartoon showing the interpretation of the basement structural and magnetic maps of Figures 1 and 2. Structural fabrics are indicated by ruled patterns and interpreted sense of displacement shown with arrows. Where the structures cross in central Kansas, complex intersecting patterns result. The fault zones form basement uplifts that divide the region into five Paleozoic depositional basins.

Regional Comparison

A similar intersection of NW-SE and NE-SW wrench fault zones was documented as occurring in the Paradox evaporite basin of southeastern Utah by Baars (1966) and Warner (1978). Warner cited evidence that the NE trending Colorado Lineament displays sinistral strike-slip offset, originating at about 1.7 Ga (middle Precambrian time). Baars and Ellingson (1984) further documented evidence that the NW trending Olympic-Wichita Lineament is a dextral wrench-fault zone of 1.6 to 1.7 Ga origin. As in Kansas, the intersecting wrench-fault zones appear to displace each other, forming a conjugate set. Stevenson and Baars (1986) interpreted the fault-bounded Paradox basin to be a large pull-apart basin of Pennsylvanian age, formed by reactivation along the NW Olympic-Wichita Lineament, with extension facilitated by the NE Colorado Lineament (Fig. 4). The fault-bounded Uncompahgre uplift to the northeast supplied vast quantities of clastic sediments to the adjoining basin.

Although having originated in middle Precambrian time, Baars (1966) documented strong evidence that movement along the Olympic-Wichita Lineament was rejuvenated in Late Cambrian, Late Devonian, Early Mississippian, and Middle Pennsylvanian time. Reactivation in Late Devonian time created shallow marine fault blocks upon which offshore sand bars formed and became petroleum reservoirs. Mississippian reactivation formed structurally controlled shoaling conditions that fostered development of Waulsortian banks, which upon dolomitization became excellent petroleum reservoirs (Baars 1966). Further reactivation in Middle Pennsylvanian time again caused subtle shoals along the southern shelf of the Paradox basin that localized the development of algal bioherms that have produced prolific amounts of petroleum (Baars & Stevenson 1982).

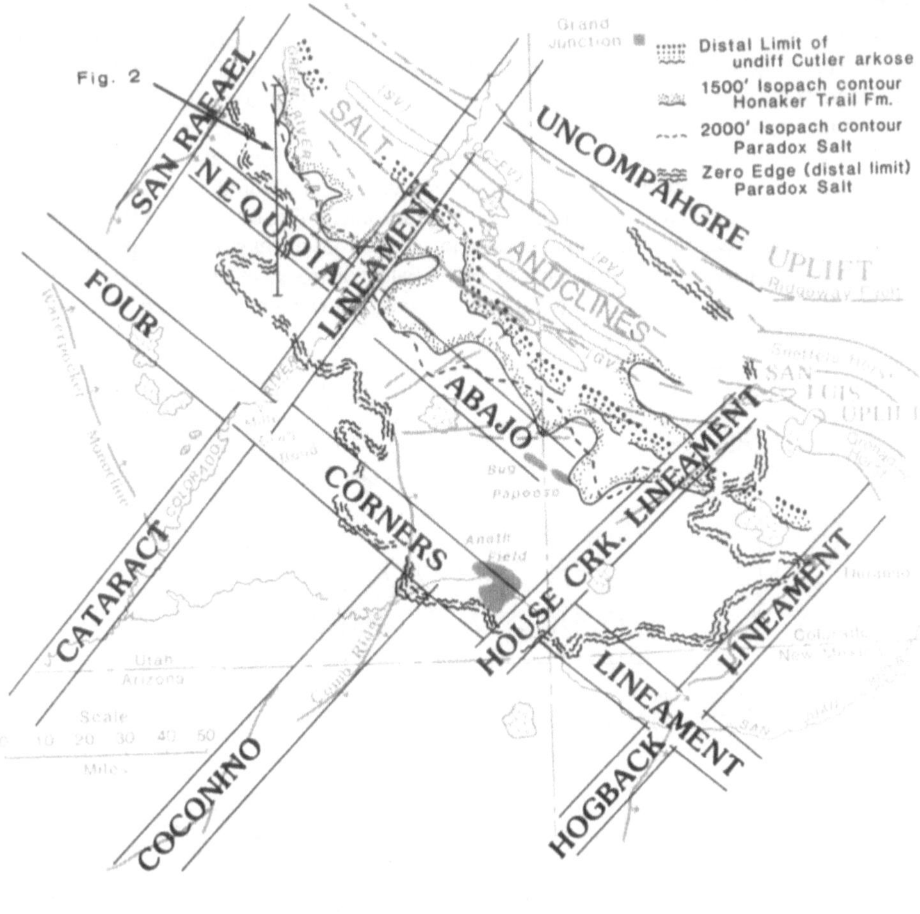

Figure 4. Tectonic map of the Paradox basin, southeastern Utah and southwestern Colorado from Stevenson and Baars (1986). NW fault zones, indicated as basement "lineaments," are parts of the Olympic-Wichita Lineament, and NE trends lie along the Colorado Lineament (see text). The Paradox evaporite basin of Pennsylvanian (Lower Carboniferous) age was interpreted to be a pull-apart basin. Compare this regmatic basement fabric with the regional structural pattern of figure 6.

Regional Setting

The Olympic-Wichita Lineament, the key element in the origin of the Paradox basin, extends to the northwest at least across Utah, and has been interpreted (Baars 1976) to be an extension of the Olympic-Wallowa Lineament of Wise (1969). It can be traced to the southeast into the fault complex of the Southern Oklahoma aulacogen, and perhaps beyond (Baars 1976). The composite magnetic anomaly map of the United States (U.S.G.S., 1982, from Hinze & Braile 1988, Plate 1B) clearly confirms these relationships. Thus, the Olympic-Wichita Lineament is a continental-scale structural feature interpreted to play a major role in the basement architecture of the North American craton (Figs. 5 & 6).

The Pennsylvanian Paradox pull-apart basin lies along the Olympic-Wichita Lineament and is complementary to a major fault-bounded uplift. In like fashion, the Pennsylvanian Anadarko and Arkoma basins of Oklahoma are intimately related to the same structural lineament, but lie in mirror-image to the Paradox basin. The uplifted sources of voluminous clastic sediments in southern Oklahoma occur to the south and the deep structural basins are to the north of the fault zone (Fig. 6). The deep basins of central Oklahoma are bounded generally to the north by a structurally controlled shallow shelf that lies along the Bourbon arch-Central Kansas fault complex that in Kansas lies sub-parallel to the Olympic-Wichita Lineament (Fig. 6). It is easy to interpret a close relationship between these Oklahoma basins to adjacent basement structures; perhaps these are also pull-apart basins (?) As in the Paradox basin, Paleozoic rocks of the Kansas shelf of the Oklahoma basins show a long history of Paleozoic tectonic rejuvenations and prolific petroleum production from these strata.

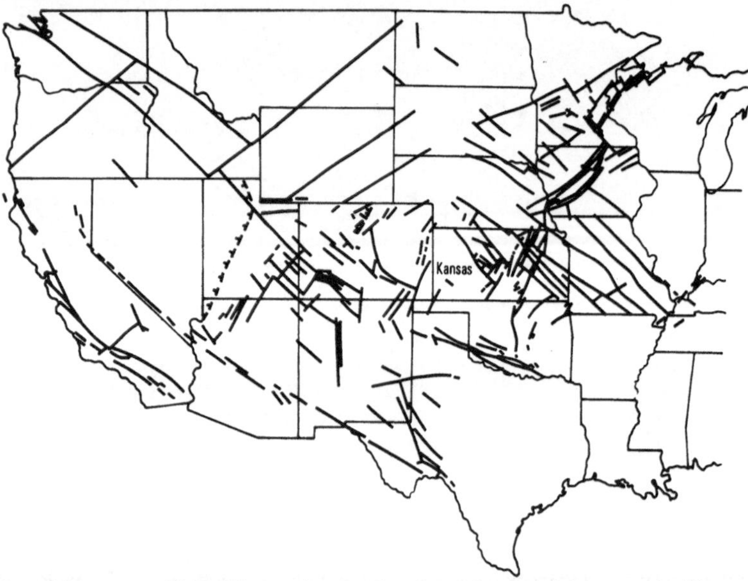

Figure 5. Map of the western United States showing locations of some of the most significant fault zones and lineaments in the Precambrian basement. Faults in the northern Midcontinent, north of Kansas, are generalized from Sims (1990). Where sense of displacement has been determined, northwesterly structures are dextral and northeasterly structures are sinistral. In most cases, complex fault zones are shown as single lines.

Northern Midcontinent Basement Structure

Precambrian basement structures of the northern Midcontinent are shown on Figure 5. To the north and east of Kansas the structure has been generalized from Sims (1990). As previously discussed, the Midcontinent Rift System and the Central Plains Orogen are the principal structural features linking Kansas to the craton. It is noteworthy that the structural fabric of the northern Midcontinent is, like the margins of the craton to the south and west, a network of NW-SE and NE-SW fault zones, many of which have been interpreted to be strike-slip structures (Sims 1990). Thus, it appears that the basement structural fabric of the North American craton comprises several suborthogonal, fault-bounded blocks.

Figure 6 is a highly generalized map showing the relationships of Kansas basement structures to others of the Midcontinent. Late Paleozoic fault-related basins are superimposed on the map to indicate their probable associations with basement structure. The interpretation shown on Figure 6 presents the MRS as a series of left-stepping en echelon fault zones as they appear to have occurred prior to extensional rifting. If this interpretation is correct, the MRS may consist of three originally distinctive compressional structures, later apparently joined by clockwise rotation of their associated crustal blocks, followed by late Precambrian relaxation of stress now exemplified by extension along the rift zone.

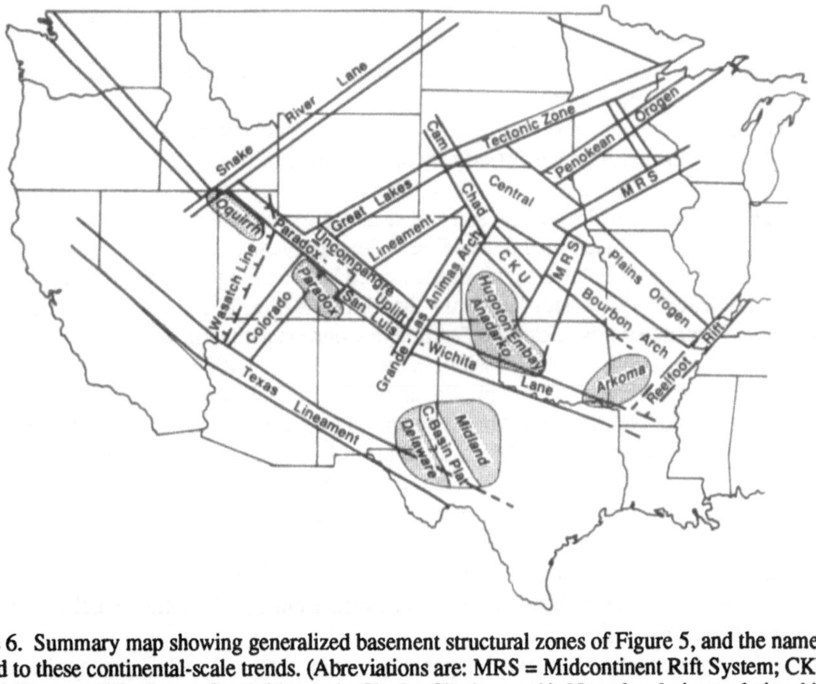

Figure 6. Summary map showing generalized basement structural zones of Figure 5, and the names applied to these continental-scale trends. (Abreviations are: MRS = Midcontinent Rift System; CKU = Central Kansas uplift; Cam = Cambridge arch; Chad = Chadron arch). Note the obvious relationships of late Paleozoic depositional basins (Oquirrh, Paradox, Anadarko, and Arkoma basins) to the basement structural fabric. These are all fault-bounded basins of Pennsylvanian-Permian age formed by rejuvenation of basement structures during the Ancestral Rocky Mountains orogeny.

Such an interpretation of the MRS would explain the apparent age and structural discrepancies along the rift. Sims (1990, p. 5) depicts the northern rift as "...a medial horst of basalt-rhyolite flows and local overlying sedimentary basins that is flanked by red-beds, which compose clastic wedges along the margins of the rift." He (Sims 1990) assigns an age of 1,000-1,200 Ma for basalts within the rift, or "about 1,100 Ma."

In contrast, the MRS in Kansas consists of numerous fault blocks (Figs. 1 & 2), some of which contain predominantly basalts and others of arkosic redbeds; others contain composite lithologies. Texaco recently drilled the Poersch #1 well within the MRS in southeastern Washington County, Kansas (sw sw sec. 31, T.5S., R.5E.), drilling and coring 8454 feet of Precambrian rocks. The upper 4583 feet of Precambrian rock consisted of predominantly basalt and gabbro, with a few thin beds of arkose. The lower 3871 feet was predominently arkose with minor amounts of basalt and gabbro; the well bottomed in arkose (Berendsen et al. 1988). K-Ar dates by three different laboratories ranged from 587 to 800 Ma for a gabbro near the top of the Precambrian to 837 to 1,021 Ma for the deepest basalt encountered (Berendsen et al. 1988). Thus, rocks drilled in the deep test were considerably younger than the 1,100 Ma date for the northern MRS. Concern within the Kansas Geological Survey regarding the viability of these dates stems more from the younger age of the rocks in Kansas than of the quality of the dating process.

Discussion

A possible interpretation of the sequence of tectonic events that molded the basement of the Midcontinent U.S.A. may be summarized as follows:

1) The crystalline basement was fractured to form orthogonal crustal blocks by NE and NW oriented faults in latest Archean or Early Proterozoic time (Sims 1990) by N-S compressional strain.
2) NE strike-slip faults developed with sinistral sense of displacement; NW faults were dextral. Master fault zones included the NE trending Colorado-Great Lakes tectonic zone, the Sierra Grande-Las Animas-Penokean trend, and the three left-stepping en echelon segments of the present-day MRS. NW master faults are the Texas-Walker Lane, the Olympic-Paradox-Wichita Lineament, and the Chadron-Cambridge-Bourbon arch trends.
3) Continued compression, or transpression, in the time frame 1,700 to 1,600 Ma started stick-slip dextral movement on the Olympic-Paradox-Wichita fault zone, and consequent rotation of the large crustal blocks of the Midcontinent. Thus, the apparent sinistral movement of faults that now offsets the MRS in southern Minnesota and southeastern Nebraska respectively resulted from dextral rotation of crustal blocks.
4) By about 1,300 Ma most of the transpressional strain was being released along the Texas Lineament-Walker Lane tectonic zone.
5) The compressional stress field began to relax by 1,100 Ma and extension and volcanic activity along the MRS slowly worked southward from the Canadian Shield, ending in Kansas in latest Proterozoic time.
6) By Cambrian time faults of the MRS apparently were locked and healed, perhaps because of the binding effects of basic igneous flows and intrusions. There is no evidence in Kansas to indicate that there was post-Proterozoic activity along the MRS except where faulting was not associated with late igneous activity, such as the Nemaha uplift-Humboldt fault zone and the southeastern Central Kansas uplift.

7) Minor intermittent activity along master faults continued throughout the lower Paleozoic, as indicated by subtle facies relationships in the sedimentary sequence.
8) By the beginning of Middle Pennsylvanian time, however, renewed intense tectonic activity, especially along the Olympic-Paradox-Wichita fault zone, created the major uplifts of the Ancestral Rocky Mountains and the complimentary deep Arkoma, Anadarko, Paradox, and Oquirrh basins. Faults of the Central Kansas-Bourbon arch and Nemaha uplifts were reactivated, forming the northern shelves of the Oklahoma basins. Kluth and Coney (1981) blamed this tectonic convulsion on the collision of the South American-African plate with North America. However, it must be emphasized that this late Paleozoic event only reactivated existing basement structures.

Conclusions

The highly complex fault patterns of Kansas fit well with the continental-scale regmatic structure of the North American craton. When viewed at a continental scale, fault trends in Kansas are seen as segments of large-scale structural lanes that comprise the regional basement fabric. Although these structures originated in late Precambrian time, they were reactivated repeatedly throughout the Phanerozoic, affecting depositional and erosional patterns that localized petroleum reservoirs. Fault-controlled basement uplifts in Kansas comprise the complex northern shelves of the Oklahoma Pennsylvanian-age basins, and are thus instrumental in controlling petroleum emplacement.

References

Baars, D.L. 1966. Pre-Pennsylvanian paleotectonics; Key to basin evolution and petroleum occurrences in Paradox basin, Utah and Colorado. American Association of Petroleum Geologists Bulletin **50**, 2082-2111.

_____ 1976. The Colorado Plateau aulacogen, key to continental-scale basement rifting. 2nd International Conference on Basement Tectonics Proceedings, 157-164.

_____ and Ellingson, J.A. 1984. Geology of the western San Juan Mountains. In, Brew, D.C. (ed.) Field Trip Guidebook: 37th Annual Meeting Rocky Mountain Section Geological Society of America, 1-45.

_____ and Stevenson, G.M. 1982. Subtle stratigraphic traps in Paleozoic rocks of Paradox basin. In, Halbouty, M.T., Deliberate search for the subtle trap: American Association of Petroleum Geologists Memoir **32**, 131-158.

Berendsen, P., & Blair, K.P. 1986. Subsurface structural maps over the Central North American rift system (CNARS), central Kansas, with discussion. Kansas Geological Survey Subsurface Geology Series **8**, 16 p.

Berendsen, P., Borcherding, R.M., Doveton, J., Gerhard, L., Newell, K.D., Steeples, D., & Watney, W.L. 1988. Texaco Poersch #1, Washington County, Kansas - Preliminary geologic report of the pre-Phanerozoic rocks. Kansas Geological Survey Open-File Report 88-22.

Dolton, G.L., & Finn, T.M. 1989. Petroleum geology of the Nemaha uplift, Central Mid Continent. U.S. Geological Survey Open-File Report 88-450D, 39 p.

Kluth, C.G., & Coney, P.J. 1981. Plate tectonics of the Rocky Mountains. Geology **9**, 10-15.

Hinze, W.J., & Braile, L.W. 1988. Geophysical aspects of the craton; U.S. In, Sloss, L.L. (ed.) Sedimentary cover - North American Craton; U.S.: Boulder, Colorado, Geological Society of America, The Geology of North America, v. **D-2**, p. 5-24.

Sims, P.K. 1990. Precambrian basement map of the northern Midcontinent, U.S.A. United States Geological Survey Map I-1853-A.

Stevenson, G.M., & Baars, D.L. 1986. The Paradox: A pull-apart basin of Pennsylvanian age. In, Peterson, J.A. (ed.) Paleotectonics and sedimentation in the Rocky Mountain region, United States: American Association of Petroleum Geologists Memoir **41**, 513-539.

Warner, L.A. 1978. The Colorado Lineament: A Middle Precambrian wrench fault system. Geological Society of America Bulletin **89**, 161-171.

Wise, D.U. 1963. An outrageous hypothesis for the tectonic pattern of the North American Cordillera. Geological Society of America Bulletin **74**, 357-362.

THE SIERRA ALTA DE SAN LUIS A CASE OF REGMAGENIC CONTROL OF GOLD MINERALIZATION.

H G.L. BASSI
CIRGEO — Ramirez de Velasco 847
(1414) Buenos Aires
Argentina

ABSTRACT. The metallogeny of the Sierra Alta de San Luis is controlled by a network of pre-intrusive and pre-volcanic fractures striking NNE and WNW supporting the hypothesis of a regmagenic network in southern South America.

A gold district is located within the crystalline basement of the Sierras Pampeanas, a complex of Early Paleozoic metamorphics intruded by Paleozoic granitoids and pegmatites related to tungsten and beryllium deposits. A pre-Tertiary peneplain was uplifted during the Andean Orogeny and broken into blocks. A volcanic event took place, yielding trachyandesitic rocks associated with a gold-silver mineralization with related lead, zinc, copper and arsenic. The district shows extensive hydrothermal alteration which is both linear and disseminated. The linear type produces very poorly defined vein-like mineral occurrences. Extensive disseminated alteration is the source of important detrital gold deposits.

As a whole the alteration conforms with a NNE and WNW metallogenic network reflecting the crustal fracturing system.

Introduction

In 1988 the author proposed the presence of a crustal fracturing network which controlled in a supraregional and integral way the geological events in southern South America.

This paper ratifies the WNW and NNE trends of that network, acting as a metallogenetic and pan-geologic control on a mining district located in the central part of Argentina.

The district contains auriferous Cenozoic mineralization, with outstanding secondary ore deposits and abundant minor primary ones. An earlier Paleozoic metallogeny is revealed especially of tungsten, and beryllium, mica, quartz and feldspar. Both cycles, although separated in time, could be geologically related. It is reavealing that the Paleozoic tungsten ore deposits of N-S trend (more than 15 km long) are interrupted at the border of the district, exactly where they are intersected by Tertiary auriferous veins on a WNW trend.

Geology

In an area of only 80 km^2 the geological components characterizing the whole province of San Luis are exposed. that is crystalline basement and Tertiary volcanics (Fig. 1).

M. J. Rickard et al. (eds.), Basement Tectonics 9, 211–222.

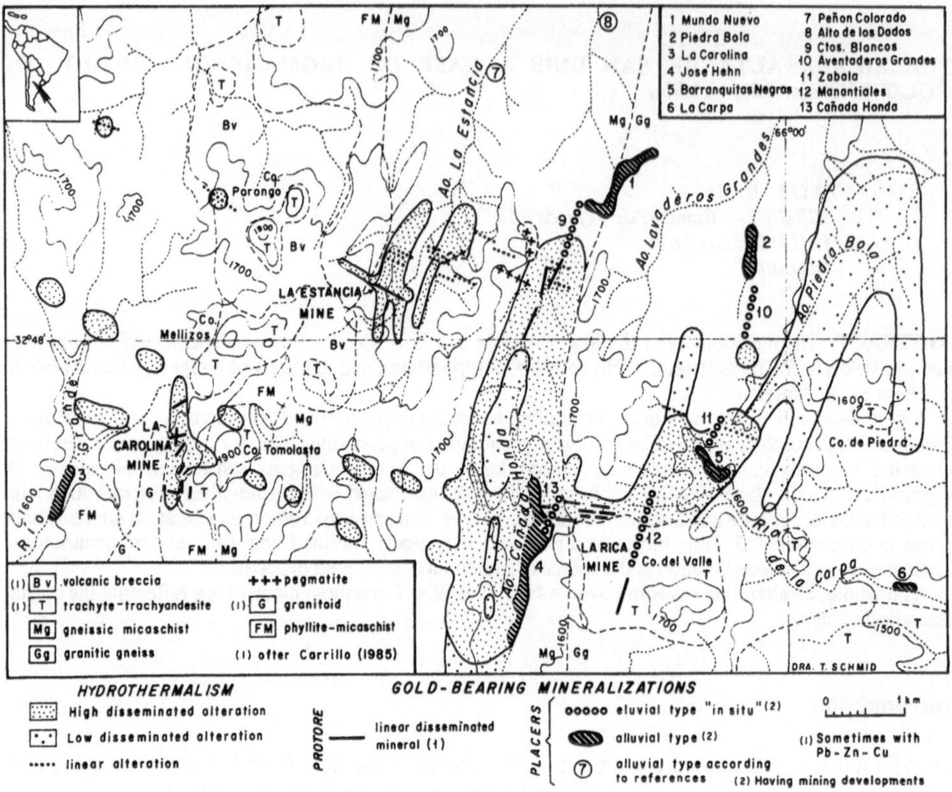

Figure 1. Schematic geology and Cenozoic mineralization in the Sierra Alta de San Luis.

CRYSTALLINE BASEMENT

The area is composed dominantly of metamorphic rocks, with a major granitoid body and pegmatite occurrences of variable extent. The metamorphic rocks show, from west to east, a gradual increase in metamorphic grade; they are grouped into three southerly elongated blocks, Western, Central and Eastern, with different petrographic characteristics.

The Western Block extends beyond the western border of the district and continues eastwards for about 5 km. This block includes the NNE lineament of volcanic cones, Cerro Tomolasta, Cerro Mellizos and Cerro Porongo. It is made up of phyllites, with subordinate intercalations of mica schist and mica-quartz schist and in places of quartzite, particularly on the eastern edge of the block.

The main mineralization at the La Carolina mine is associated with a lenticular bed of pure quartzite, interbedded between phyllite and mica-quartz schist, 15m in maximum width and about 150m long. The quartzite is massive, consisting essentially of polygonal granoblastic quartz, and passes transitionally into mica-quartz schist. The hydrothermal alteration of these rocks is of two types. Tourmalinization is tentatively assigned to the Paleozoic metallogenetic cycle, whereas silicification occurred during the Tertiary cycle. The latter is expressed as bands of secondary quartz, showing an apparent schistosity, or increase in quantity of quartz grains.

The Central Block between 3 km and 4 km wide extends from the aforementioned volcanic lineament to slightly east of the Cañada Honda. The rock is gneissic mica schist with granoblastic clear layers alternating with lepidoblastic dark ones and rare interstratified amphibolite. In the La Estancia mine the proportion of granoblastic to lepidoblastic bands is two to one, with a thickness between 20 cm and 10 cm respectively.

The Cenozoic hydrothermal alteration of the granoblastic layers of the La Estancia mine shows a total alteration of the microcline to sericite and clay, advanced silicification, altered biotite and abundant powdery pyrite. Scheelite, corresponding to the Paleozoic metallogeny, has also been observed. Within the lepidoblastic layers, which do not hold feldspars, the argillitization and sericitization are less evident, but the rock contains, instead, abundant cubes of pyrite.

The Eastern Block extends from the Cañada Honda to beyond the eastern border of the district. This Block is characterized by complex structures in the metamorphic rocks, with abundant folds of varying shapes, intensity and trends. The main rock type is a banded coarse grained granitic gneiss showing granoblastic-cataclastic texture.

Regionally cataclasis increases from the Central Block to the Eastern Block, but an intense zone has been identified at the summit of the divide of the Cerritos Blancos area. Here the granoblastic bands of the gneissic micaschist are remarkedly coarse grained and resemble the granitoid of La Carolina.

La Carolina is a syntectonic granitoid body, about 1 km thick, exposed in the Western Block near its contact with the Central Block. It extends from the Cerro Tomolasta south for several kilometres. The contacts with the metamorphic country rock are sharp and in places faulted (e.g. the Eastern border, near the Tomolasta hill). The granitoid body contains xenoliths of metamorphic rocks similar in schistosity, trend, dip and lithology to those of the country rock.

In the Central and Eastern blocks there are many pegmatitic bodies, generally concordant with the schistosity, with thicknesses between a few centimetres and 20m. Discordant ones as at Cerritos Blancos stand out. This body, between 3m and 4m thick, crops out on both sides of the main road, is sub-horizontal and covers about 4 hectares; it is now being mined for quartz and feldspar. Five hundred metres to the south there are similar bodies plunging between 10° and 15° to the north and north-east. Farther south they are parallel to the schistosity, trending NE and dipping E at about 70°. The change in dip of the pegmatites from sub-horizontal to subvertical in addition to the evidence of cataclasis on the dividing summit mentioned above, suggests the possibility of an intrusive body of the Carolina type at depth. Such a body, acting as a possible source for the pegmatitic intruded through early fractures which may have been reactivated in the Tertiary and utilized by the intermediate extrusions.

In this metamorphic setting we also find amphibolites, which are concordant with schistosity and nearly always of reduced length. In the Central Block they have thicknesses between 5m and 30m, whereas in the Eastern block some may reach thicknesses considerably larger, commonly in the cores of folds. The rock is composed of hornblende, augite and abundant opaque minerals in a granoblastic texture with slight schistosity. In the WNW zones of linear alteration the hornblende appears replaced by epidote, quartz, titanite and calcite. Originally, these amphibolites were interpreted as products of metamorphic processes affecting zonal ultra-basic bodies (Horok 1973), but now they are considered as metamorphosed basic complexes (Merodio et al. 1978).

TERTIARY VOLCANICS

The volcanics are by far the largest exposures of the Tertiary effusions in the Province of San Luis. There are cones and domes forming high hills, with abrupt walls, standing out clearly against the Crystalline Basement. They are aligned in a WNW trend known as a Morro-Tomolasta regional run. In detail two alignments 3 km apart with that trend can be distinguished. The northern one is formed by the Porongo and de Piedra Hills, and the southern one by Tomolasta and del Valle hills. There is also a NNE alignment over 7 km, along which the Tomolasta, Mellizos and Porongo are disposed. Further, a second alignment in this direction is outlined by the Cerro de Piedra and Sololosta (not included in Fig.1).

These four reticulated volcanic alignments determine the distribution of the Tertiary metalliferous mineralization and reproduce, in detail, the supraregional scheme of crustal fracturing (Bassi 1988). These effusions are formed of trachyte-trachyandesite (Böckmann 1948) and volcanic breccias which sporadically surround the extrusive bodies (Carrillo 1985). The rocks of the Tomolasta-Porongo run are porphyric with about 25% phenocrysts of either andesine (trachyandesite) or sanidine (trachyte), with hornblende or biotite (Carrillo 1985). Hydrothermal alteration has altered the phenocrysts to chlorite, sericite and calcite, which is also disseminated in the groundmass. According to Böckmann (op. cit.) the volcanic hills of the region are cumulo-volcanic type, except the dome Canutal an Los Pájaros, which is a neck.

TECTONIC FRAMEWORK AND STRUCTURE

Metamorphics with a subvertical to easterly dipping schistosity are the most widespread rocks in the region. According to Kilmurray (1977) these were formed by the Devonian F2 deformation phase. Their metamorphic grade increases eastward and three blocks (Western, Central and Eastern), elongated in the direction of schistosity, can be defined.

A granitoid body appears to have been tectonically emplaced along the boundary between the first two blocks. Its origin is assigned to Kilmurray's G3 cycle. The existence of another granitoid body, not exposed, is inferred from the intensive migmatization and pegmatite injection within the outcropping metamorphics; it should be emplaced at the stream divide, striking E-W.

During the Tertiary the district was affected by volcanic effusions following two WNW lineaments one 3 km apart from the other. They give rise to a regional run more than 50 km long extending up to the Morro. This regional fracture system corresponds to the reactivation of a primogenial cycle belonging to the Gondwana Regmagenic Network (Bassi 1988).

The tectonic scheme for the district, including metallogenic events is proposed in Table 1.

Cenozoic Metallogeny

Even though the primary gold mineralization was Tertiary, the more important deposits, from an economic point of view, are the secondary ones which formed in the Quaternary.

PRIMARY ORE DEPOSITS

The auriferous mineralization of the Sierra Alto de San Luis are not vein-like occurrences, but the result of hydrothermal alteration along ill-defined fractures, concordant or transverse to schistosity. That is why none of the mines of the district have continuity of mineralization in a single structure and are nearly always disconnected sections. Sporadically high ore grade occurs where veins cross. These (locally named 'cruceros') are of restricted length and commonly rich in free gold.

The main occurrence of the La Carolina mine is an exception, developed in the west wall of a lenticular quartzite horizon with a maximum thickness of 15m and not more than 150m long, interbedded between phyllite to the west and mica-quartz schist to the east (Fig. 2). The vein-like body has a maximum exposed width of 2.50m (one sample yielded 1 g Au/t). In some instances the workings of the area extend towards the quartzitic horizon as a pit, indicating the auriferous fluids penetrated the quartzite wall rock. A sample from the central quarry yielded also an average grade over 14m thickness of 1 g Au/t, indicating an auriferous alteration halo around the vein-like structure. The inset of Figure 2 shows a band of strong alteration of up to 20m wide which includes the quartzitic horizon.

Paleozoic	Cambrian Subcycle	E-W Fractures (F1. Kilmurray)	Preintrusive path of the inferred granitoid (Blancos Hill) Paths for the development of the drainage of the Paleozoic peneplain (relics observed in the present morphology
	Carboniferous Devonian Subcycle	NNE * Fractures (Kilmurray's F 2-3)	Preintrusive path of the concordant Cycle granitoid of La Carolina. Paths of the Caledonian metallogeny (Angelelli 1984) precursors for the Andine blocking.
Tertiary Cycle	NNE * Subcycle & disseminated	Alignment of Volcanics	Tomolasta — Canutal run Sololasta — Valle — de Piedra run
		Lineal impregnation Carpa belts hydrothermal alteration.	Carolina, Estancia, Cañada Honda and
	WNW *	Alignment of Volcanics	Porongo — de Piedra run Tomolasta — del Valle run
		Linear and disseminated hydrothermal alteration	C° de Piedra and C° Tomolasta belts

Table 1. Tectonic scheme for the Sierra Alta de San Luis.
(* belonging to the Gondwana Regmagenic Network).

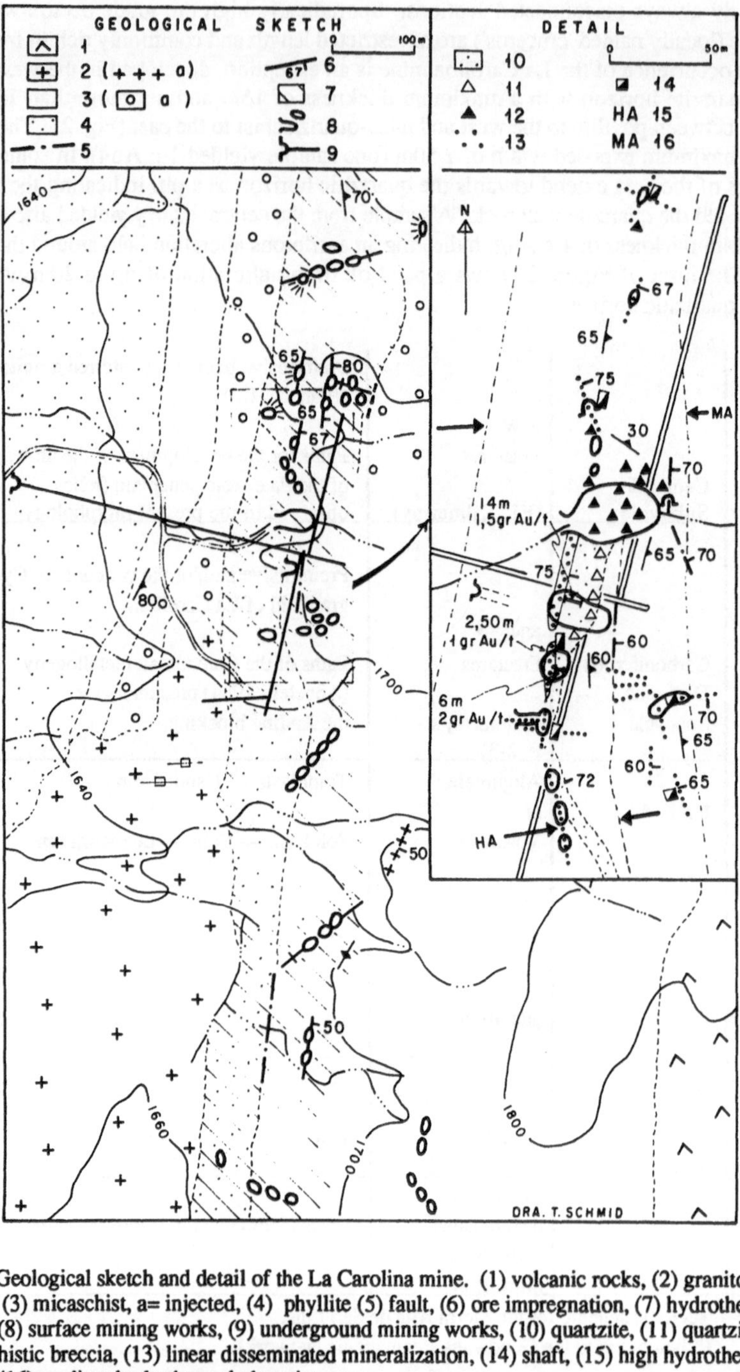

Figure 2. Geological sketch and detail of the La Carolina mine. (1) volcanic rocks, (2) granitoid, a= pegmatite, (3) micaschist, a= injected, (4) phyllite (5) fault, (6) ore impregnation, (7) hydrothermal alteration, (8) surface mining works, (9) underground mining works, (10) quartzite, (11) quartzitic breccia, (12) micashistic breccia, (13) linear disseminated mineralization, (14) shaft, (15) high hydrothermal alteration, (16) medium hydrothermal alteration.

The rest of the vein-like mineralizations cropping out in this mine are of variable trend and short length and are distributed along a moderately altered band more than 1000m long and up to 150m wide. The underground mining, apparently completely performed by West Mining Co. between 1884 and 1889 (Gerez 1934) consisted of a crosscut of 380m and a 400m gallery along the trend. According to Gerez — probably the last one who had access to these workings — only a low percentage of the reserve was exploited. The mean grade obtained by Gerez, from 56 samples taken in the non-collapsed working was 4 g Au/t. Consequently, considering modern mining methods, this ore deposit ought to be explored as a low-grade disseminated gold prospect.

Examples of mineralization that are discordant with schistosity are the La Estancia and La Rica mines. The latter, now inaccessible, comprises a group of short, sub-parallel, vein-like occurrences, distributed in an E-W belt 800m long and 200m wide. La Estancia constitutes a case of mineralization lodged in a single fracture, observable on the surface for more than 500m and surveyed underground for a length of 200m. In spite of the persistence of the fault development, the mineralization of interest is only found in certain areas of the run. It occurs in subvertical ore shoots generated by the intersection of the fracture with transverse belts of hydrothermal alteration, concordant with the schistosity of the metamorphic country rock (Fig. 3). The mean grade of the adit over the minimum width of exploitation, of about 1m: 1.4 g Au/t, 62 g Ag/t, 3.1 %Pb, 3.8 %Zn and 1.1 %As (sampling Lapidus 1952). In the exploited ore shoot (probably up to 60m deep), located to the west and near the road, one sample obtained at the mouth of the shallow shaft yielded, for the 1m thick vein: 7 g Au/t, 240 g Ag/t and 0.4 %As and the wall-rock yielded 1 grAu/t, 36 g Ag/t and .02 %As. From the underground mining morphology and the transverse alteration observed on the surface, two more ore shoots of minor importance are predicted to be present along the run. Consequently we prefer to consider these vein-like occurrences of the district as 'linear disseminated-mineral zones', meaning — a hydrothermal alteration with variable mineralization and intermittent run.

Apart from the known mines, linear disseminated-mineral zones are observed near the headwaters of the Cañada Honda creek. Mineralized zones 1-2m wide crop out intermittently along the water course for almost two kilometres, characterized by ferruginous schist in places with fresh pyrite. This linear disseminated mineralization is accompanied by a zone of wall-rock alteration 50m wide containing disseminated iron oxides. One sample of the pyritic schist yielded 1 g Au/t, 2 g Ag/t and traces of As. North of the Cerritos Blancos eluvial placer this alteration disappears. Another similar impregnation is exposed on the western foot of the Cerro del Valle, covered in the north by the Manantiales eluvial placer. The mineralization of these linear alteration zones consists of gold and silver associated with pyrite and often accompanied by Pb, Zn, Cu and As sulphides. In La Estancia mine, pyrite is the predominant sulphide, in places replaced by marcasite, and in decreasing order, sphalerite, galena and arsenopyrite.

A clear relationship between gold and arsenic in the ore of La Carolina and La Estancia mines is observed. This relationship should be useful in a regional strategic resource evaluation.

Hydrothermal Ateration and Minerogenetic Belts In contrast to the limited magnitude of the vein-like occurrences, the disseminated and linear hydrothermal alterations are particularly widespread. They affect the development of the secondary auriferous occurrences and define minerogenetic belts.

The linear alterations are either concordant or discordant with schistosity. The concordant zones are difficult to identify especially in areas of disseminated alteration. On the other hand, the transverse linear alterations are easily seen as WNW striking topographic highs 10-15m wide approximately 1.5m above their surroundings. They are visible in La Estancia mine area where the

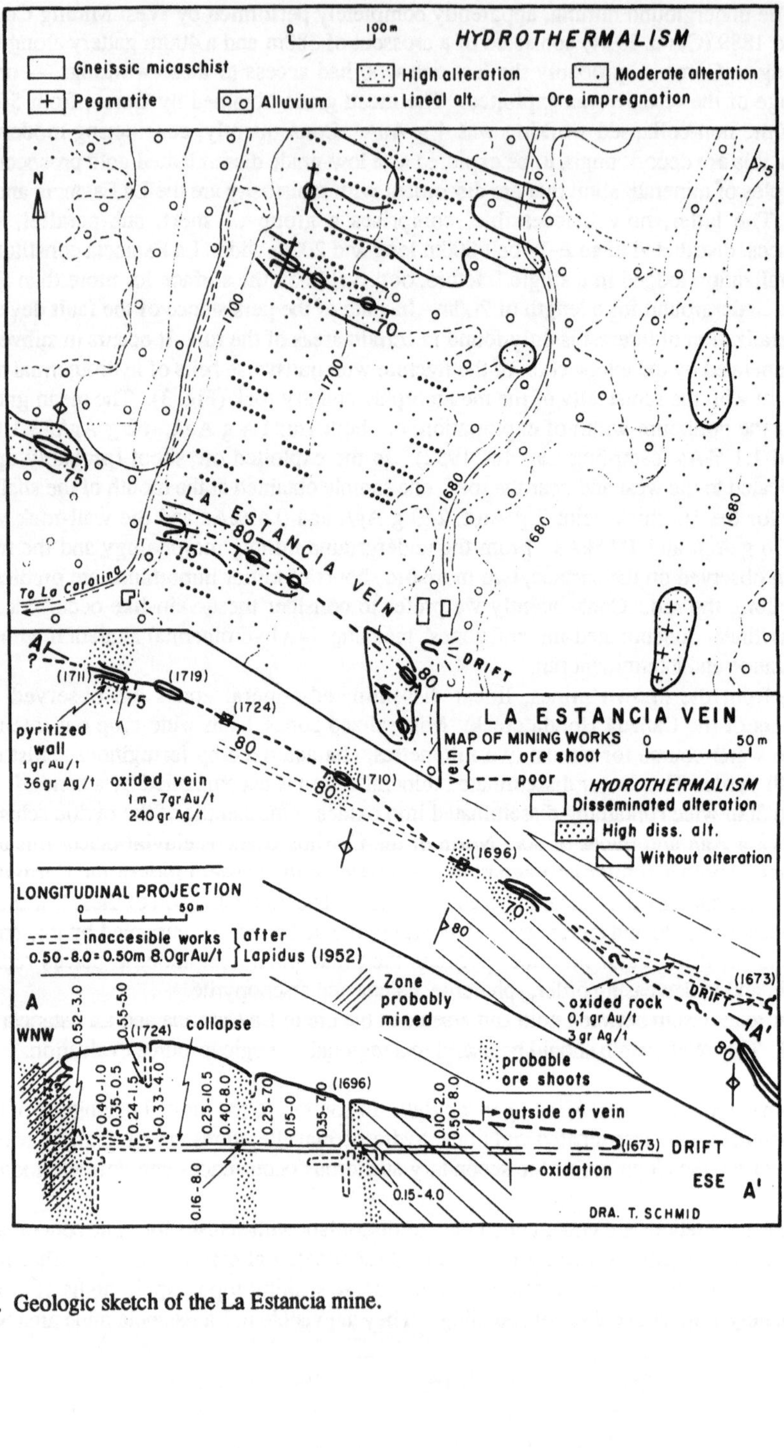

Figure 3. Geologic sketch of the La Estancia mine.

ridges are about 50m apart from one another. The altered schist is distinguished by a slight argillation, iron oxides and silicification, made evident by its erosive competence, but difficult to detect because of the abundance of quartz in those rocks.

The disseminated hydrothermal alteration is also panoramically observed because of its lighter colouring, with a yellowish limonitic tinge. The latter suggests the presence in the primary zone of disseminated pyrite, which is associated with gold in this region. The dissemination alteration is distributed in the district as belts of significant thicknesses and lengths which we interpret as associated with regional minerogenetic lineaments. They have NNE and WNW trends (Fig. 4) which agree with the suggested Gondwana Regmagenic Network (Bassi 1988).

In the N20°E direction, four belts, from west to east: known as Carolina, Estancia, Cañada Honda and Carpa have been determined. The first and the third stand out due to their particular structural position and metalliferous importance. The Carolina belt, situated in the Western block of metamorphic rocks includes the granitoid body, the Cerro Tomolasta-Cerro Porongo volcanic run and the La Carolina mine together with their disseminated alteration and hydrothermal metamorphism which surrounds the Cerro Porongo. The Estancia belt includes the mine and two alteration areas of the Maray sector, containing as other belts concordant bands of more intense alteration, which carry the ore shoots of the main vein already described. Towards the North, where the belt is crossed by the provincial route, a preliminary sample taken out along 70m yielded 0.1 g Au/t, 12 g Ag/t and 0.003% As. The Cañada Honda belt located in the Central block, near the western border includes the largest disseminated alteration area in the district, forming the

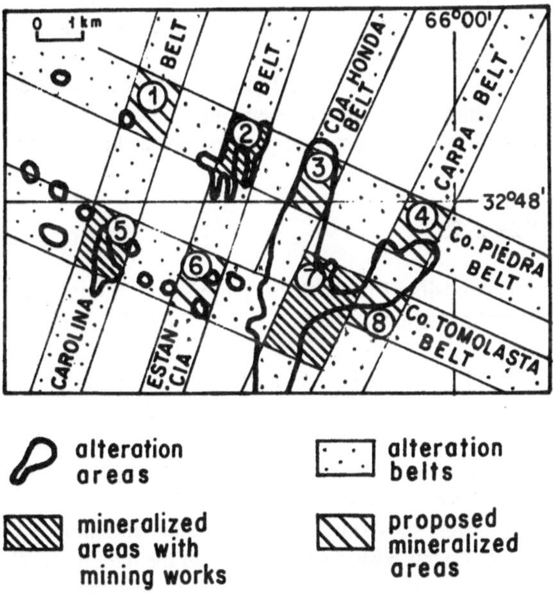

Figure 4. Scheme interpreting the distribution of the Tertiary hydrothermalism and the resulting mineralized areas in the Sierra Alta de San Luis. (1) C° Porongo, (2) Estancia, (3) Cerritos Blancos, (4) Zabala, (5) Carolina, (6) Maray, (7) Rica, (8) Manantiales.

principal gold source of the detrital ore deposits. It contains the Cerritos Blancos and Cañada Honda alluvial type placers and also the La Rica mine.

La Carpa belt includes the well mineralized northern contact (Rosello & Garcia 1983) of the andesitic-trachyandesitic body of the Cerro del Valle, exposed mainly in the basin of the Rio de la Carpa headwaters. This basin trends NNE, is about $2km^2$, and coincides with a zone of intense alteration which facilitated erosion. The La Carpa belt contains an exposed linear disseminated-mineral zone which gave rise to the Manantiales eluvial-type placer, and possibly two other zones, now covered by sediments, which gave rise to the formation of the Zabala and Aventaderos Grandes eluvial placers.

The Cerro de Piedra and Cerro Tomolasta belts have been identified with WNW trending disseminated alteration. The former to the north, includes the alteration area of the western foot of the Cerro Porongo, the La Estancia mine, the linear disseminated-mineral zone of Cañada Honda and the alteration zone which was the source for the Aventaderos Grandes eluvial placer. The Cerro Tomolasta belt connects the La Carolina and Rica mines and a group of minor areas of hydrothermalism.

The NNE and WNW alteration sets form a network. The areas of intersection of the two trends coincide with known mineralized occurrences with the exception of Cerro Porongo and Maray. Only Carolina, Estancia and Rica have workings on the primary occurrences. The NNE-WNW network is similar on a smaller scale to the supraregional crustal fractures controlling the large lineaments. The principle on which the hypothesis of regmatic network for southern South America has been based initially is the preferential trends of the tectonic orientations (Bassi 1988). These have been been locally identified in some mines of that large region. With them a parallelism between the areal distribution of the mineralization and the dominant orientation of the megafractures was established.

In the Sierra Alta de San Luis this supraregional pattern is manifested. Consequently this framework of intersecting sets with its minerogenetic interpretation could serve as a basic tool for the development of a primary mineral prospecting programme and as a starting point in prospecting for placers.

SECONDARY ORE DEPOSITS

The secondary auriferous deposits of the district comprise traditional alluvial-type placers, and some eluvial-type placers with unusual features.

Eluvial-type Placers The eluvial placers are auriferous accumulations formed by the in-situ weathering of linear disseminated-mineral zones, removal of the gangue and accumulation of free gold without any lateral displacement. They are called "in situ eluvial-type placers". Such a process requires a relatively flat terrain and a vertical primary original occurrence, so that the gold can concentrate directly over its original source.

Following Tertiary metallogenesis, oxidation and secondary enrichment processes took place in a near-surface environment. Decomposition and erosion of the light material followed, and finally a barren overburden of eolian sediments, (which also constitutes the overburden of the alluvial-type placers) was emplaced. The overburden thickness is not more than 4m, judging by the ancient workings and the exploratory holes made by Ahlers in the Cerritos Blancos eluvial placer (Burmeister 1934-35).

Five eluvial placers have been identified, all of them in the eastern half of the district. Cerritos Blancos and Cañada Honda deposits are related to the minerogenetic belt of Cañada Honda.

Aventaderos Grandes, Zabala and Manantiales deposits are associated with the Carpa belt. The Cerritos Blancos and Aventaderos Grandes are located on the northern slope, near major drainage divides. Erosion of these in situ deposits gave rise to the Mundo Nuevo and Piedra Bola alluvial placers, respectively. They have been mined from about the middle of last century with great success, and supported an important small village and commercial trading station (Burmeister, 1934-35). The Cerritos Blancos eluvial placer is about 600m long and originated from the linear disseminated-mineral zones of the upper stretch of the Arroyo Cañada Honda. The impregnation to the south of the divide was eroded by the stream and its gold contributed, quite probably, to the José Hehn alluvial placer. The Cañada Honda eluvial placer is located on the western end of La Rica mine. In 1944, 0.40m of the bed rock (representing the eluvial placer) together with 0.30m of the alluvial placer yielded 60% of the gold of the whole with a grade around 6 g Au/m^3 (Bassi 1948). The Zabala and Manantiales eluvial placers are located in the headwaters of the La Carpa river basin. The former has evidence of mining over about 400m length and the latter over about 800m.

The grade of these eluvial placers depends, not only upon the gold concentration of the original primary mineralization but on the depth of the eroded column. In the case of Zabala and Manantiales, which are located at the bottom of the basin, superimposed on a Paleozoic drainage system identifiable on its borders, the eroded column is 60m deep. With a grade of only 1 g Au/t and 1m of width of its primary mineralization, the gold accumulated could be in the order of 24 g Au/m^3 per metre of thickness of eluvium.

Alluvial-type Placers Deposits are distributed on both watersheds. The northern one, with a lower slope, contains large shallow placers, whereas the southern one contains narrow deposits of greater thicknesses and grade (Bassi 1948).

The alluvial placer here named José Hehn — in memory of one of its more persistent exploiters — originated in the first place by several contributions within the basin catchment of the Cañada Honda stream, from the linear impregnation of the headwaters of the gulch, also responsible for the Cerritos Blancos eluvial placer, and then, from the erosion of La Rica mine and the Cañada Honda eluvial placer, which underlies the alluvium. Finally, from the supply of the erosion of practically 3 km^2 of intense hydrothermal alteration which, assuming only 0.05 g Au/t, would theoretically represent 40 tons of gold per 100m of the eroded column.

Similarly, though with an auriferous contribution coming from a third of the area, it is worth emphasizing the primary source of the headwaters basin of the La Carpa river. It is probable that the gold content of the eroded column could be higher than that of the basin of the Cañada Honda stream, judging from a greater hydrothermal alteration grade.

The La Carolina alluvial placer could be in third place, with respect to the two mentioned above, as its source for gold is the extensive alteration including the primary ore deposit. Probably the volcanic breccia which surrounds Porongo hill, drained by the same catchment network, would also contain an obscured hydrothermal alteration in the tuffs. In which case the alluvial accumulation situated upstream from the supply of the Carolina stream, could also be an auriferous placer.

Conclusions

The metallogeny of the Sierra Alta de San Luis is controlled by a NNE-WNW net of pre-intrusive and pre-effusive fractures, part of the regmagenic network of southern South America.

Although it has relatively few workable primary auriferous deposits, it shows important hydrothermal areas, quite probably everywhere accompanied by disseminated gold — constituting a very important source of alluvial gold.

The gold mined from the placers, up to the present about one or two tons, represents only a small proportion of the potential reserves of the alteration-erosion association.

Therefore the region is considered prospective for the development of an integrated programme of alluvial exploration based on tectonic principles.

Acknowledgements

Thanks are due to the Consejo Nacional de Investigaciones Cientificas y Técnicas and to the Centro de Investigaciones en Recursos Geológicos for their grant.

References

Angelelli, V. 1984. Ore mineral deposits of Argentina (Spanish). Comisión Investigatión Científica Provincia de Buenos Aires, V. 1 & 2, La Plata.

Bassi, H.G.L. 1948. The auriferous alluvial type placers of La Carolina-Carpa River zone, Provincia de San Luis (Spanish). Revista Asociación. Geológica Argentina 3, 5-53.

----1988. Hypotheses concerning a regmagenic network controlling metallogenic and other geologic events in the South American Austral Cone. Geologische Rundschau 77, 491-511.

Böckmann, S.E. 1958. Study of the Tertiary Volcanoes of the Province of San Luis (Spanish). Unpublished Doctoral Thesis, Buenos Aires University.

Burmeister, C. 1934 & 35. The gold of the Sierra de San Luis (Spanish). Revista Minera 6, 65-120 and 7, 8-41.

Carrillo, R. 1985. Tertiary vulcanism between the Tomolasta and Redondo hills (Spanish). CIRGEO, unpublished report, Buenos Aires.

Gerez, J.M. 1934. Preliminary report on the sampling of the La Carolina gold mine (Spanish). Secretary of Energy and Mining — Argentina — unpublished report.

Horak, M. 1973. Mining reservation area No 7, basic rocks (Spanish). Dirección General de Fabricaciones Militares, portfolio p.704-11, unpublished, Buenos Aires.

Kilmurray, J.O., & Dalla Salda, L. 1977. Structural and petrological characteristics of the Central and South region of the Sierra de San Luis (Spanish). Obra del Centenario del Museo de La Plata 4, 167-178.

Lapidus, A. 1952. Report on the La Estancia mine, San Luis (Spanish). Secretaría de Minería, unpublished report, Buenos Aires.

Merodio, J.C., Dalla Salda, L., & Rapela, C.W. 1978. Preliminary geochemical and petrological study of the San Francisco basic body region, Provincia de San Luis (Spanish). Revista de la Asociación Geológicia Argentina 33, 122-138.

Rosello, E.A., & García, H.H. 1983. Geology of the Cerro del Valle and its adjacent auriferous occurrences area (Spanish). 2nd Economic Geological National Congress, San Juan, Argentina 2, 617-630.

TAR PAVEMENT RIFT - TRANSFORM TECTONIC MODEL AND SOME EXAMPLES IN NATURE

M.B. KATZ
Department of Applied Geology
University of New South Wales
P.O. Box 1
Kensington N.S.W. 2033
Australia

ABSTRACT. Tar-pavement structures have been observed to geometrically resemble rift-transform fault systems. These consist of a series of spaced, dilatant fractures or rifts connected up by linear arrays of en echelon cracks or 'transform fault' lines. The geometrical relationships in these tar pavement structures resemble tectonic rift-rift offsets and transform faults described in nature and experiment. These structural models are applied to aspects of the tectonic evolution of the East African rift and Rhine graben, as well as the southern portion of the Australian New England Orogen.

Introduction

An understanding of continental rift-rift offsets often invokes the oceanic rift-transform fault model (Illies 1972; Courtillot et al. 1974; Fairhead 1980; Illies 1981; Elmohandes 1981; Katz 1987). The extension in the rift is accommodated by high-angle shear transform faulting, which in a continental setting is characterised by a zone of en echelon faults (Illies 1972). These transform, shear-initiated en echelon faults have been described as either low-angle (~10°) Riedel shears or higher angle (~45°) fissures or tension gashes, to the transform direction (Courtillot et al. 1974). Similar features are observed in a tar pavement model described below where the shear transform zone analogues are characterised by low- angle (10°-20°) to higher angle (40°-50°), often sygmoidal, en echelon cracks. Most cracks are dilatant and this would indicate that both shear and tensional forces are operating (Hancock 1972).

Tar Pavement Models

TRANSFORM LINES

On a gently inclined tar-pavement sidewalk in Sydney Australia, a series of spaced open fractures or rifts, oriented perpendicular to the sidewalk slope direction, are connected up by lines of sygmoidal, en echelon cracks or transform faults (Fig.1). The sidewalk rifts are parallel or offset. Parallel rifts are connected up by a complex line of shear (Fig. 1). Offset rifts are connected by either sinistral or dextral, en echelon 'transform' gash arrays (Figs. 2 & 3). The 'transform' gash arrays are tapered cracks with sygmoidal form, that represent either tension cracks or Riedel shears or a combination of both. The angle of these cracks to the transform line direction is about 10°-30°

M. J. Rickard et al. (eds.), Basement Tectonics 9, 223–231.

and can increase to about 40°-50° as the line of rifting is approached (Figs. 2 & 3). The low-angle (10°-30°) cracks are usually less dilatant and are similar to Riedel shears (Fig. 2). The higher angle (40°-50°) cracks are tension gashes. The fact that most of the 20°-30° cracks are dilatant (Fig. 2) would suggest that both shear and tensional forces are operating (Hancock 1972). The lower angle, less dilatant cracks are more commonly developed in the middle of the transform line and become higher angled and more dilatant as the rift is approached (Fig. 3). The rift itself joints the transform line by a prominent high-angle bend.

Figure 1. Up slope view of tar pavement showing parallel and offset rifts connected by en echelon transform lines, Sydney, Australia.

RIFTS

The tar-pavement rifts commonly have a zig zag, en echelon geometry (Figs. 2 & 3). which also implies tensional and shear controls (Freund & Merzer 1976) One such feature is shown in Figure 4. A sinistral, en echelon transform line is connected to a dextral en echelon rift line. An interpretative resolution of the shear forces suggests that the extension is not operating normal to the rift axis but at some oblique angle, thus the rift is a transtensional feature.

TRANSFORM LINE-RIFT TRANSITION ZONE

When the transform lines of en echelon cracks meet the rifts there are progressive changes in the crack geometry and orientation (Figs. 1, 2, 3 & 4):

1) The transform lines display a bend or fork into the rift (virgation, Courtillot et al. 1974)
2) The en echelon cracks change in their orientation from 10°-20° to the transform direction, to 30°-40° as they bend or fork into the rift.
3) The en echelon cracks become more dilatant as they approach the virgation.
4) Where the rift connects into the transform line of en echelon cracks a conspicuous rift bifurcation is developed (Fig. 5).

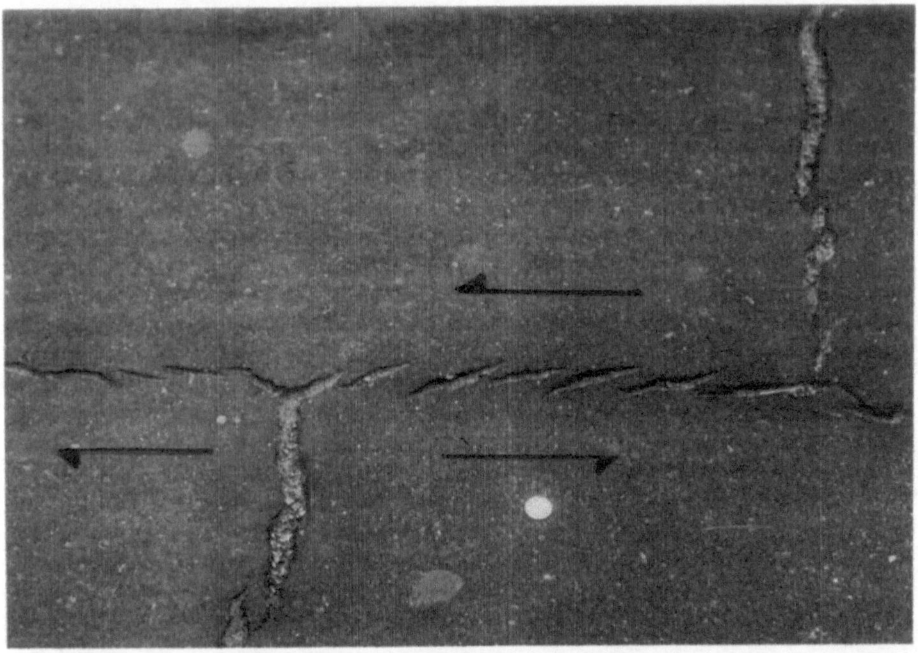

Figure 2. Tar-pavement right-hand offset rifts connected up by a sinistral en echelon, slightly sygmoidal (S), cracks at an angle of about 20° to the transform-line direction, to 30° as it bends or forks into the rift. Dextral en echelon zone to the left displays a lower angle of about 10° and the cracks are not as dilatant as the cracks in the sinistral zone. Rifts have en echelon shear features and can be considered transtensional structures. Twenty Cent piece (2.8cm) for scale.

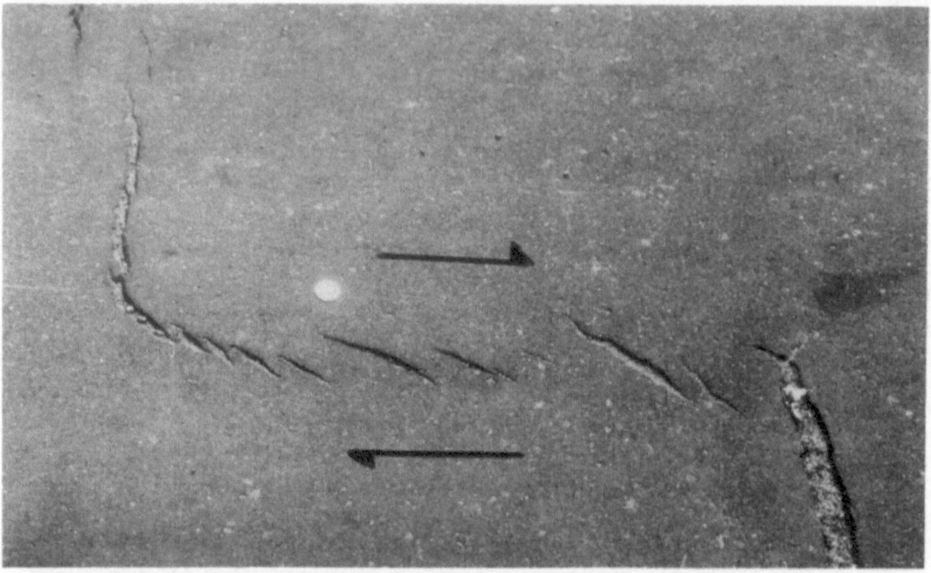

Figure 3. Tar-pavement left-handed offset rifts connected up by a dextral, en echelon, slightly sygmoidal (Z), cracks at an angle of about 20°-25° to the transform-line direction to 40° as it bends or forks into the rift. Middle of zone less dilatant. Rifts have en echelon shear features and can be considered transtensional structures. Twenty Cent piece (2.8cm) for scale.

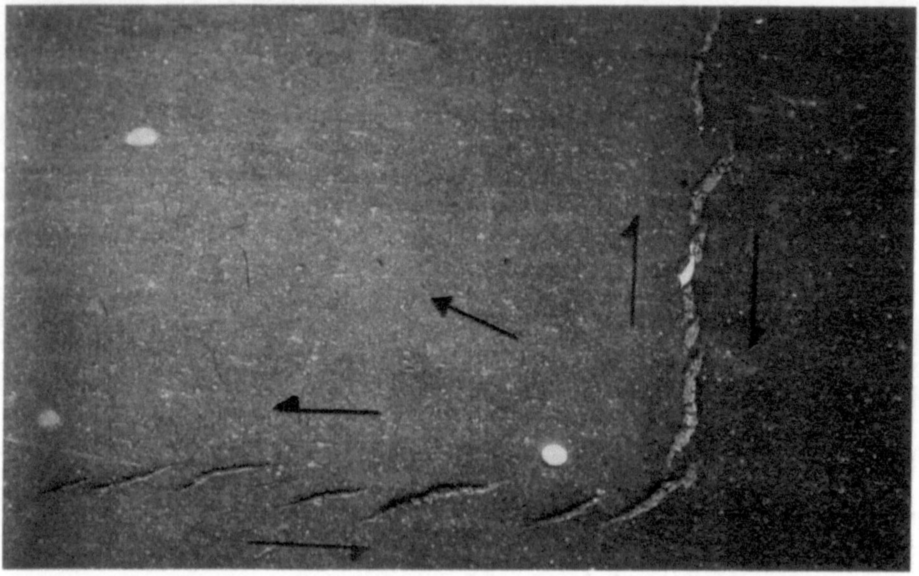

Figure 4. Tar-pavement sinistral en echelon transform line connecting up with a dextral en echelon rift line which implies oblique extension or transtension. Twenty Cent piece (2.8cm) for scale.

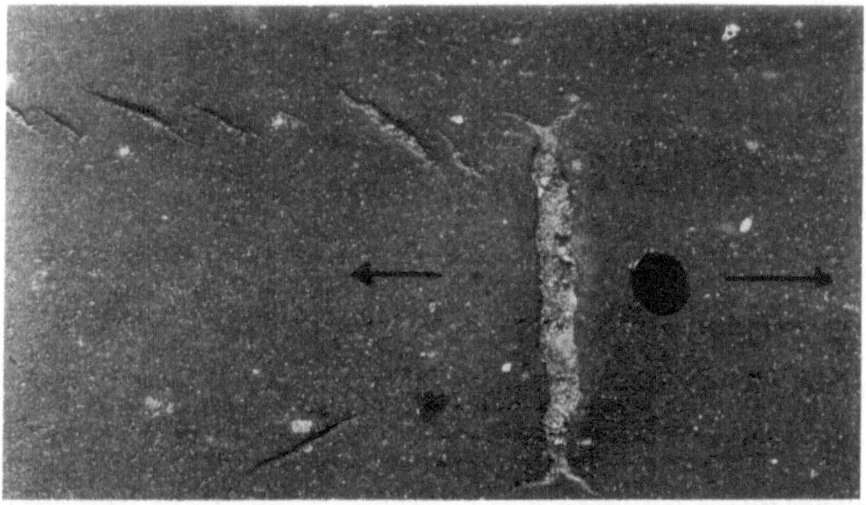

Figure 5. Doubly terminated rift marked by rift bifurcations as the rift passes into transform lines of en echelon cracks.

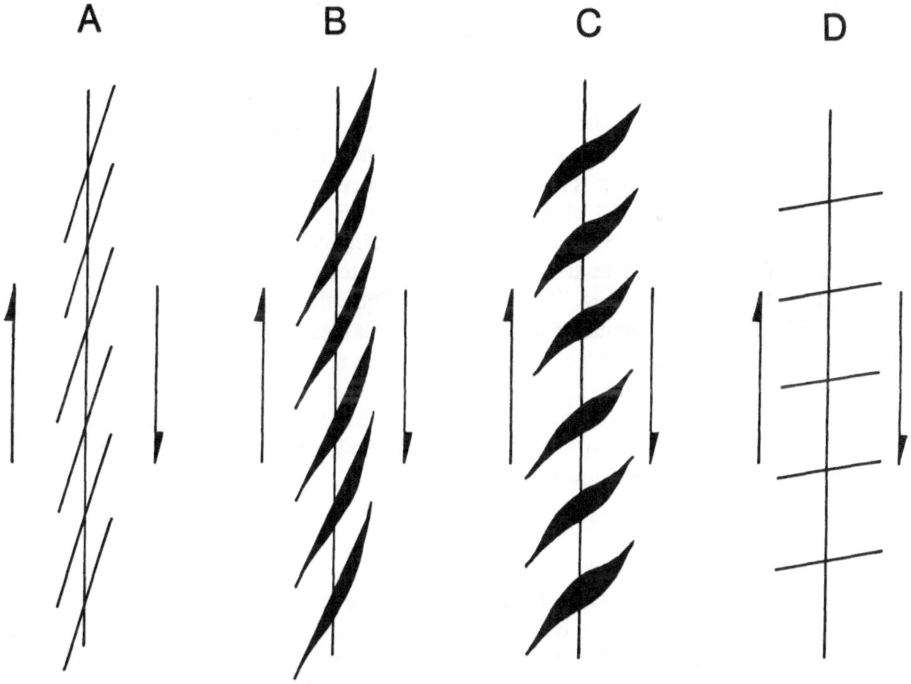

Figure 6. Range of dextral transform-line structures found in tar pavement and experimental clay models of Elmohandes (1981) A). Lowest angle (10°) Riedel shear cracks B). Low-angle (20°) transtensional cracks C). High-angle (45°) tension cracks D). Highest angle (80°) conjugate Riedel shear cracks

Discussion

The features described in tar-pavement rifts—transform lines have been observed in similar clay model experiments (Courtillot et al. 1974; Elmohandes 1981). Low-angle (10°) Riedel shear cracks, higher angle (20°-30°) transtensional cracks and high-angle (40°-50°) tensional cracks are observed in both clay and tar-pavement models. Conjugate Riedel shears which are developed at very high angles (80°-90°) to the transform lines have been noted in clay experiments (Elmohandes 1981) although they were not noted in the tar pavement. A range of Riedel shears, transtensional and tensional structures can be observed along these transform lines (Fig. 6). In geological terms the Riedel shears are faults and the transtensional and tensional structures can be fault troughs, or local grabens, and if deep seated, may tap magma.

Natural Examples

EAST AFRICAN RIFT

The geological-tectonic development of the East African rift system has been described by McConnell (1972). Rift-transform interpretations have been given by Mohr (1976) and Katz (1987). An en echelon transform fault has been suggested for the connection between the Kenya rift and the Pangani rift of Tanzania (Fairhead 1980). On structural tectonic maps (McConnell 1972), and space imagery (Mohr 1976), the Kenyan rift on its southern extension bifurcates towards the west-southwest and east-southeast (Fig. 7, compare with Fig. 2). To the west-southwest McConnell (1972) has the rift following the Lake Eyasi structure as it strikes into the Tanzanian craton. Mohr (1976) shows these WSW lineaments discontinuously penetrating the Tanzanian craton up to the Lake Tanganyika rift (Fig. 7). In contrast, the ESE trends are not clear, for this transitional bifurcated area is mainly covered by an easterly chain of Cenozoic volcanics which extends to the Pangani rift of northern Tanzania (Fig. 7). This volcanic chain which includes Mount Kilimanjaro, has been interpreted as being controlled by SE-NW striking en echelon fissures developed by a dextral transform fault which connects the Kenyan rift with the Pangani rift (Fairhead, 1980) (Fig. 7 and Fig. 3). The relation of the Kenya rift and the Tanganyika rift and other segments of the West rift is not clear. A series of discontinuous, en echelon WSW-ENE lineaments of the Lake Eydsi trend are recognized by Mohr (1976) as far as Lake Tanganyika and beyond. It is conceivable that these WSE-ENE structures could also be the expression of a wide east-west transform line connecting the Kenyan rift with the Lake Tanganyika rift (cf. Fig. 7, Figs. 2 & 4). The zig zag pattern of both the Kenya and Tanganyika rifts (Fig. 7) may thus be explained by transtensional processes (see Fig. 4).

RHINE GRABEN RIFT

The Rhine graben - Bresse graben right handed offset has been interpreted by Illies (1972, 1981) to be a wide zone of oblique sinistral transform faulting which consists of en echelon Riedel and conjugate Riedel faults (Fig. 8). The rift valley itself is a sinistral transtensional feature which has an internal Riedel shear pattern. The northern part of the Rhine graben shows a marked bifurcation, to the northwest into the Lower Rhine rift and to the northeast into the Hessen depression. Both forks are characterised by Cenozoic volcanism.

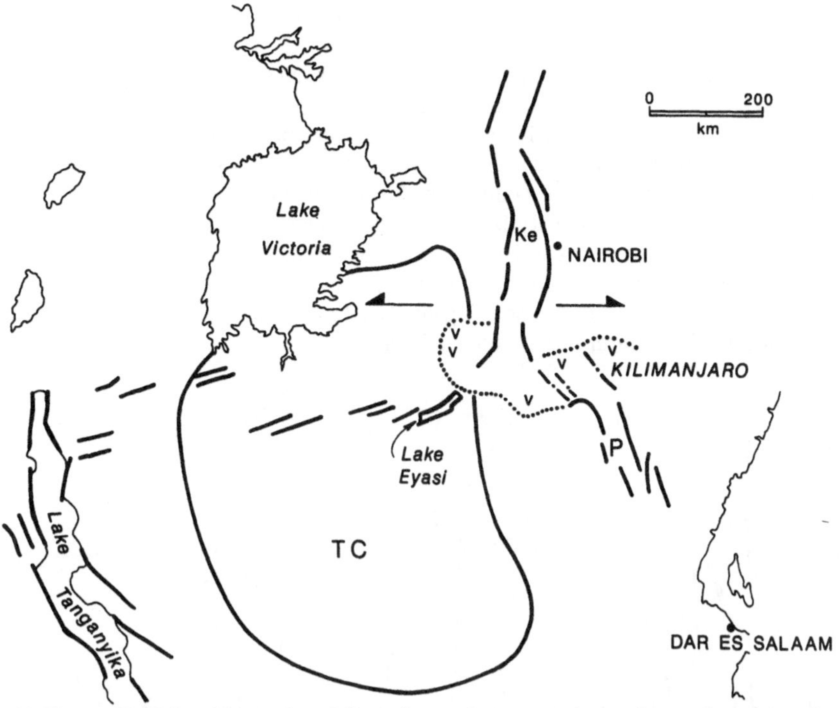

Figure 7. Kenya rift (Ke) and its southern bifurcation, to the east-southeast along a dextral transform line connecting to the Pangani rift (P) (Fairhead, 1980) and possibly to the west-southwest along a sinistral transform line connecting to the Tanganyika rift. Note extensive volcanics in the bifurcation area. TC - Tanzanian craton, V - volcanic field.

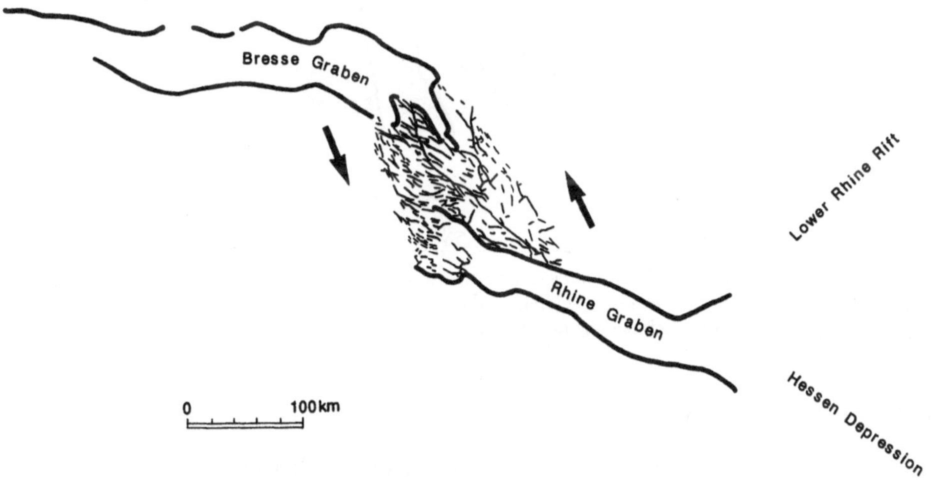

Figure 8. Rhine graben - Bresse graben transform zone and northern bifurcation into the Lower Rhine rift and Hessen depression marked by volcanic fields (after Illies 1972).

SOUTHERN PART OF THE TAMWORTH TROUGH, N.S.W., AUSTRALIA

Although the Tamworth trough is not a rift structure sensu stricto it has geometrical and structural features that can be modelled using a rift- transform line mechanism (Fig. 9). The NNW trending Tamworth trough bends to the ESE on its southward extension into a zone of complicated faulting with local igneous activity (Roberts and Engel 1987). This zone is geometrically similar to a dextral transform zone which would connect the Tamworth trough to some as yet unknown structure offset to the east. Fault analysis to the north also suggests dextral transtension (Katz 1986). Faults trending southeast can be considered as dextral Riedel-type faults, and those trending north-northeast are the sinistral conjugate Riedels. The Barrington Tops granodiorite could be interpreted, using this model, as being emplaced in a NW trending tensional gap at the Tamworth trough ESE transform-line virgation (Fig. 9).

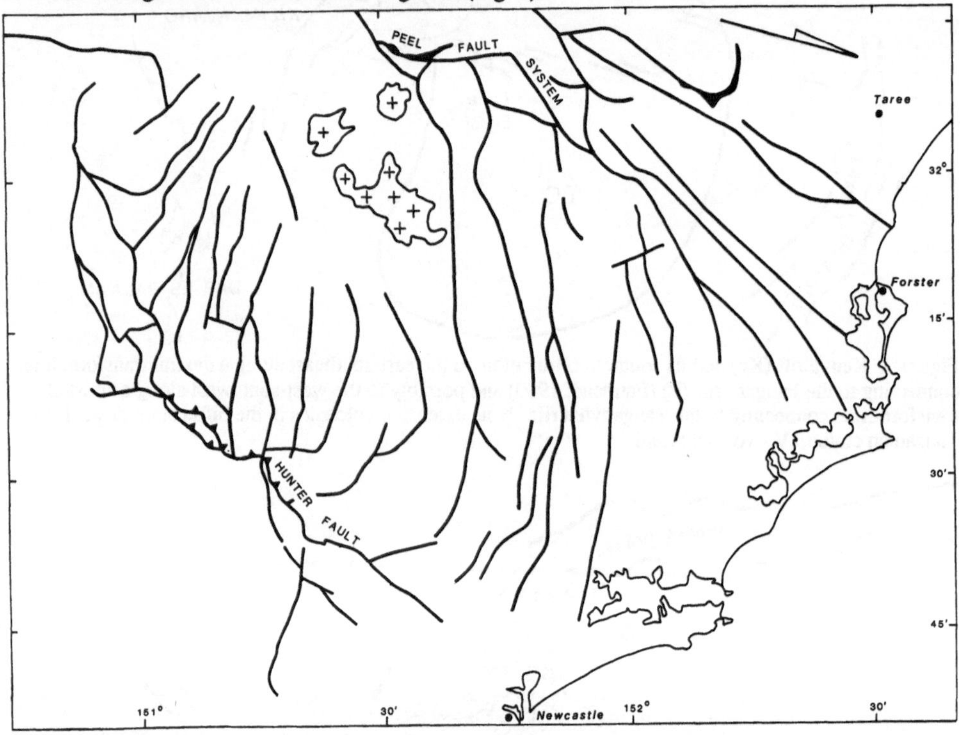

Figure 9. Southern Tamworth trough structures (after Roberts and Engel, 1987) interpreted as a dextral ESE transform line made up of SE Riedel and NNE conjugate Riedel faults. Barrington Top granodiorite (+) emplaced in a NW trending tensional gap developed at the Tamworth trough virgation.

Conclusions

Tar-pavement rift-transform fault relationships are similar in geometry and orientation to experimental clay models and can be utilized to model continental rift-transform tectonics as seen in the East African rift, Rhine graben and Tamworth trough. The transform line is made up of en echelon Riedel shears, transtensional and tensional structures each with their own orientation to the transform direction and dilatational expression. The rifts can also be considered as transtensional features suggesting oblique extension. The transitional area between the rift and the transform line is a preferred dilatant zone for the intrusion of volcanics and the emplacement of igneous bodies.

References

Courtillot, V., Tapponnier, P., & Varet, J. 1974. Surface features associated with transform faults. A comparison between observed examples and an experimental model. Tectonophysics **24**, 317-329.

Elmohandes, S.E. 1981. The Central European graben system: rifting initiated by clay modelling. Tectonophysics **73**, 69-78.

Fairhead, J.D. 1980. The structure of the cross-cutting volcanic chain of northern Tanzania and its relation to the East African rift system. Tectonophysics **65**, 193-208.

Freund, R., & Merzer, A.M. 1976. The formation of rift valleys and their zigzag fault pattern. Geological Magazine **113**, 561-568.

Hancock, P.L. 1972. The analysis of en echelon cracks. Geological Magazine **109**, 269-276.

Illies, J.H. 1972. The Rhine graben rift system—plate tectonics and transform faulting. Geophysical Surveys **1**, 27-60.

Illies, J.H. 1981. Mechanism of graben formation. Tectonophysics **73**, 249-266.

Katz, M.B. 1986. Tectonic analysis of the faulting at Woodsreef Asbestos Mine and its possible relationship to the kinematics of the Peel Fault. Australian Journal of Earth Sciences **33**, 99-105.

Katz, M.B. 1987. East African rift and northeast lineaments: continental spreading - transform system? Journal African Earth Sciences **6**, 103-107.

McConnell, R.B. 1972. The geological development of the Rift System of eastern Africa. Geological Society of America Bulletin **83**, 2549-2572.

Mohr, P.A. 1976. ENE trending lineaments of the African Rift System. Proceedings 1st International Conference New Basement Tectonics, Utah Geological Association Publication **5**, 327-334.

Roberts, J., & Engel, B.A. 1987. Depositional and tectonic history of the southern New England Orogen. Australian Journal of Earth Sciences **34**, 1-20.

NORTH-SOUTH LINEAMENTS IN THE BRITISH ISLES: A 1500MYR RECORD OF REACTIVATION

R.S. HASZELDINE
Department of Geology and Applied Geology
University of Glasgow
Glasgow G12 8QQ
Scotland
U.K.

ABSTRACT. The British Isles is a classical area for the development of geological concepts, and although it has a reputation of detailed complexity, many simple linear features can also be observed. Following Russell's (1968) identification of large-scale crustal "geofractures" locating major ore deposits in Ireland, a re-examination of the large-scale sedimentary, igneous and tectonic features has been conducted of the British Isles and its surrounding continental shelf. Very long N-S geofracture lineaments are identifiable. Some linears show geological activity at the present day and some can be identified back to the mid-Proterozoic. Such linears control the locations of features as diverse as:- modern earthquakes, Palaeocene volcanic centres, the giant petroleum province of the Viking Graben, 5 km deep Carboniferous sedimentary basins, Carboniferous sedimentary-exhalative base-metal giant ore deposits, Silurian post-orogenic granite plutons, and a syndepositional mid-Proterozoic graben fault. These major features have a persistent N-S orientation (360±10°), suggesting that an extensional E-W stress field has been reactivated many times. Linears can also be identified with E-W, NW-SE, and NE-SW orientations, but are not discussed here in detail, for their control on economic mineral or hydrocarbon accumulations seems to be less dominant. The cause of this stress field and its constant orientation are not identifiable with certainty. One feasible hypothesis proposed by Carey (1988) suggests that the earth has expanded, so that the British Isles, in common with many continents, remains in a constant orientation with respect to true north, resulting in constancy of lineament orientation through time. The earth's rotation may also be producing SW-sinistral and SE-dextral shears, which result in a persistent E-W tension.

Introduction

The British Isles are a part of the world which have received intense geological investigation since the 1700's. Consequently, geological mapping now provides precise locations of surface boundaries, and there is a long history of resource exploitation, ranging from coal, base-metals, ironstone and refractory clays to oil, gas, and now gold. One role of a geologist is to understand, and then predict, the location of economic resource deposits. Thus a geologist needs to use predictive models which, in some aspect reflect the way the geological world works, and also strip away confusing detail to leave the more fundamental controls apparent. In 1968, Russell published an hypothesis that major sedimentary exhalative base-metal ore deposits in Ireland were controlled by N-S linear "geofractures". Although never "popular", his hypothesis has remained intact to the present day, without being seriously challenged, and has accumulated several predictive successes, for the locations of newly discovered deposits (including Navan, Europe's largest Zn supplier) have relentlessly fallen on geofractures (Russell & Haszeldine 1991). It has

M. J. Rickard et al. (eds.), Basement Tectonics 9, 233–246.

also become apparent to me that such linear features can be used successfully to locate the position of many other geological features in the British Isles. In other countries, linears control the location of some important geological features and ore deposits. For example Chukwu-Ike & Norman (1977) defined N-S alignments of tin-bearing granites in Nigeria. In Australia, O'Driscoll (1986) has located many ore deposits on lineaments with a variety of orientations. In these countries, it is obvious that lineaments have been active through long periods of geological time. Consequently a search was made of British geology to ascertain if Russell's geofractures were

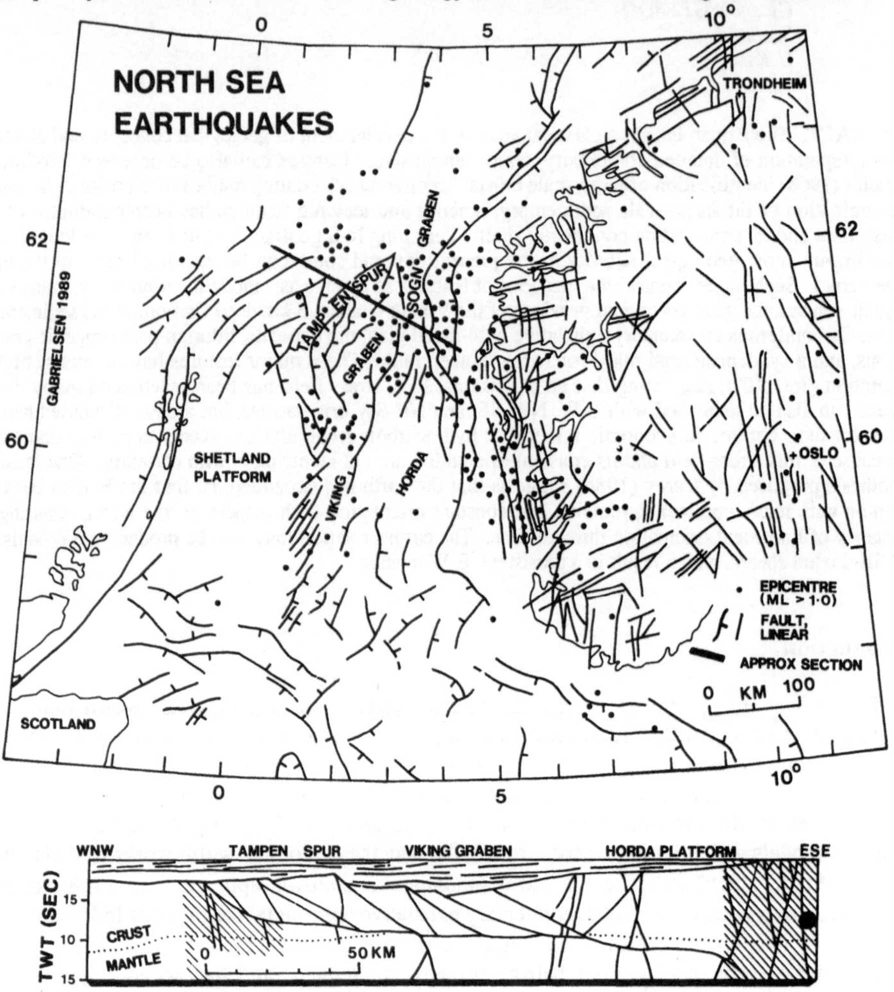

Figure 1. Map of the northern North Sea with a partial cross-section (on a different horizontal scale) derived from interpretations of seismic reflection data. Note the abundance of earthquake foci on the western Norwegian coast and around the Sogn Graben—which is a present day seafloor low. Many foci (but by no means all!) fall on or close to N-S offshore faults mapped from seismic data, or onshore linears mapped by fieldwork and from satellite images. On the cross-section, a recent deep earthquake epicentre falls vertically below the interpretation of major N-S normal faults on the Horda Platform west of Norway. Seismic interpretation and map redrawn and simplified from Gabrielsen (1989).

restricted to the Lower Carboniferous, or if they too were manifestations of longer-lived lineaments. This paper, together with Haszeldine (1988), Haszeldine & Russell (1987) and Russell & Haszeldine (1991) outlines some of the evidence to support the existence of long lived lineaments in the classical geological territory of the British Isles, and also makes the inference that their activity, over geologically long timespans, must be controlled by a planet-wide process which has maintained a constant orientation, such as earth rotation and expansion.

Lineament activity through time

PRESENT DAY EARTHQUAKES

Although the British Isles, and northwestern European shelf are usually considered to comprise a stable continental plate, there is a good record of present day earthquake activity. The vast majority of such earthquakes are of small magnitude, but events of magnitude 5, 6 or 7 in the past 10,000 years can be deduced from sedimentary evidence, such as submarine slumps offshore of southwestern Norway and an associated tsunami deposit in eastern Scotland (Long et al.1989). Direct recordings of earthquakes are well known in the North Sea (Havskov et al.1989). Gabrielsen (1989) has noted that epicentres from some of the larger earthquakes lie at some 23 km down in the North Sea offshore of southwestern Norway (Fig. 1), and lie vertically below the surface expression of N-S faults mapped by seismic reflection. Gabrielsen (1986, 1988), in common with many oil companies in the North Sea, shows that these same faults are part of a N-S array which controlled sedimentation in the petroleum basins. Where these extend laterally onshore into the crystalline crust, the faults or shear zone-fault complexes have a very steep dip. From seismic reflection (Gabrielsen 1986) and sedimentological evidence, these were growth faults during rift sedimentation, were reactivated as normal planar faults during thermal subsidence and compaction of the basin, and from earthquake epicentre evidence are still active faults at the present day.

On the onshore British Isles, Davenport et al. (1989) have shown that glacial lake sediments younger than 12,000bp have been deformed by an earthquake of magnitude 7 in Glen Roy, Scotland. The fault producing the earthquake was inferred to be N-S in orientation, and lies on the northward extension of the Loch Lomond/Tyndrum geofracture (Russell 1972, Haszeldine 1988) defined from geological evidence and present-day topography.

PALAEOCENE VOLCANICS

During the early Tertiary, the British Isles experienced its final phase of continental rupture associated with the opening of the present North Atlantic ocean. As part of this, a group of alkali-basalt lavas were erupted, intruded by large basaltic volcanoes, and cut by basalt dykes. These volcanoes are located on the western seaboard of Scotland, and the volcanoes have been eroded to expose their gabbro and granite roots. Macintyre et al. (1975) compiled K-Ar dates from these Tertiary volcanic centres together with their locations (Fig. 2). On a local (10km) scale, it is apparent that the volcanoes of Skye, Rhum, Ardnamurchan, Mull and Arran lie on earlier NE-SW structures conventionally assigned to the Caledonian Orogeny. On a larger (100km) continental scale, Macintyre et al.(1975) showed that the volcanoes from the Faeroe Islands in the north, through western Scotland, to Lundy Island in the south, lie on or close to a NNW-SSE line. The K-Ar dates range from 62-53 Ma, showing that the major volcanic activity on this line was

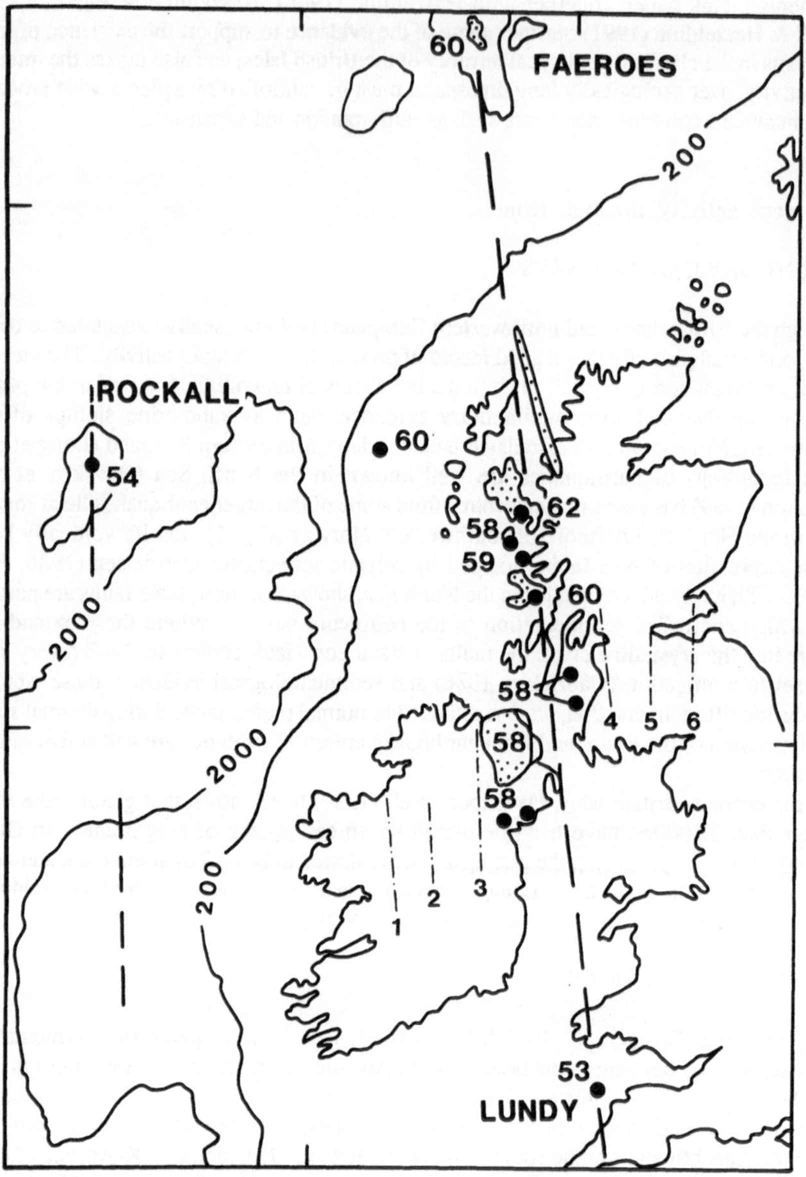

Figure 2. The Tertiary volcanoes west of Britain are K-Ar dated, and formed in two spasms, predominantly at 59Ma, then reactivated at 52 Ma (error on each date 1-4Ma). Their centres and the line of a 1.1km-wide Tertiary dyke fall close to a N-S (353°) line. Redrawn from Macintyre et al. (1975). Lines 1-6 refer to Russell's (1972) geofractures. Lines through Rockall and to the south are magnetic lineations.

geologically short in time. An exceptionally long and wide Tertiary dyke off the northwestern Scottish mainland (Ofoegbu & Bott 1985) runs sub-parallel to Macintyre et al.'s N-S line. The extension of this line to southeast Greenland on a map of pre-Atlantic refitted continents suggests that this line also intersects the thickest sequence of Palaeocene volcanics on the Greenland coast (Macintyre et al. 1975).

UPPER JURASSIC GRABEN

In the late 1960's and early 1970's, the North Sea became known as a world-class hydrocarbon province, following seismic-reflection surveys and successful exploration drilling by a number of companies (Brennand et al. 1990). Northern North Sea reserves alone currently rank it as the thirteenth largest petroleum basin in the world (Selley 1985). The major hydrocarbon source rock is the Upper Jurassic Kimmeridge Clay, and most reservoirs in the northern North Sea are Middle or Upper Jurassic sandstones (with about 30% of the reserves in Tertiary and other sediments). The deposition of these Jurassic sediments, together with the structural traps which locate the oilfields, are all controlled by the Viking Graben and its subsidiary fault terraces and flanks in the

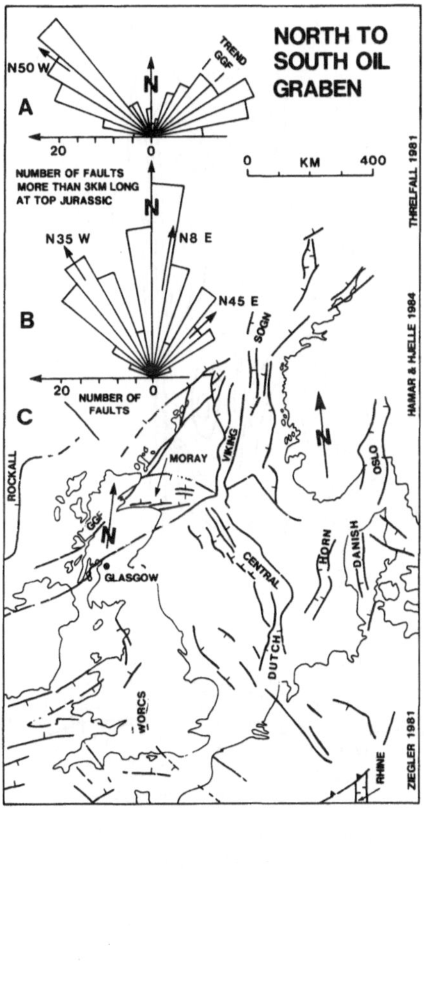

Figure 3. A) Shows the azimuths of the number of faults more than 3km in length from the top Jurassic level of reflection seismic data in the Inner Moray Firth (Threlfall 1981).
B) Shows the azimuths of faults more than 3km long between the Shetland Islands and the centre of the Viking Graben (Threlfall 1981). This northern North Sea is the world's 13th largest petroleum basin (Selley 1985).
C) Is a sketch map (after Ziegler 1981, Hamar & Hjelle 1984) showing patterns of large faults around the Viking Graben, which is a structurally dominant N-S giant petroleum basin 10km deep (Hospers & Ediriweera, 1991). Palaeogeographic evidence (Stow et al. 1982) shows the N-S faults on its western side to form syndepositional submarine fault scarps during its Late Jurassic rifting phase. Also shown are the N-S Graben of Rockall, Sogn, Oslo, Danish, Dutch, Worcester and Rhine.

East Shetland Basin. The Viking Graben is a N-S graben (Fig. 3) which was controlled by subsidence along N-S growth faults in the Upper Jurassic (Brown 1990), by planar normal faulting in the mid-Jurassic as sediments compacted over deeper structures (Badley et al. 1988), and by the same syndepositional growth faults in the Lower Jurassic, Triassic (Glennie 1990) and possibly even in the Carboniferous and Devonian (Haszeldine & Russell 1987). Magnetic, seismic-reflection and borehole data show that the entire fill to the Viking Graben and adjacent basins has a maximum thickness of 10km (Hospers & Ediriweera 1991); Jurassic and younger basin fill to the present day exceeds 5km. This Graben cuts across earlier basement structures of the Ordovician-Silurian Caledonian Orogeny and is here considered to be a N-S lineament structure initiated during the Devonian, but most obviously active during the Mesozoic.

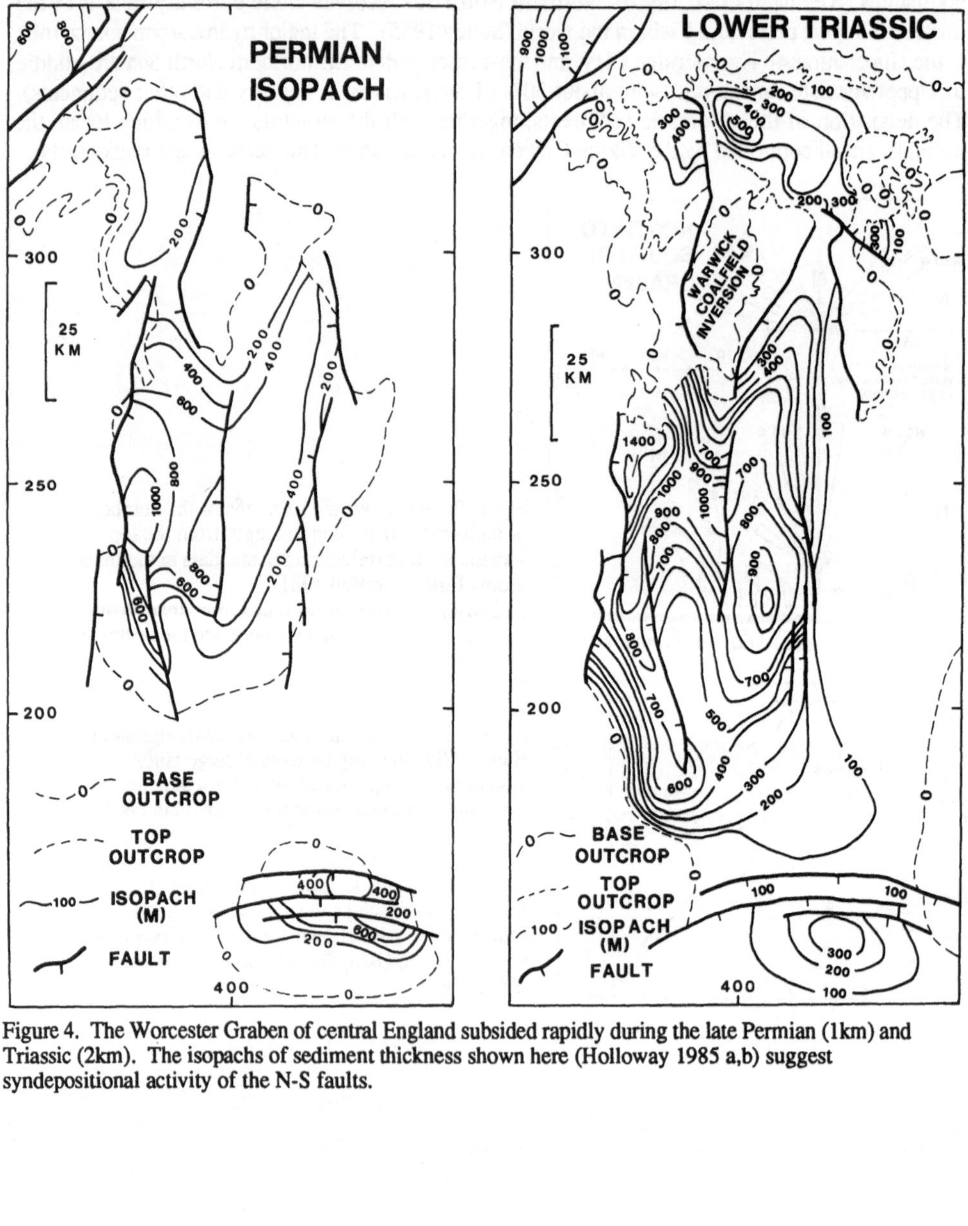

Figure 4. The Worcester Graben of central England subsided rapidly during the late Permian (1km) and Triassic (2km). The isopachs of sediment thickness shown here (Holloway 1985 a,b) suggest syndepositional activity of the N-S faults.

TRIASSIC GRABEN

Many oil company geologists working in and around the UK like to consider the Triassic as the start of tectonic patterns which controlled hydrocarbon accumulations (Glennie 1990). The Worcester Graben of central England is a good example, for this graben shows no definite expression in the Carboniferous (although an earlier basin in this location may, in fact, have been inverted and eroded). During the Permian, some 1km of subsidence took place along newly initiated N-S faults. In the Triassic these new faults became very well defined and localised more than 2km of Graben subsidence, cutting across earlier NW-SE and NE-SW structures of the Caledonian Orogen in this "Midlands Microcontinent" (Fig. 4).

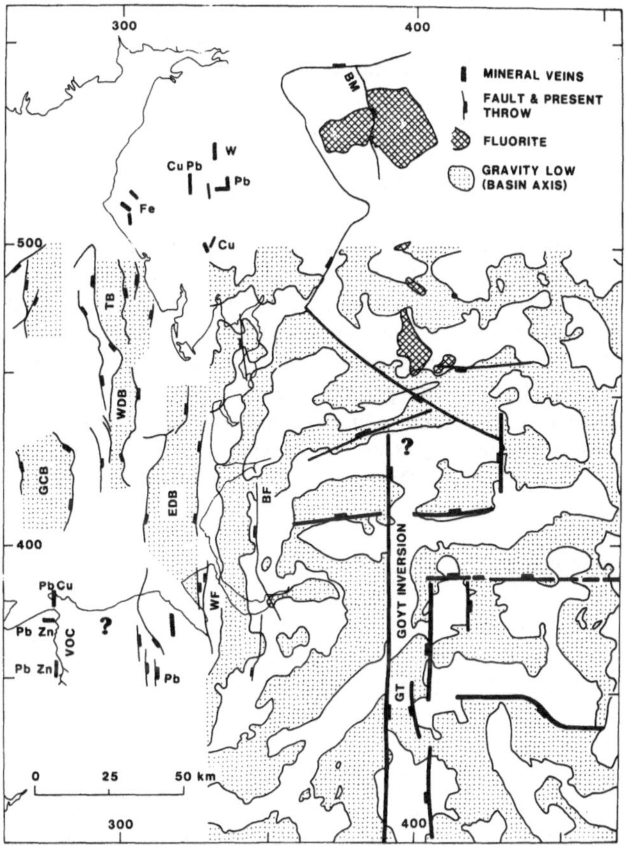

Figure 5. Carboniferous N-S graben (dots) were syndepositional. Their centre axes (not fault-bounded edges) sometimes align with major ore deposits (after Russell & Haszeldine 1991), see text. Geographical abbreviations:- BF (Boundary Fault), BM (Burtreeford Monocline), EDB (East Deemster Basin), GCB (Godred Croven Basin), GT (Goyt Trough), TB (Tynwald Basin), WDB (West Deemster Basin), WF (Woodchurch Fault), VOC (Vale of Conway).

CARBONIFEROUS GRABEN AND ORE DEPOSITS

During the Lower Carboniferous, the British Isles experienced an unusually intense episode of E-W tension, which led to fracturing of the crust and the formation of the world class sedimentary exhalative ore deposits in Ireland (Russell 1968). These formed by convective circulation of huge water volumes through fractured crust (Russell 1978). The locations of many of these major ore deposits fall on N-S "geofracture" lineaments (Russell 1968), which cut across the NE-SW Caledonian basement of Ireland, and have a spacing of approximately 45km (Fig. 2). Using gravity residual processing across onshore north central England, Lee (1988) discovered Carboniferous sedimentary fills to 3km deep faulted graben (Fig. 5). These too have an exact N-S orientation and a regular spacing of 40km. Directly offshore, Jackson et al. (1987) used boreholes and seismic to define Permian and Triassic syndepositional graben which continue the same pattern (Fig. 5, EDB, WDB, TB, GCB). Haszeldine (1988) and Russell & Haszeldine (1991) point out that these graben, and the ore-deposit geofractures, have the same orientation and spacing—and there are several cases where large base-metal vein deposits fall exactly on the extrapolated axis of a N-S graben (Fig. 5).

SILURIAN-DEVONIAN POST-OROGENIC GRANITES

Post-dating the final ductile deformation of the Caledonian Orogeny, a suite of calc-alkaline "newer granites" was intruded throughout the British Isles during the late Silurian and early Devonian. These are well exposed at the surface in Scotland, but are sediment covered in England, Wales and most of Ireland. Even a casual inspection of the geological map of Scotland enables arrays of N-S, E-W, NE-SW and NW-SE pluton elongations to be defined (Fig. 6). In several well-mapped cases, the internal zonation of the pluton parallels the outcrop pattern—showing that the outcrop today is no 'freak chance' of glaciation. Once again, these pluton arrays cross-cut the earlier structural grain of the Caledonian Orogen. Note also that parallel pluton alignments and elongations can be defined in the mid-Ordovician "younger basic" gabbros (Fig. 6), suggesting that the same stress field was reactivated.

MID PROTEROZOIC GRABEN

The cynic will naturally suggest that the N-S geological patterns described above are chance alignments, or a reactivation of an end-Caledonian fracture pattern. To counter this, a deliberate search was made to ascertain if N-S linears could be detected in the oldest rocks in Britain.

Northwest Scotland has a basement of Archaean Lewisian gneiss, unconformably overlain by Stoer Group redbeds, which are overlain with angular unconformity by Torridonian Group redbeds (comprising the Diabeg, Applecross and Aultbea Formations), overlain with planar unconformity by Lower Cambrian marine quartzites (Anderton et al. 1979). The two redbed sequences are not well dated, but here are thought to be mid-and late- Proterozoic respectively by their good fit onto palaeomagnetic polar wandering curves (Piper 1988). On the western seaboard of Scotland an outlier of the Cambrian quartzite occurs at Achiltiebuie, with a N-S strike, and is down faulted to the west by the Coigach Fault. This occurs adjacent to the stratigraphically lowest Stoer Group sediments exposed at Enard Bay (Gracie & Stewart 1967). This fault must have a throw between 1 and 3km to produce the observed field relations (Stewart 1988b), for the main outcrop of the Cambrian quartzite is some 20km to the east, where it dips regionally 20/120°. This

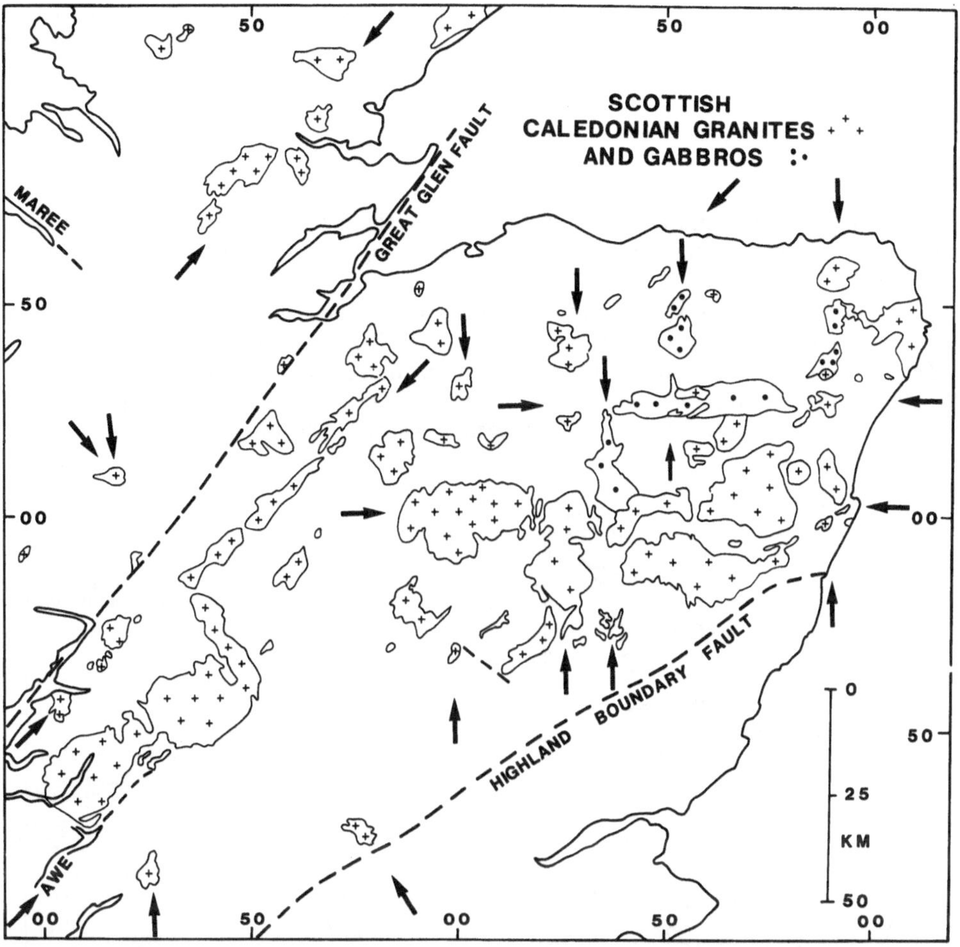

Figure 6. Late Silurian - Early Devonian post orogenic calc-alkaline "newer granites" (crosses) intruded the Scottish Highlands during 410-390 Ma, post-dating the Caledonian Orogeny. N-S, W-E, NW-SE and NE-SW trends are apparent in their elongations (arrows), and in the elongations of the syn-metamorphism and syn-deformation "younger basic" gabbros (dots) of 460-480Ma. Some of these alignments appear to cross the Great Glen Fault. Significantly, both the mid-Ordovician and Lower Devonian plutons show parallel linear trends, suggesting that the same stress field was reactivated .

Coigach Fault can be traced onshore and offshore to run N-S for some 100km (Johnstone & Mykura 1989). To achieve the present-day field and facies relations across this fault (Fig 7), it must have:-

1. Acted as a normal fault after deposition of the Cambrian quartzite.
2. Have been a normal fault after deposition of the mid-Proterozoic Torridon Group but before deposition of the Cambrian.
3. Been active during deposition of the mid-Proterozoic Stoer Group sediments (Stewart 1988a).

Thus the same N-S fault was reactivated at least three times throughout a 1Ga history, and is here considered a lineament of the same class as the Palaeozoic, Mesozoic and Tertiary structures described above.

Discussion

I have attempted to demonstrate that N-S faults were major controls on graben formation, ore deposits and volcanics throughout 1,500 Ma of British geologic history, and are still active today. Two remarkable features of this observation are that:-

1) the stress field producing such fractures was constant in orientation, and that
2) this stress field was spasmodically active over such a long period of time.

The simplest stress field to activate the lineament in each case was one of E-W tension. According to current plate tectonic theory, the British Isles were widely separated parts of a Proterozoic supercontinent (Piper 1988). The component parts of the British Isles ruptured

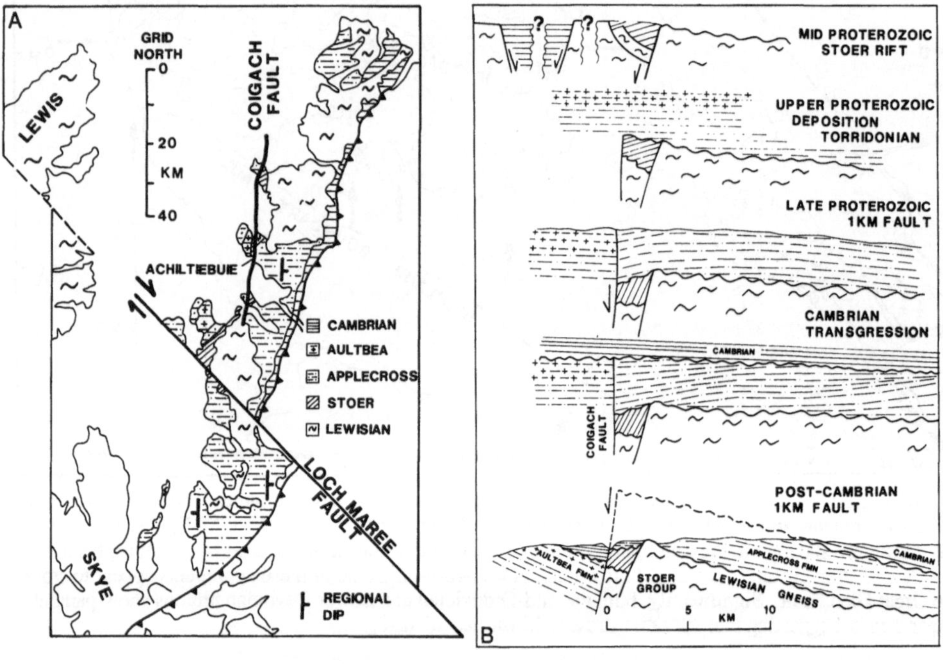

Figure 7. A) The N-S Coigach Fault of NW Scotland in map view (after Johnstone & Mykura 1989). I infer here a dextral SE displacement along the Archaean Loch Maree Fault, some of which postdated deposition of the Stoer Group (Stewart 1988a), and some may also have postdated deposition of the Aultbea Formation. This reconstruction aligns the Stoer Group across the Loch Maree Fault, and importantly, aligns the N-S Coigach Fault with a N-S Lewisian gneiss dome.

B) Cross-section of the Coigach Fault at Achiltiebuie, from author's fieldwork. The present-day exposed geometry is shown at the bottom. Qualitative sketch reconstructions of fault movements suggest several episodes of faulting:- in the post-Cambrian; late Proterozoic; mid-Proterozoic. This is the oldest undeformed N-S fault linear in Britain.

separately off this supercontinent, and were brought together for the first time by large-scale ranscurrent faulting during closure of the Iapetus Ocean and the Caledonian Orogeny. During and after suturing together, the British Isles rotated roughly 80 degrees clockwise as it drifted northwards since the Ordovician (Smith et al. 1981).

Numerous explanatory options are possible, but basically there are three main alternatives:-

1) Either the stress field must rotate with the continental fragments, or the primordial fractures are reactivated by stress fields from very many directions,
2) New fractures are propagated into the continental fragments from older cratonic areas,
3) Perhaps the stress field was constant and the magnetic field of the earth rotated with respect to true north. The first option must invoke a completely unknown mechanism for producing tension, and so is not favoured. The second option still does not explain why such fractures should remain constant in orientation, and so is rejected. The third option remains a viable possibility, although unpopular amongst the majority of geologists. Carey (1988) has replied to his palaeomagnetic critics, and pointed out that the palaeomagnetic evidence cited in favour of continental drift can be equally interpreted to support a magnetic pole moving with respect to true north on an expanding globe.

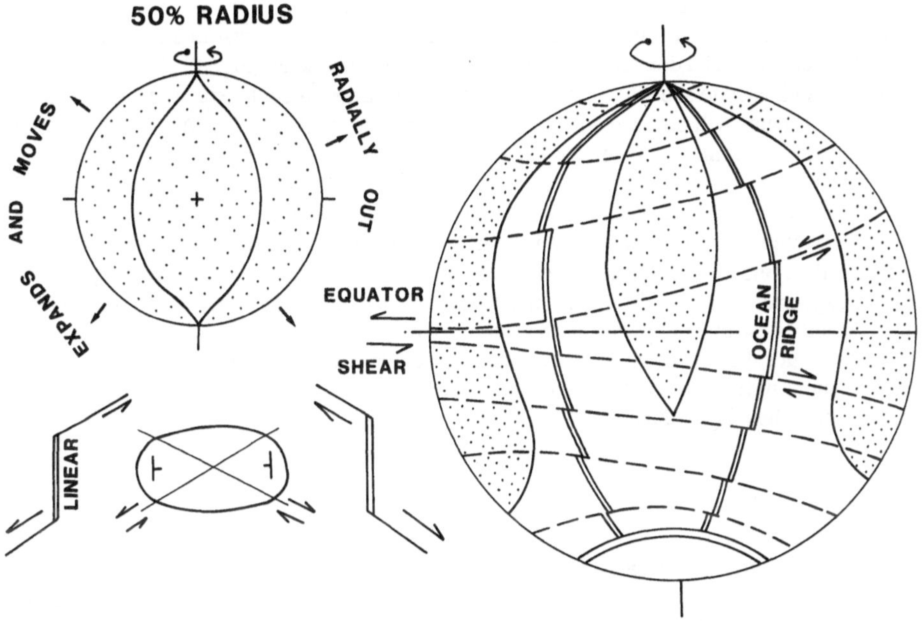

Figure 8. Earth expansion, from 55% of the present radius (Owen 1983) can be considered as the movement of continents radially away from the core, rather like petals of a tulip opening. Such motion would keep continental N-S orientations constant with respect to true north, and hence lineament orientations would remain constant through time. The persistent reactivation of N-S lineaments suggests a persistent E-W tension. This may be produced as one consequence of the Earth's rotation, for equatorial continents and oceans must have a faster angular velocity than high latitude continents and oceans. This produces a sinistral SW shear and a dextral SE shear, which translate into E-W tension. Superimposed onto this pattern is a sinistral shear between the northern and southern hemispheres, due to the slower rotation of the northern hemisphere (Carey 1988), and the slow northern drift of continents (Wegener 1924).

If Carey (1988) is correct in inferring that the earth is expanding, then continents will remain in a constant orientation relative to true north. On Carey's earth, plate tectonics has not translated continents laterally across its surface during the past 1500 Myr, but continents have moved radially outwards from the centre, and so *appear* to have moved laterally. An apparent northwards motion is due to greater expansion in the southern hemisphere, which is evidenced in the magnetic polarity patterns of the ocean floor in the southern Pacific (Owen 1983).

On Carey's earth, there is still the possibility that the British Isles in particular are an in a unique setting, in that the opening of the present North Atlantic Ocean can be viewed as radiating from the Alaskan Oroclin (Carey 1988, Figs. 15, 16, 17), so that tension in the British Isles has remained exactly E-W. The evidence of intra-Caledonian N-S and E-W gabbro plutons in northeast Scotland (Fig. 6), and of a Proterozoic N-S fault (Fig.7) is crucial here in extending the evidence of the same stress field back well before the opening of the present North Atlantic Ocean. This is easily explained if the Caledonian Orogeny does not represent closure of the wide Iapetus Ocean (or indeed any ocean), and the mega-continent has ruptured apart for the first time to form the present Atlantic. Otherwise it is difficult to explain why the post Caledonian rifting of re-assembled continental fragments should mimic a pre-Caledonian tension so precisely.

Whilst the earth rotates, the oceanic and continental crust at the equator will rotate with a faster angular velocity than the crust nearer the poles. This may produce a sinistral-shear couple in the northern hemisphere and a dextral-shear in the southern hemispheres (Fig. 8). Carey (1988) has also provided evidence to show that the asymmetric distribution of continents in the northern and southern hemispheres will produce an equatorial sinistral shear. Combining this with a long term northward drift of continents (Wegener 1924) relative to the equator, produces a spiral sinistral NE-SW shear in the northern hemisphere. On a stress ellipse, such a shear equates to an E-W tensional stress (Fig. 8). Thus such a shear and its consequent E-W tension will remain constant in orientation relative to the continents and to true north. Spasms of tension (and activity along lineaments) then reflect changes in the rotational velocity of the earth's crust, and may relate to spasms of expansion of the earth's radius on a radially expanding earth where ocean is created but not significantly destroyed. Such spasms of activity may yet be shown to correlate with epeiric episodes of basin subsidence and relative "eustatic" sea-level changes. No doubt there are many other suggestions of how lineaments can be reactivated with the same orientation during 1500 Myr. I look forward to hearing of them.

Conclusions

1. N-S faults in the British Isles and around its continental shelves are still active today.
2. Such faults have controlled the siting of major volcanoes, petroleum basins, sediment-filled graben, ore deposits and granites for the last 1500 Ma.
3. E-W, NE-SW and NE-SW faults have also been persistently active during this timespan, but only localise major geological sites (such as a volcano) where they are cross-cut by a N-S linear.
4. This E-W tensional stress field, has been spasmodically active in the British Isles for at least 1500Ma, and so the geographical orientation of the British Isles may have been constant with respect to true north. One explanation for this is that the stress is driven by earth's rotation, and the stress orientation is kept constant if the earth is expanding, in contrast to conventional plate tectonics.

Acknowledgements

For guidance and experienced discussions I thank M J Russell (Glasgow), S.W. Carey (Hobart), and H.G. Owen (British Museum). Clark Fenton (Glasgow) showed me Gabrielsen's work, and many past and present colleagues and students have shown me their 'pet' N-S linear. Thanks to all at the 9th IBT conference for considering these ideas; my attendance was funded by the Royal Society of London and the University of Glasgow.

References

Anderton, R., Bridges, P.H., Leeder, M.R., & Sellwood, B.W. 1979. A dynamic stratigraph of the British Isles. Allen and Unwin, London, 301p.

Badley, M.E., Price, J.D., Dahl, C.R., & Agdestein, T. 1988. The structural evolution of the northern Viking Graben and its bearing upon extensional modes of basin formation. Journal of the Geological Society of London **145**, 455-472.

Brennand, T.P., Van Hoorn, B., & James, K.H. 1990. Historical review of North Sea exploration. In, Glennie, K.W. (ed.) Introduction to the petroleum geology of the North Sea 3rd edn. Blackwell Scientific, Oxford, 1-33.

Brown, S. 1990. Jurassic. In, Glennie, K.W. (ed.) Introduction to the petroleum geology of the North Sea 3rd edn. Blackwell Scientific, Oxford, 219-254.

Carey, S.W. 1988. Theories of the earth and universe, a history of dogma in the earth sciences. Stanford University Press, U.S.A. 413p.

Chukwu-Ike, I.M., & Norman, J.W. 1977. Mineralized crustal failures shown on satellite imagery of Nigeria. Transactions, Institution of Mining and Metallurgy **86**, B55-B57.

Davenport, C.A., Ringrose, P.S., Becker, A., Hancock, P., & Fenton, C. 1989. Geological investigations of late and post glacial earthquake activity in Scotland. In, Gregersen, S., & Basham, P.W. (eds) Earthquakes at North Atlantic passive margins. Kluwer Academic, 175-194.

Gabrielsen, R.H. 1986. Structural elements in graben systems and their influence on hydrocarbon trap types. In, Spencer A.M. et al. (eds) Habitat of Hydrocarbons on the Norwegian continental shelf. Graham & Trotman, 55-60.

Gabrielsen, R.H. 1989. Reactivation of faults on the Norwegian continental shelf and its implications for earthquake occurrence. In, Gregersen, S., & Basham, P.W. (eds) Earthquakes at North Atlantic passive margins. Kluwer Academic, Netherlands, 67-90.

Glennie, K.W. 1990. Outline of North Sea history and structural framework. In, Glennie, K.W. (ed.) Introduction to the petroleum geology of the North Sea 3rd edn. Blackwell Scientific, Oxford, 34-77.

Gracie, A.J., & Stewart, A.D. 1967. Torridonian sediments at Enard Bay, Ross-shire. Scottish Journal of Geology **3**, 181-194.

Hamar, G.P., & Hjelle, K. 1984. Tectonic framework of the More basin and the northern North Sea. In, Spencer, A.M. et al. (eds) Petroleum geology of the North European margin. Graham & Trotman, p 349-358.

Haszeldine, R.S. 1988. Crustal lineaments in the British Isles; their relationship to Carboniferous basins. In, Besley B.M., & Kelling, G. (eds) Sedimentation in a synorogenic basin complex. Blackie, Glasgow, 53-68.

Haszeldine, R.S., & Russell, M.J. 1987. The late Carboniferous northern North Atlantic Ocean: implications for hydrocarbon exploration from Britain to the Arctic. In, Brooks, J., & Glennie, K.W. (eds) Petroleum geology of NW Europe. Graham & Trotman, 1163-1175.

Havskov, J., Lindholm, C.D., & Hansen, R.A. 1989. Temporal variations in North Sea seismicity. In, Gregersen, S., & Basham, P.W. (eds) Earthquakes at North Atlantic passive margins: neotectonics and postglacial rebound. Kluwer Academic, Dordrecht, Netherlands, 413-427.

Holloway, S. 1985a. The Permian. In, Whittaker, A. (ed.) Atlas of onshore sedimentary basins in England and Wales. Blackie, Glasgow, 26-30.

Holloway, S. 1985b. Triassic. In, Whittaker, A. (ed.) Atlas of onshore sedimentary basins in England and Wales. Blackie, Glasgow, 31-36.

Hospers, J., & Ediriweera, K.K. 1991. Depth and configuration of the crystalline basement in the Viking Graben area, northern North Sea. Journal of the Geological Society of London, **148**, 261-266.

Jackson, D.I., Mulholland, P., Jones, S.M., & Warrington, G. 1987. The geological framework of the East Irish Sea Basin. In, Brooks, J., & Glennie, K.W. (eds) Petroleum geology of northwest Europe. Graham & Trotman, London, 191-203.

Johnstone, G.S., & Mykura, W. 1989. The northern highlands of Scotland 4th edn. HMSO, London, 219p.

Lee, A.G. 1988. Carboniferous basin configuration of central and northern England modelled using gravity data. In, Besley, B.M., & Kelling, G. (eds) Sedimentation in a synorogenic basin complex. Blackie, Glasgow, 69-84.

Long, D., Dawson, A.G., & Smith, D.E. 1989. Tsunami risk in northwestern Europe: A Holocene example. Terra Nova **1**, 532-537.

Macintyre, R.M., McMenamin, T., & Preston, J. 1975. K-Ar results from W Ireland and their bearing on the timing and siting of Thulean magmatism. Scottish Journal of Geology **11**, 227-250.

O'Driscoll, E.S.T. 1986. Observations of the lineament-ore relation., Philosophical Transactions Royal Society of London (Ser.A.) **317**, 195-218.

Ofoegbu, C.O., & Bott, M.H.P. 1985. Interpretation of the Minch linear magnetic anomaly and of a similar feature on the shelf north of Lewis by non-linear optimization. Journal, Geological Society of London **142**, 1077-1087.

Owen, H.G. 1983. Atlas of continental displacement 200 million years to the present. Cambridge University Press, 160p.

Piper, J.D.A. 1987. Palaeomagnetism and the continental crust. Open University Press, U.K. 434p.

Russell, M.J. 1968. Structural controls of base-metal mineralization in Ireland in relation to continental drift. Transactions, Institution of Mining and Metallurgy **77**, B117-128.

Russell, M.J. 1972. North-south geofractures in Scotland and Ireland. Scottish Journal of Geology **8**, 75-84.

Russell, M.J. 1978. Downward excavating hydrothermal cells and Irish type ore deposits: importance of an underlying thick Caledonian prism. Transactions, Institution of Mining and Metallurgy **87**, B168-171.

Russell, M.J., & Haszeldine, R.S. 1991. Accounting for geofractures. In, Bowland, M. et al. (eds) The Irish minerals industry - a review of the decade. Publication Irish Association for Economic Geology, Geological Survey, Dublin. (in press).

Selley, R.C. 1985. Elements of petroleum geology. Freeman, New York, 449p.

Smith, A.G., Hurley, A.M., & Briden, J.C. 1981. Phanerozoic paleocontinental world maps. Cambridge University Press, U.K. 102p.

Stewart, A.D. 1988a. The Stoer Group. In, Winchester, J.A. (ed.) Late Proterozoic stratigraphy of the north Atlantic region. Blackie, Glasgow 97-103.

Stewart, A.D. 1988b. The Torridon Group. In, Winchester, J.A. (ed.) Late Proterozoic stratigraphy of the north Atlantic region. Blackie, Glasgow 104-112.

Stow, D.A.V., Bishop, C.D., & Mills, S.J. 1982. Sedimentology of the Brae oilfield North Sea: fan models and controls. Journal of petroleum geology **5**, 129-148.

Threlfall, W.F. 1981. Structural framework of the central and northern North Sea. In, Illing, L.V., & Hobson, G.D. (eds) Petroleum geology of the continental shelf of north-west Europe. Heyden, London, 98-103.

Wegener, A. 1924. The origin of continents and oceans. (English translation by W.A. Skerl). Methuen, London.

Ziegler, P.A. 1981. Evolution of sedimentary basins in north-west Europe. In, Illing, L.V., & Hobson, G.D. (eds) Petroleum geology of the continental shelf of north-west Europe. Heyden, London, 3-39.

BASEMENT-COVER RELATIONSHIPS IN OROGENIC BELTS

M.J. Rickard
Geology Department
Australian National University
GPO Box 4
Canberra ACT 2601
Australia

ABSTRACT: Basement Tectonic Conferences have long been preoccupied with relationships between structures in cratons and their platform covers, ie intraplate tectonics. There are, however, many instances where basement influences, especially the 'buttress-anvil' effect of the craton edge, play a major role in determining structural development in the orogenic belt itself. Some examples, mainly from the Appalachians, of plan and section patterns are outlined.

Introduction

In concluding these proceedings this brief paper presents a contention that 'basement tectonics' should not be regarded exclusively as an intraplate phenomenon as suggested in the introductory keynote address. Many features of the structural geometry of orogenic belts may be directly or indirectly related to the influence of buried basement and especially to the buttress or anvil edge effect of continent (craton) margins.

This is not a new proposition, but in many cases the descriptions and explanations of these structures emphasize the orogenic processes such as thrusting, complex folding or metamorphism and underestimate or ignore the role of the basement. There are many general examples in works on subduction and collision tectonics (eg Coward & Ries 1986, McClay & Price 1981) and indeed the 8th International Basement Tectonics Conference had as its major theme a comparison of ancient and Mesozoic continental margins (Bartholomew et al. in Press).

The influence on cratons of events in orogenic belts has long been recognized, but the reverse effect less so. Using the Appalachians as an example, Rodgers (1987) has described chains of reactivated basement uplifts within the craton, but marginal to the orogenic belt. These have located thrust-faulted monoclines in the platform cover during orogenic deformation. Within the orogenic belt the continental edge has been recognized as a major factor in the development of sedimentary basins and orogenic structures (eg Sheridan 1974, Cook & Oliver 1981). Price & Hatcher (1983) have shown that steep gravity gradients mark the edge of the buried craton in both the Appalachians and Canadian Cordillera. Reactivation of buried basement during orogenic activity has been invoked to explain chains of gneiss domes (eg Currie 1983) and the location of break thrusts (eg St Julien & Hubert 1975, Lillie 1985). A now classical study by Brace (1958) illustrated the mutual interaction of basement and cover within the Appalachian fold belt.

In spite of these works, and similar ones in other orogenic belts, the major influence of basement as a factor in the generation of orogenic structures has been generally neglected. In this paper a preliminary attempt is made to survey the tectonic influence of basement on the structure of

M. J. Rickard et al. (eds.), Basement Tectonics 9, 247–254.

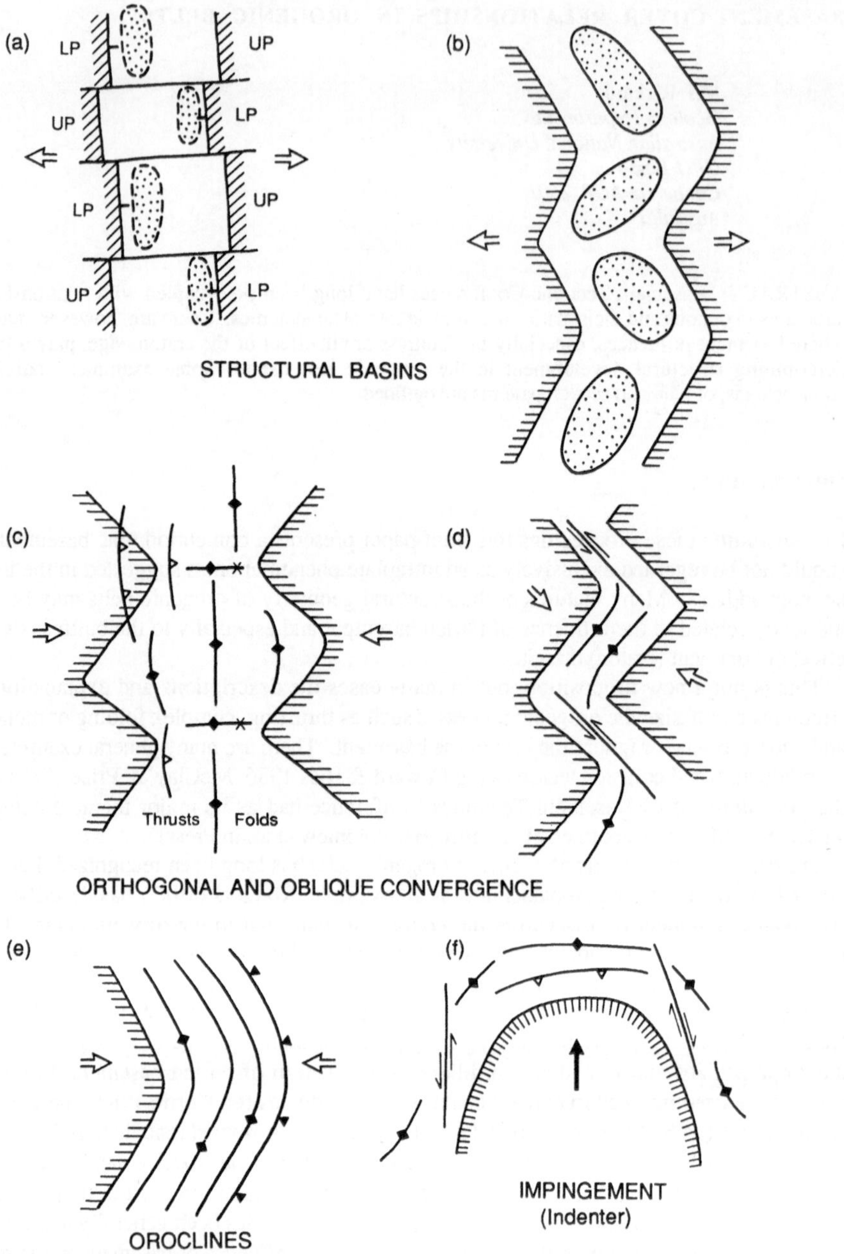

Figure 1. Basement effects on orogens, in plan.
(a) Structural basins at lower and upper plate margins (b) Structural basins in salients of plate margins
(c) Orthogonal and (d) Oblique convergence of plate margins affects major fold and fault patterns
(e) Curving fold trends (oroclines) formed by plate convergence or subduction (f) Impingement of basement into orogen (indenter effect).

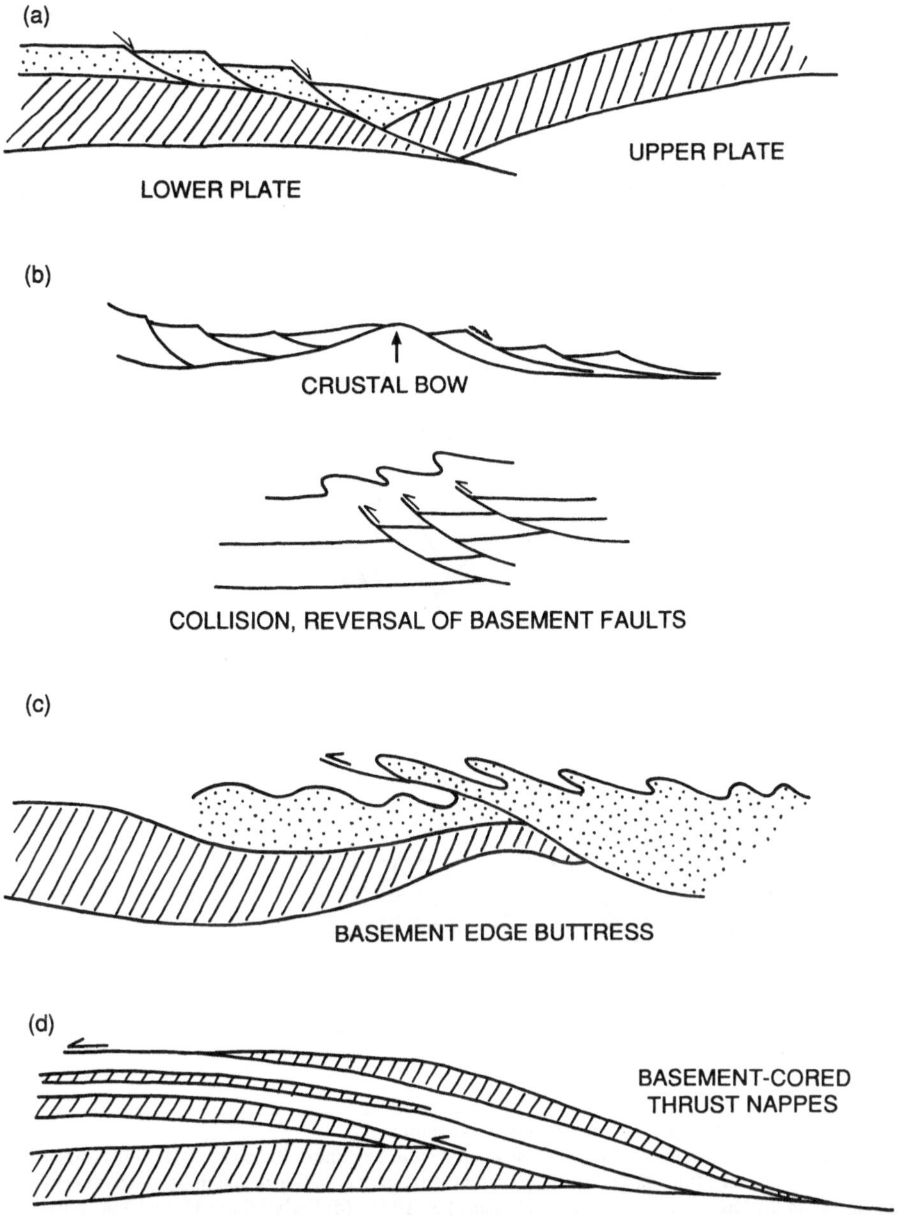

Figure 2. Basement effects on orogens in section.
(a) Lower and upper plate basement geometries influence cover deformation (b) Basement faults reverse on collision and affect overlying strata during orogenic phase (c) Basement edge acts as 'buttress-anvil' and controls position of major thrusts (d) Fault reversal allows development of basement-cored thrust nappes.

orogenic belts. Figures 1 and 2 set out, in cartoon form, the various types of interaction in plan and section.

Basement effects in plan

The shape of the fractured plate margin has a profound control on the geometry of the initial sedimentary basins and their subsequent deformation, basins depending for their shape on the

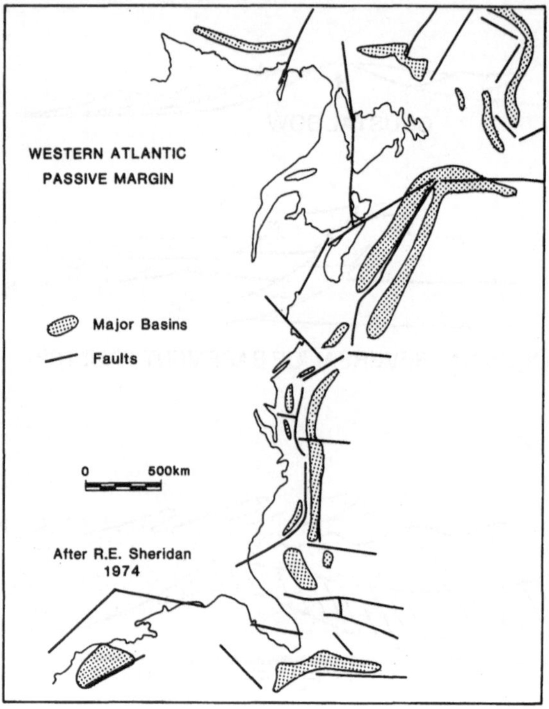

Figure 3. Structural control of sedimentary basins by craton-margin shape and basement-induced fault patterns (after Sheridan 1974).

Figure 4. Basement control of orogenic structures — examples:
(a) & (b) Appalachians of southern Canada — development in plan (after Rickard 1991): folds (diamonds), thrusts (triangles) cleavage (thin dashed lines) basement (stippled); (c) Wrench and thrust development with oblique closure in the Quebec re-entrant (after Doolan et al. 1982); (d) Early Middle Ordovician collision in the Quebec Appalachians with obduction and thrusting located at the buried craton-margin edge (after St Julien & Hubert 1975); (e) Cartoon section across areas a & b in southern Quebec Appalachians for the Devonian (after Rickard 1991): HT = Hinesburg Thrust, SA = Sutton Anticline, SBS = Sargent Bay Syncline; S1, S1x, S2, S1D are foliations.

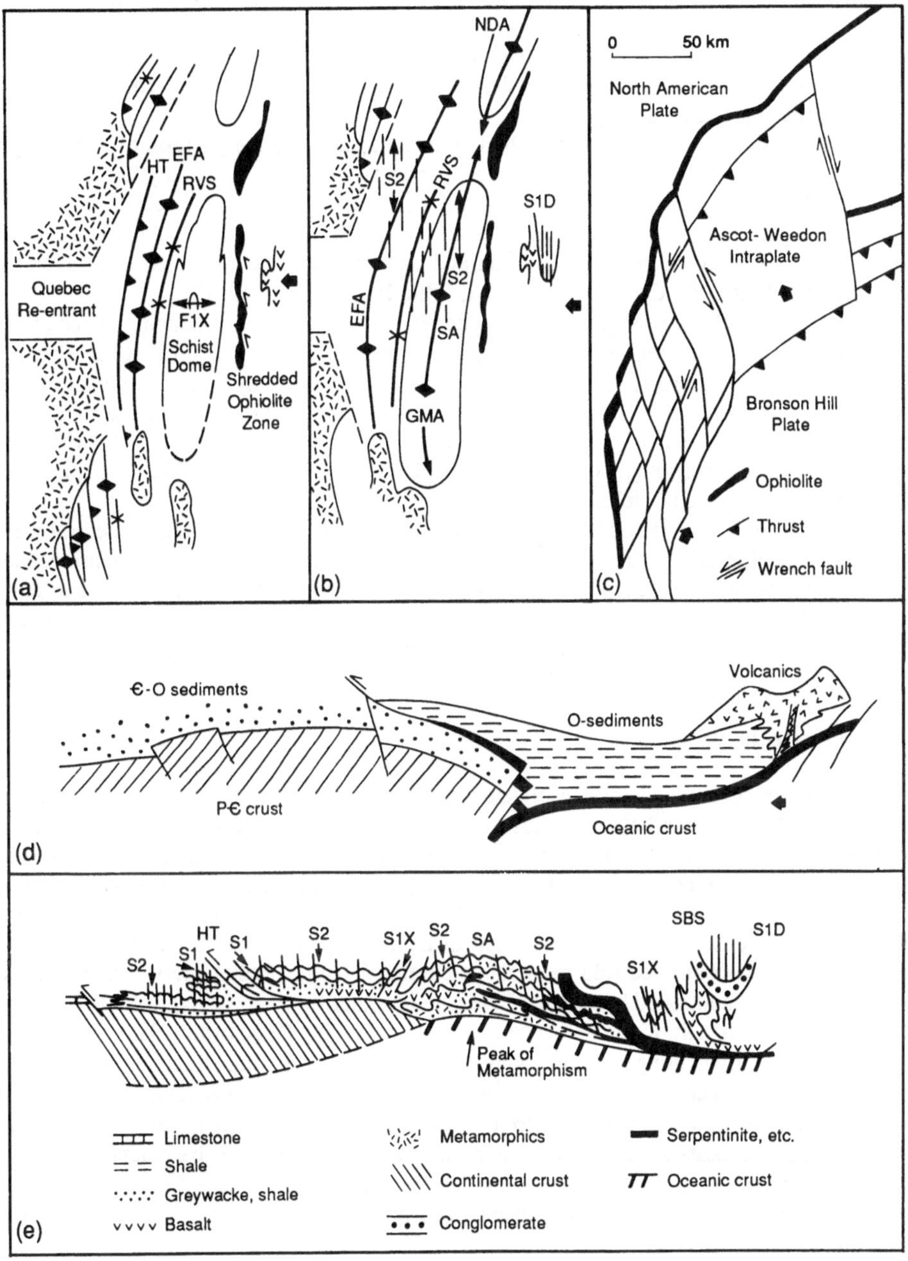
HT
EFA
RVS
Quebec
Re-entrant
F1X
Schist
Dome
Shredded
Ophiolite
Zone
(a)
NDA
S2
RVS
S1D
EFA
S2
SA
GMA
(b)
0 50 km
North American
Plate
Ascot- Weedon
Intraplate
Bronson Hill
Plate
Ophiolite
Thrust
Wrench fault
(c)
Є-O sediments
Volcanics
O-sediments
PЄ crust
Oceanic crust
(d)
HT
S1
S2
S1X
S2
SA
S2
SBS
S1D
S2
S1
S1X
Peak of
Metamorphism
Limestone
Shale
Greywacke, shale
Basalt
Metamorphics
Continental crust
Conglomerate
Serpentinite, etc.
Oceanic crust
(e)

salients and recesses of the fractured craton margin (Fig. 1a, b & 3). Basement highs and faults propagate from the fractured margins into the developing basins (Sheridan 1974, Norris & Carter 1982).

The subsidence history of the basin may well depend on whether the basin sits over an upper- or lower-plate margin (Lister et al. 1986) and these may alternate across basement-controlled transfer faults along the craton margin (Fig. 1a).

During closure, these boundary controls may in turn influence the position of metamorphic zones in thicker sections, fold trends from basin to basin, cross folds, and many faults (Fig. 1c, d & 4a, b). Salients and recesses in the orogenic belt may reflect changes, especially triple junctions, in the original plate margin (Rankin 1975, Thomas 1977). The original junction between continent and oceanic crust may be reflected in a marked upwarp and a concentrated zone of thrusts (Schuepbach & Vail 1980, Cook & Oliver 1981, Lillie 1985, Rickard 1991).

With orthogonal closure, original craton salients may collide first, producing zones of high compression and development of strong thrusts, curving fold trends, and cross folds (Fig. 1c & 4a) (Thomas 1977, Rickard 1991). Thrust complexes may develop by ramping over and around basement massifs (Hatcher Jr. 1983). Salient-bounding basement uplifts may cause rotation and imbrication of thrust sheets (Craddock et al. 1988). With oblique closure (Fig. 1d) the buried basement would set up strong zones of shear (Doolan et al. 1982), offset fold trends, and even cause cleavage transection of folds (Fig. 4b, c).

Plate closure by either subduction or collision, may cause simple curving structural trends (oroclines) wrapping around continental nuclei (Marshak 1988) (Fig. 1e). More intense versions occur where plate collision involves large-scale imprignment of basement cratons and lateral 'extrusion' of the deforming metasedimentary welt (Fig. 1f). Well-known examples occur in northwest Spain (Matte 1968) and with the Indian-Tibet collision (Tapponnier et al. 1986).

Basement effects in section

The buried craton, being made of consolidated continental crust, is more rigid than the overlying sediments of the orogenic belt, and thus has a marked effect on structural profiles (Fig. 2). Basement has a primary control of thick-skinned and thin-skinned structural styles. The differences in architecture of upper and lower plates (Lister et al. 1986), one with a gentle upwarp and the other with collapsed listric faults (Fig. 2a), would be expected to influence the overlying structures in fold belts. The former with simple disharmonic folds above decollement slide zones, the latter with fault-controlled fold belts above basement faults. Sedimentary-facies packages, themselves perhaps controlled by original basement irregularities, may have greater influence on orogenic structures than the direct influence of basement faulting (Woodward 1988).

Crustal thinning by faulting and up-bowing of margins provides a pre-structured basement that can react with the cover during orogenic closure (Fig. 2b). Wiltschko & Eastman (1983) have shown with computer models that irregularities in rigid basement disturb regional stress trajectories and concentrate stresses so that thrust ramps develop. Furthermore, basement uplifts may indirectly locate thrusts in the overlying zone (eg Winslow 1981). These reactivated basement faults extend upwards with reverse movement (Jackson 1980) and may even penetrate into over-riding nappes (Thomas 1983).

The strongest structural influence is the 'buttress-anvil' effect of the craton edge, which during compression is likely to control the location and propagation of major thrust zones and breaks in structural geometry (Fig. 2c). There are numerous examples of such geometry in many orogenic belts, either as a direct 'buttress' effect (eg in the Appalachians, St Julian & Hubert 1975, Doolan et al. 1982, Lillie 1985, Rickard 1991— see Fig. 4d & e; in the Canadian Cordillera, Brown 1981, Price 1981, Price & Hatcher 1983) or by reactivation and reversal of earlier basement faults at the craton margin (eg in the Appalachians, Hatcher Jr. 1983, Stanley & Ratcliffe 1985; and in the Zagros Mountains, Alawi 1980). The craton buttress may even locate rising metamorphic domes, slide nappes, and later structural zones (eg Rickard 1991, Fig. 4e).

It is likely that, with intense deformation, pre-existing craton-margin normal faults may be reversed to develop major basement-cored nappe sheets (Fig. 2d) as in the Norwegian Calidonides (Cuthbert et al. 1983, Rickard 1985). Indeed, it is difficult to envisage how basement slicing can occur by any other mechanism.

Epilogue

I have drawn attention to the fact that many orogenic structures may be attributed to basement control; thus the scope of 'basement tectonics' should be widened to embrace the role of buried cratons in orogenic belts. Hopefully this topic will become a major theme for future symposia on International Basement Tectonics.

References

Alawi, M. 1980. Tectonostratigraphic evolution of the Zagrosides of Iran. Geology **8**, 144-149.

Bartholomew, M.J., Hyndman, D.W., Mogk, O.W., & Mason, R. (in press). Characterization and comparison of ancient and Mesozoic continental margins: Proceedings of the Eighth International Conference on Basement Tectonics, Kluwer, Dordrecht.

Brace, W. 1958. Interaction of basement and mantle during folding near Rutland, Vermont. American Journal of Science **256**, 241-256.

Brown, R.L. 1981. Metamorphic complex of SE Canadian Cordillera and relationship to folding. In, McClay, K.R., & Price, N.J. (eds) Thrust and Nappe Tectonics. Geological Society of London Special Publication **9**, 463-473.

Cook, F.A., & Oliver, J.E. 1981. The Late Precambrian-Early Paleozoic continental edge in the Appalachian orogen. American Journal of Science **281**, 993-1008.

Coward, M.P., & Reis, A.C. (eds) 1986. Collision Tectonics. Geological Society of London, Special paper **19**, 415p.

Craddock, J.P., Kopania, A.A., & Wiltschko, D.V. 1988. Interaction between northern Idaho-Wyoming thrust belt and bounding basement blocks, central western Wyoming. In, Schmidt, C.J., & Perry, W.J. Jr. (eds) Interaction of Rocky Mountains foreland and Cordilleran thrust belt. Geological Society of America, Memoir **171**, 333-351.

Cuthbert, S.J., Harvey, M.A., & Carswell, D.A. 1983. A tectonic model for the metamorphic evolution of the Basal Gneiss Complex, Western South Norway. Journal of metamorphic Geology **1**, 63-90.

Doolan, B.L., Gale, M.H., Gale, P.N., & Hoar, R.S. 1982. Geology of the Quebec Re-entrant: possible constraints from early rifts and the Vermont-Quebec Serpentinite Belt. In, St. Julian, P., & Béland, J. (eds) Major structural zones and faults of the Northern Appalachians. Geological Association of Canada Special Paper **24**, 661-692.

Hatcher Jr., R.D. 1983. Basement massifs in the Appalachians: their role in deformation during the Appalachian orogenies. Geological Journal **18**, 255-265.

Jackson, J.A. 1980. Reactivation of basement faults and crustal shortening in orogenic belts. Nature **283**, 343-346.

Lillie, R.J. 1985. Tectonically buried continent/ocean boundary, Ouachita Mountains, Arkansas. Geology **13**, 18-21.

Lister, G.S., Etheridge, M.A., & Symonds, P.A. 1986. Application of the detachment fault model to the formation of passive continental margins. Geology **14**, 246-250.

Marchak, S. 1988. Kinematics of orocline and arc-formation in thin skinned orogens. Tectonics **7**, 73-86.

Matte, Ph. 1968. La structure de la virgation hercynienne de Galice (Espagne). Géologie Alpine **44**, 157-280.

McClay, K.R., & Price, N.J. (eds) 1981. Thrust and Nappe Tectonics. Geological Society of London, Special Publication **9**, 539p.

Norris, R.J., & Carter, R.M. 1982. Fault-bounded blocks and their role in localizing sedimentation and deformation adjacent to the Alpine fault, southern New Zealand. Tectonophysics **87**, 11-23.

Price, R.A., & Hatcher, R.D. Jr. 1983. Tectonic significance of similarities in the evolution of the Alabama-Pennsylvania Appalachians and the Alberta-British Columbia Canadian Cordillera. In, Hatcher, R.D. Jr., Williams, H., & Zietz, I. (eds) Contributions to the tectonics and geophysics of Mountian Chains. Geological Society of America Memoir **158**, 149-160.

Price, R.A. 1981. The Cordilleran foreland thrust and fold belt in the southern Canadian Rocky Mountains. In, McClay, K.R., & Price, N.J. (eds) Thrust and Nappe Tectonics. Geological Society of London Special Publication **9**, 427-448.

Rankin, D.W. 1975. The continental margin of eastern North America in the southern Appalachians: The opening and closing of the proto-Atlantic ocean. American Journal of Science **275A**, 298-336.

Rickard, M.J. 1985. The Surnadal synform and basement gneisses in the Surnadal-Sunndal district of Norway. In, Gee, D.G., & Sturt, B.A. (eds) The Caledonide Orogen — Scandinavia and related areas. Wiley, 485-498.

Rickard, M.J. 1991. Stratigraphy and Structural Geology of the Cowansville-Sutton-Mansonville Area in the Appalachians of Southern Quebec. Geological Survey of Canada Paper 88-27.

Rodgers, J. 1987. Chains of basement uplifts within cratons marginal to orogenic belts. American Journal of Science **287**, 661-692.

Schuepbach, M.A., & Vail, P.R. 1980. Evolution of outer highs on divergent continental margins. In, Continental Tectonics — studies in geophysics. US National Academy of Science, Washington D.C., p50-61.

Sheridan, R.E. 1974. Altantic Continental Margin of North America. In, Burke, C.A., & Drake, C.L. (eds) The geology of continental margins. Springer-Verlag, New York 391-407.

St. Julien, P., & Hubert, C. 1975. Evolution of the Taconian Orogen in the Quebec Appalachians. American Journal of Science **275A**, 337-362.

Stanley, R.S., & Ratcliffe, N.M. 1985. Tectonic synthesis of the Taconian Orogeny in western New England. Geological Society of America Bulletin **96**, 1227-1250.

Tapponnier, P., Peltzer, G., & Armijo, R. 1986. On the mechanics of the collision between India and Asia. In, Coward, M.P., & Ries, A.C. (eds) Collision Tectonics. Geological Society of London, Special Publication **19**, 115-157.

Thomas, W.A. 1977. Evolution of Appalachian-Ouachita salients and recesses from re-entrants and promontories in the continental margin. American Journal of Science **277**, 1233-1278.

Thomas, W.A. 1983. Basement-cover relations in the Appalachian fold and thrust belt. Geological Journal **18**, 267-276.

Wiltschko, D.V., & Eastman, D. 1983. Role of Basement warps and faults in localizing thrust fault maps. In, Hatcher, R.D. Jr., Williams, H., & Zietz, I. (eds) Contributions to the tectonics and geophysics of mountain chains. Geological Society of America, Memoir **158**, 177-190.

Winslow, M.A. 1981. Mechanisms for basement shortening in the Andean foreland fold belt of southern South America. In, McClay, K.R., & Price, N.J. (eds) Thrust and Nappe Tectonics. Geological Society of London Special Publication **9**, 513-528.

Woodward, N.B. 1988. Primary and secondary basement controls on thrust sheet geometries. In, Schmidt, C.J., & Perry, W.J. Jr. (eds) Interaction of Rocky Mountains foreland and Cordilleran thrust belt. Geological Society of America, Memoir **171**, 353-366.

PART III

TITLES OF PAPERS READ AT CONFERENCE

PART III TITLES OF PAPERS READ AT CONFERENCE

(* denotes full paper published in this volume)

Structure of the Australian craton and cover basins

Rutland, R.W.R.* Basement tectonics in Australia: An introductory perspective.

Williams, P.R. & Duggan, M.B. Tectonics of the Yilgarn Craton, Western Australia: Vertical versus horizontal models in the Eastern Goldfields.

Whitaker, A. Magnetic lineament and deformation patterns in the southern Yilgarn Block.

White, S.H. & Muir, M.D. Role of Archaean shear zones in the evolution of mobile zones surrounding the Yilgarn Craton.

Byrne, D.R. & Harris, L.P.* The structural elements of the Northampton Block, Western Australia.

Korsch, R.J., Wake-Dyster, K.D., O'Brien, P.E., Finlayson, D.M. & Johnstone, D.W.* Geometry of Permian to Mesozoic sedimentary basins in eastern Australia and their relationship to the New England Orogen.

Zhou, B., Mills, K.J. & Liu, S.F. The boundary between the Tasman Fold Belts and the Australian Craton: A reappraisal from studies of mafic rocks.

Wellman, P. Siting of sedimentary troughs along the boundaries between geophysical domains, Tasman Orogenic System.

Little, T.A., Holcombe, R.J., Gibson, G.M., Silwa, R. & Dobos, S.K. Late Palaeozoic tectonics of the North D'Aguilar Block, southeast Queensland.

Krokowski, J. & Olissoff, S.* The Pine Creek Shear Zone north of Pine Creek (Northern Territory) structural evolution and experimental studies.

Shaw, R.D., Korsch, R.J., Wright, C., Goleby, B.R. & Collins, C.D.N.* Thrust tectonics in Central Australia based on the Arunta-Amadeus Seismic Reflection Profile.

Liu, S.F. & Flemming, P.D. Tectonic models for the Cambrian Kanmantoo Group in southeast Australia.

Braun, J. & McQueen, H. 'Devon-shear' or basement tectonic evolution and basement development in Central and Western Australia in the late Palaeozoic.

Basement structures of Continental Regions

Bankwitz, P. Basement-cover relations in Central Europe.

Thomas, M.D. A plate tectonic framework for the Trans-Hudson Orogen, Canadian Shield, based on gravity and magnetic anomaly patterns.

Haszeldine, R.S.* North-south British Isles linears; Proterozoic to present.

Alsinawi, S.A. & Al-Banna A.S.* An E-W transect section through central Iraq.

Sindi, H.O.* Basement tectonism of the Arabian-Nubian Dome.

Edgell, H.S.* Basement tectonics of Saudi Arabia as related to oid field structures.

Kodama, K.* Structural analysis of the basin by Virtual Basement Displacement (V.B.D.) method.

Parker Gay Jr., S. Basement control of oil and gas traps: More common than we thought?

Adams, D.P.M. & Henley, R.W. Basement-linked syn-volcanic faulting and high-grade epithermal gold mineralization at Bimurra, northeast Queensland.

Carlson, M.P. Genetic relationships between the Precambrian basement and Phanerozoic tectonics, Midcontinent region, North America.

Baars, D.L.* Conjugate basement rift zones in Kansas, Midcontinent, USA.
Gibson, G.M. High-grade metamorphic rocks of the Tuhua Orogen, Western New Zealand: Lower crustal analogues of the Lachlan Fold Belt, SE Australia.
Katz, M.B.* Tar pavement rift - transform tectonic model and some examples in nature.
Jöns, H.P. Basement tectonics on Mars.

Structural patterns and Mineral Deposits

O'Driscoll, E.S.T.* Elusive trails in the basement labyrinth.
Bassi, H.G.L.* The Sierra Alta de San Luis, Argentina, South America: A case history of regmagenic control of gold metallogeny.
Muir, M.D. Structural controls on the East Alligator Rivers uranium field, Northern Territory, Australia.
White, S.H., Muir, M.D. & Smith, C.B. Basement reactivation and mineralization, Kimberley area, northwestern Australia.
Hodgson, R.A. & Kvet, R. Global fracture systems: mapping, analysis and economic utilization.

Techniques for analysing basement structures

Nash, C.R.* Factors affecting the acquisition of structural data from remotely sensed images of eastern Australia.
Brown, R., Fitzgerald, P. & Gleadow, A. Applying apatite fission-track analysis to tectonics: Examples from southern Africa, south eastern Australia and Antarctica.
Dunlap, W.J. Kinematic modeling and balancing of high-strain thrust systems.
Creasy, J. Synthesis of multiscale remote sensing interpretations for defining basement/cover structures.

Structural patterns in Oceanic crust

Palmer, J., Sempéré, J-C., Christie, D., Phipps Morgan & Shor, A. Seafloor spreading in the Australian-Antarctic Discordance.
Bostrom, R.C. Relation of ocean-floor structures to the Australian continental margin: Seasat images.

Epilogue

Rickard, M.J. Basement-cover relationships in orogenic belts.

Proceedings of the International Conferences on Basement Tectonics

1. R. Mason (ed.): *Basement Tectonics 7.* Proceedings of the Seventh International Conference on Basement Tectonics, held in Kingston, Ontario, Canada, August 1987. 1992 ISBN 0-7923-1582-0
2. J. Bartholomew (ed.): *Basement Tectonics 8.* 1992 (*in prep.*)
3. M. J. Rickard, H. J. Harrington and P. R. Williams (eds.): *Basement Tectonics 9 – Australia and Other Regions.* Proceedings of the Ninth International Conference on Basement Tectonics, held in Canberra, Australia, July 1990. 1992 ISBN 0-7923-1559-6

KLUWER ACADEMIC PUBLISHERS – DORDRECHT / BOSTON / LONDON